GENETIC ALGORITHMS AND ENGINEERING DESIGN

**WILEY SERIES IN ENGINEERING DESIGN
AND AUTOMATION**

Series Editor
HAMID R. PARSAEI

GENETIC ALGORITHMS AND ENGINEERING DESIGN
Mitsuo Gen and Runwei Cheng

GENETIC ALGORITHMS AND ENGINEERING DESIGN

MITSUO GEN
RUNWEI CHENG
Ashikaga Institute of Technology
Ashikaga, Japan

A Wiley-Interscience Publication
JOHN WILEY & SONS, INC.
New York • Chichester • Brisbane • Toronto • Singapore • Weinheim

Library of Congress Cataloging in Publication Data:

Gen, Mitsuo, 1944–
    Genetic algorithms and engineering design / Mitsuo Gen, Runwei
Cheng.
       p.   cm. — (Wiley series in engineering design and automation)
    Includes index.
    ISBN 0-471-12741-8 (alk. paper)
    1. Industrial engineering–Mathematical models. 2. Genetic
algorithms. 3. Mathematical optimization. I. Cheng, Runwei.
II. Title. III. Series.
T56.24.G46   1996
670′.285′51–DC20   96-8437

Printed in the United States of America

10 9 8 7 6 5 4 3 2 1

To Our Fathers, Mothers, and Families

PREFACE

Genetic algorithms are powerful and broadly applicable stochastic search and optimization techniques based on principles from evolution theory. In the past few years, the genetic algorithm community has turned much of its attention toward the optimization problems of industrial engineering, resulting in a fresh body of research and applications that expands previous studies.

This book is intended as a text covering the central aspects of genetic algorithms and their applications to difficult-to-solve optimization problems inherent in industrial engineering and manufacturing systems design. The book is designed as a self-study guide for professionals or as a textbook for the undergraduate- or graduate-level students in the departments of industrial engineering, management science, operations research, computer science, and artificial intelligence in colleges and universities. The book may also be useful as a comprehensive reference text for system analysts, operations researchers, management scientists, engineers, and other specialists who face the challenging difficult-to-solve optimization problems inherent in industrial engineering/operations research. Although there have been several very good books on genetic algorithms published in recent years, this book is the first book that provides readers with a comprehensive view and a unified treatment of the state of the art of genetic algorithms in the application of industrial engineering/operations research.

We have summarized the results of related genetic algorithm studies of recent years (from early 1992 to the present) and have adopted an algorithmic approach to reorganize all selected materials, but we do not present an in-depth discussion of current theory because genetic algorithms are still far from maturity. We hope that readers may be benefited from this approach by learning many useful and necessary techniques when using genetic algorithms to solve real-world problems in industrial engineering/operations research.

As we know, most professional articles on genetic algorithms are difficult for beginners because every author has a different representation approach as well as a unique use of terms and symbols. In this book, all selected previous works were systematically rearranged, terms and symbols were unified, and mistakes were corrected. The emphasis is on readability, understandability, and clarity of exposition. All algorithms are explained in intuitive, rather than highly technical, language with illustrative figures and numerical examples.

The prerequisites for a course using this book can be relatively modest because the mathematics has been kept at a relatively elementary level. Most of the material requires no mathematics beyond introductory college level. For each problem, this book briefly introduces its background, history, definition, formulation, and recent research progress in its solution techniques other than genetic algorithms, while the emphasis is placed on recent research results that apply genetic algorithms to obtain solutions. Written with the self-contained approach, this book also needs much fewer prerequisites to the domain knowledge of each topic.

This book has been organized to provide great flexibility in the selection of topics. Chapter 1 through Chapter 3 provide the fundamental knowledge for readers. The remaining chapters can be covered nearly independently, except that they all use the basic material represented in the first three chapters. The plan of this book is schematically described in the diagram on page ix.

Chapter 1 gives a tutorial on the basic concept of genetic algorithms. A revised general structure of genetic algorithms is described which can better reflect the current practice of genetic algorithm methods. A thorough discussion on encoding problems will provide readers with some helpful insight on how to design a suitable chromosome representation for a real-world problem. A comprehensive survey of selection strategies is also given with the point of view of the algorithmic approach in this book.

As we know, many industrial engineering design problems are very complex in nature and are therefore difficult to solve with conventional optimization techniques. In recent years, genetic algorithms have received considerable attention regarding their potential as a novel optimization technique. Chapter 2 discusses the applications of genetic algorithms to solve unconstrained optimization, nonlinear programming, stochastic programming, goal programming, and interval programming. Chapter 3 explains how to solve combinatorial optimization problems—including the knapsack problem, the quadratic assignment problem, the minimum spanning tree problem, the traveling salesman problem, and the film-copy deliver problem—with genetic algorithms. All techniques developed for them are applicable to other combinatorial problems. In particular, the traveling salesman problem is usually used as a paradigm for combinatorially difficult problems, and many new ideas have been tested on it. Because many difficult-to-solve engineering optimization problems are characterized as either constrained optimization or combinatorial optimization problems, these three chapters provide a foundation of genetic algorithms for readers' further studies of interest in these fields.

Chapter 4 shows how to solve reliability optimization problems, including the problem of a redundancy system with several failure modes, the problem of both redundant units and alternative design, the problem with fuzzy goal and

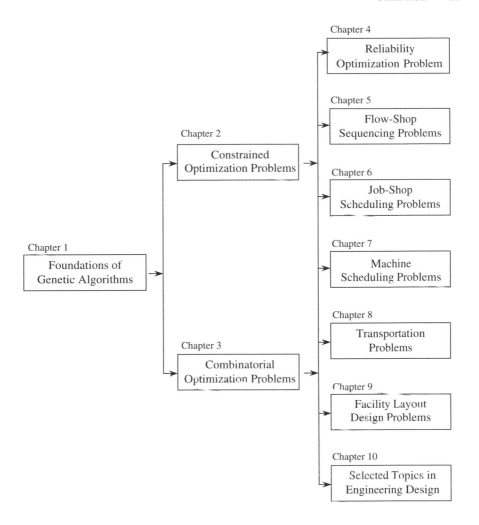

Chapter 4
Reliability
Optimization Problem

Chapter 5
Flow-Shop
Sequencing Problems

Chapter 2
Constrained
Optimization Problems

Chapter 6
Job-Shop
Scheduling Problems

Chapter 1
Foundations of
Genetic Algorithms

Chapter 7
Machine
Scheduling Problems

Chapter 8
Transportation
Problems

Chapter 3
Combinatorial
Optimization Problems

Chapter 9
Facility Layout
Design Problems

Chapter 10
Selected Topics in
Engineering Design

fuzzy constraints, the problem with interval coefficients, and the problem with redundant mixing components.

Chapter 5 describes how to solve flow-shop sequencing problems. Chapter 6 gives a review of recent literature on solving job-shop scheduling problems using genetic algorithms. Job-shop scheduling problems are among the most difficult types of combinatorial optimization and have captured the recent interest of genetic algorithm researchers. In this chapter, we concentrate on the representation schemes proposed for job-shop scheduling. The studies on genetic algorithm/job-shop scheduling provide invaluable help with regard to the constrained combinatorial optimization problems. All of the techniques developed for job-shop scheduling may be useful for other scheduling problems in modern flexible manufacturing systems and other combinatorial optimization problems.

Chapter 7 discusses how to apply genetic algorithms to single- and multiple-machine scheduling problems with nonregular measures. The machine scheduling problem is a rich and promising field of research with applications in manufacturing, logistics, computer architecture, and so on.

Chapter 8 investigates how to solve transportation problems, including bicriteria linear transportation problems, bicriteria solid transportation problems, and fuzzy multicriteria solid transportation problems. The techniques developed in this chapter can be generalized into other multicriteria optimization problems.

Chapter 9 shows how genetic algorithms can be successfully applied to various machine layout and facility layout problems. Facility layout design has been the subject of an interdisciplinary interest for over two decades. Because of the combinatorial nature of the facility layout problem, the genetic algorithm method is one of the most promising approaches for solving practical size layout problems.

Chapter 10 discusses some selected topics in industrial engineering design, including resource-constrained project scheduling problems, fuzzy vehicle routing and scheduling problems, location-allocation problems, obstacle location-allocation problems, and production plan problems. Most of them are more complex then those discussed in previous chapters. From this material, the reader will gain a further understanding of genetic algorithms, be able to appreciate the application potential of genetic algorithms, and discover many stimulations for their own work. We hope that this book will contribute to a better understanding, advancement, and development of new applications of genetic algorithms.

We would first like to express our sincere appreciations to Dr. Hamid R. Parsaei, University of Louisville, Editor-in-Chief of *Engineering Design and Automation* and Series Editor of the Wiley Series in Engineering Design and Automation, for giving us a chance to write this book as the first in a new series. We also would like to express our sincere appreciation to Dr. Hamed K. Eldin, University of Iowa, Editor-in-Chief of *Computers and Industrial Engineering*, for giving one of the authors a chance to edit a special issue on genetic algorithms and industrial engineering, which stimulated our initial motivation to devote our efforts to the research of industrial-engineering-oriented genetic algorithms.

This book benefited from numerous discussions with our colleagues and friends. We would like to thank the following individuals for their valuable comments: Dr. C. L. Hwang and Dr. Frank A. Tillman, Kansas State University & HTX International; Dr. Inyong Ham, Pennsylvania State University; Dr. Sang M. Lee, University of Nebraska—Lincon; Dr. Way Kuo, Texas A&M University; Dr. Gary S. Wasserman, Wayne State University; Dr. Alice E. Smith, University of Pittsburgh; Dr. Zbigniew Michalewicz, University of North Carolina; Dr. David B. Fogel, Natural Selection, Inc.; Dr. John R. Koza, Stanford University; Dr. Xin Yao, Australian Defense Force Academy; Dr. Chaiho Kim, Santa Clara University; Dr. Gursel A. Suer, University of Puerto Rico—Mayaguez; Dr. Moo Young Jung, Pohang University of Science & Technology; Dr. Hawk Hwang and Dr. Jong-Hwan Kim, Korea Advanced Institute of Science & Technology; Dr. Weixuan Xu and Dr. Jifa Gu, Chinese Academy of Sciences; Dr. Zhihou Yang, Northeastern University; Dr. Suebsak Nanthavanij, Thammasat University; Dr. Katsundo Hitomi, Ryukoko University; Dr. Hideo Tanaka, Dr. Hiroshi Ohta, Dr. Hidetomo Ichihashi, and Dr. Hisao Ishibuchi, Osaka Prefecture University; Dr. Toshio Fukuda and Dr. Takeshi Furuhashi, Nagoya University; Dr. Masatoshi Sakawa, Hiroshima University; Dr. Shigenobu Kobayashi and Dr. Tomoharu Nagao, Tokyo Institute of Technology; Dr. Hirokazu Osaki, Okayama University; Dr. Andrzej Osyczka, Tokyo Metropolitan University; Dr. Genji Yamazaki, Tokyo Metropolitan Institute of Technology; Dr. Masao Mukaidono, Meiji University; Dr. Tadaaki Fukugawa, Keio University; Dr. Junzo Wadata, Osaka Institute of Technology; Dr. Shohachiro Nakanishi, Tokai University; Dr. Masayuki Matsui, Electro-Communication University; Dr. Toshinori Kobayashi, Dr. Mitsuo Yamashiro, Dr. Kenichi Ida, Dr. Yashuhiro Tsujimura, and Dr. Takao Yokota, Ashikaga Institute of Technology.

We would like to give our special thanks to Ph.D. students Gengui Zhou and Dijin Gong for their careful proofreading of the manuscript. We want to thank Dr. Baoding Liu as well as the following graduate students at Ashikaga Institute of Technology for their creative and devoted work with us during the past few years: Takeaki Taguchi, Chunhui Cheng, Yinzhen Li, Elica Kubota, Dawei Zheng, and Yingxiu Li.

It is perhaps obvious that we are indebted to many researchers who have developed the underlying concepts that permeate this text. Although far too numerous to mention, we have tried to recognize their contributions through bibliographic references at the end of this book.

It was a real pleasure working with John Wiley & Sons' professional editorial and production staff, especially Mr. Robert L. Argentieri, Executive Editor of this book series, Mr. Robert H. Hilbert, Ms. Allison Ort Morvay, and Ms. Minna Panfili.

This project was supported by the International Scientific Research Program, the Grant-in-Aid for Scientific Research (No. 07045032: 1995.4 — 1998.3) the Ministry of Education, Science and Culture, the Japanese government between Ashikaga Institute of Technology, Japan and the Institute of Policy and Management, the Chinese Academy of Sciences.

We thank our wives, Eiko Gen and Liying Zhang, and children for their love, encouragement, understanding, and support during the preparation of this book.

MITSUO GEN
RUNWEI CHENG

Ashikaga Institute of Technology
April 1996

CONTENTS

1

FOUNDATIONS OF GENETIC ALGORITHMS

1.1 INTRODUCTION

Many optimization problems from the industrial engineering world, in particular the manufacturing systems, are very complex in nature and quite hard to solve by conventional optimization techniques [458, 470]. Since the 1960s, there has been an increasing interest in imitating living beings to solve such kinds of hard optimization problems. Simulating the natural evolutionary process of human beings results in stochastic optimization techniques called *evolutionary algorithms*, which can often outperform conventional optimization methods when applied to difficult real-world problems [11, 373, 437, 453]. There are currently three main avenues of this research: *genetic algorithms* (GAs), *evolutionary programming* (EP), and *evolution strategies* (ESs). Among them, genetic algorithms are perhaps the most widely known type of evolutionary algorithms today.

Recently, genetic algorithms have received considerable attention regarding their potential as an optimization technique for complex problems and have been successfully applied in the area of industrial engineering. The well-known applications include scheduling and sequencing, reliability design, vehicle routing and scheduling, group technology, facility layout and location, transportation, and many others.

1.1.1 General Structure of Genetic Algorithms

The usual form of genetic algorithm was described by Goldberg [171]. Genetic algorithms are stochastic search techniques based on the mechanism of natural selection and natural genetics. Genetic algorithms, differing from conventional search techniques, start with an initial set of random solutions called *population*. Each individual in the population is called a *chromosome*, representing a solution to the problem at hand. A chromosome is a string of symbols; it is usually, but not necessarily, a binary bit string. The chromosomes *evolve* through successive iterations, called *generations*. During each generation, the chromosomes are *evaluated*, using some measures of *fitness* [439]. To create the next generation, new chromosomes, called *offspring*, are formed by either (a) merging two chromosomes from current generation using a *crossover* operator or (b) modifying a chromosome using a *mutation* operator. A new generation is formed by (a) selecting, according to the fitness values, some of the parents and offspring and (b) rejecting others so as to keep the population size constant. Fitter chromosomes have higher probabilities of being selected. After several generations, the algorithms converge to the best chromosome, which hopefully represents the optimum or suboptimal solution to the problem. Let $P(t)$ and $C(t)$ be parents and offspring in current generation t; the general structure of genetic algorithms (see Figure 1.1) is described as follows:

Procedure: Genetic Algorithms

> **begin**
> $t \leftarrow 0$;
>   initialize $P(t)$;
>   evaluate $P(t)$;
>   **while** (not termination condition) **do**
>   recombine $P(t)$ to yield $C(t)$;
>   evaluate $C(t)$;
>   select $P(t + 1)$ from $P(t)$ and $C(t)$;
> $t \leftarrow t + 1$;
>   **end**
> **end**

It is a modified version of Grefenstette and Baker's description [192, 287]. Usually, initialization is assumed to be random. Recombination typically involves crossover and mutation to yield offspring. In fact, there are only two kinds of operations in genetic algorithms:

1. *Genetic operations*: crossover and mutation
2. *Evolution operation*: selection

The genetic operations mimic the process of heredity of genes to create new offspring at each generation. The evolution operation mimics the process of *Darwinian evolution* to create populations from generation to generation. This description differs from the paradigm given by Holland, where selection is made to obtain parents for recombination [220].

Crossover is the main genetic operator. It operates on two chromosomes at a time and generates offspring by combining both chromosomes' features. A simple way to achieve crossover would be to choose a random cut-point and generate the offspring by combining the segment of one parent to the left of the cut-point with the segment of the other parent to the right of the cut-point.

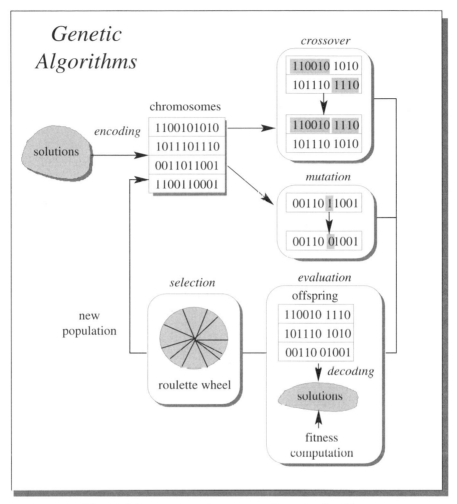

Figure 1.1. The general structure of genetic algorithms.

This method works well with the bit string representation. The performance of genetic algorithms depends, to a great extent, on the performance of the crossover operator used.

The *crossover rate* (denoted by p_c) is defined as the ratio of the number of offspring produced in each generation to the population size (usually denoted by *pop_size*). This ratio controls the expected number $p_c \times pop\_size$ of chromosomes to undergo the crossover operation. A higher crossover rate allows exploration of more of the solution space and reduces the chances of settling for a false optimum; but if this rate is too high, it results in the wastage of a lot of computation time in exploring unpromising regions of the solution space.

Mutation is a background operator which produces spontaneous random changes in various chromosomes. A simple way to achieve mutation would be to alter one or more genes. In genetic algorithms, mutation serves the crucial role of either (a) replacing the genes lost from the population during the selection process so that they can be tried in a new context or (b) providing the genes that were not present in the initial population.

The *mutation rate* (denoted by p_m) is defined as the percentage of the total number of genes in the population. The mutation rate controls the rate at which new genes are introduced into the population for trial. If it is too low, many genes that would have been useful are never tried out; but if it is too high, there will be much random perturbation, the offspring will start losing their resemblance to the parents, and the algorithm will lose the ability to learn from the history of the search.

Genetic algorithms differ from conventional optimization and search procedures in several fundamental ways. Goldberg [171] has summarized this as follows:

1. Genetic algorithms work with a coding of solution set, not the solutions themselves.
2. Genetic algorithms search from a population of solutions, not a single solution.
3. Genetic algorithms use payoff information (fitness function), not derivatives or other auxiliary knowledge.
4. Genetic algorithms use probabilistic transition rules, not deterministic rules.

1.1.2 Exploitation and Exploration

Search is one of the more universal problem-solving methods for such problems where one cannot determine *a priori* the sequence of steps leading to a solution. Search can be performed with either *blind strategies* or *heuristic strategies* [44]. Blind search strategies do not use information about the problem domain. Heuristic search strategies use additional information to guide the search along with the best search directions. There are two important issues in search strategies: exploiting the best solution and exploring the search space [46]. Michalewicz gave a comparison of hill-climbing search, random search, and genetic search [287]. Hill-climbing is an example of a strategy which exploits the best solution for possible improvement while ignoring the exploration of the search space. Random search is an example of a strategy which explores the search space while ignoring the exploitation of the promising regions of the search space. Genetic algorithms are a class of general-purpose search methods combining elements of directed and stochastic search which can make a remarkable balance between exploration and exploitation of the search space. At the beginning of genetic search, there is a widely random and diverse population and crossover operator tends to perform widespread search for exploring all solution space. As the high fitness solutions develop, the crossover operator provides exploration in the neighborhood of each of them. In other words, what kinds of searches (exploitation or exploration) a crossover performs would be determined by the environment of the genetic system (the diversity of population), but not by the operator itself. In addition, simple genetic operators are designed as general-purpose search methods (the domain-independent search methods); they perform essentially a blind search and could not guarantee to yield an improved offspring.

1.1.3 Population-Based Search

Generally, the algorithm for solving optimization problems is a sequence of computational steps which asymptotically converge to optimal solution. Most classical optimization methods generate a deterministic sequence of computation based on the gradient or higher-order derivatives of objective function. The methods are applied to a single point in the search space (Figure 1.2). The point is then improved along the deepest descending/ascending direction gradually through iterations. This point-to-point approach takes the danger of falling in local optima. Genetic algorithms perform a multiple directional search by maintaining a population of potential solutions. The population-to-population approach attempts to make the search escape from local optima. Population undergoes a simulated evolution: At each generation the relatively good solutions are reproduced, while the relatively bad solutions die. Genetic algorithms use probabilistic transition rules to select someone to be reproduced and some-

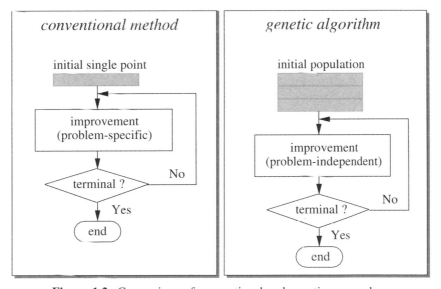

Figure 1.2. Comparison of conventional and genetic approaches.

one to die so as to guide their search toward regions of the search space with likely improvement.

1.1.4   Meta-heuristics

At first, genetic algorithms were created as a generic tool useful for many difficult-to-solve problems. Much of the early works of genetic algorithms used a universal internal representation involving fixed-length binary strings with binary genetic operators to operate in a domain-independent fashion at the level without any knowledge of the phenotypic interpretation of the strings. This universality was reflected in a strong emphasis on the design of robust adaptive systems with a broad range of applications. However, simple genetic algorithms are difficult to apply directly and successfully into many difficult-to-solve optimization problems. Various nonstandard implementations have been created for particular problems in which genetic algorithms are used as a *meta-heuristics*. Michalewicz gave his book the vivid name *Genetic Algorithms + Data Structures = Evolution Programs* to emphasize this trend. We do not restrict ourselves to the use of fixed-length binary string and binary genetic operators; to the contrary, we prefer to use a nature representation (any data structure suitable for a given problem) together with any set of meaningful genetic operators applicable to the data structure.

1.1.5   Major Advantages

Genetic algorithms have received considerable attention regarding their potential as a novel optimization technique. There are three major advantages when applying genetic algorithms to optimization problems.

1. Genetic algorithms do not have much mathematical requirements about the optimization problems. Due to their evolutionary nature, genetic algorithms will search for solutions without regard to the specific inner workings of the problem. Genetic algorithms can handle any kind of objective functions and any kind of constraints (i.e., linear or nonlinear) defined on discrete, continuous, or mixed search spaces.

2. The ergodicity of evolution operators makes genetic algorithms very effective at performing global search (in probability). The traditional approaches perform local search by a convergent stepwise procedure, which compares the values of nearby points and moves to the relative optimal points. Global optima can be found only if the problem possesses certain convexity properties that essentially guarantee that any local optima is a global optima.

3. Genetic algorithms provide us a great flexibility to hybridize with domain-dependent heuristics to make an efficient implementation for a specific problem.

Table 1.1   Explanation of Genetic Algorithm Terms

Genetic Algorithms	Explanation
Chromosome (string, individual)	Solution (coding)
Genes (bits)	Part of solution
Locus	Position of gene
Alleles	Values of gene
Phenotype	Decoded solution
Genotype	Encoded solution

1.1.6   Genetic Algorithm Vocabulary

Because genetic algorithms are rooted in both natural genetics and computer science, the terminologies used in genetic algorithm literature are a mixture of the natural and the artificial.

In a biological organism, the structure that encodes the prescription specifying how the organism is to be constructed is called a *chromosome*. One or more chromosomes may be required to specify the complete organism. The complete set of chromosomes is called a *genotype*, and the resulting organism is called a *phenotype*. Each chromosome comprises a number of individual structures called *genes*. Each gene encodes a particular feature of the organism, and the location, or *locus*, of the gene within the chromosome structure determines what particular characteristic the gene represents. At a particular locus, a gene may encode any of several different values of the particular characteristic it represents. The different values of a gene are called *alleles*.

The correspondence of genetic algorithm terms and optimization terms is summarized in Table 1.1.

1.2   EXAMPLES WITH SIMPLE GENETIC ALGORITHMS

In this section we explain in detail about how a genetic algorithm actually works, using two simple examples. We follow the approach of implementation of genetic algorithms given by Michalewicz [287].

1.2.1   Optimization Problem

The numerical example of unconstrained optimization problem is given as follows:

$$\max f(x_1, x_2) = 21.5 + x_1 \sin(4\pi x_1) + x_2 \sin(20\pi x_2)$$
$$-3.0 \leq x_1 \leq 12.1$$
$$4.1 \leq x_2 \leq 5.8$$

A three-dimensional plot of the objective function is shown in Figure 1.3.

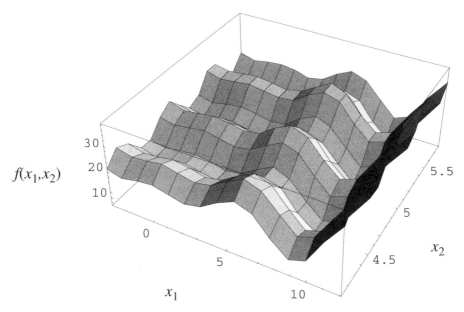

Figure 1.3. Objective function.

Representation. First, we need to encode decision variables into binary strings. The length of the string depends on the required precision. For example, the domain of variable x_j is $[a_j, b_j]$ and the required precision is five places after the decimal point. The precision requirement implies that the range of domain of each variable should be divided into at least $(b_j - a_j) \times 10^5$ size ranges. The required bits (denoted with m_j) for a variable is calculated as follows:

$$2^{m_j - 1} < (b_j - a_j) \times 10^5 \leq 2^{m_j} - 1$$

The mapping from a binary string to a real number for variable x_j is straightforward and completed as follows:

$$x_j = a_j + decimal(substring_j) \times \frac{b_j - a_j}{2^{m_j} - 1}$$

where $decimal(substring_j)$ represents the decimal value of $substring_j$ for decision variable x_j.

Suppose that the precision is set as five places after the decimal point. The required bits for variables x_1 and x_2 is calculated as follows:

$$(12.1 - (-3.0)) \times 10,000 = 151,000$$

$$2^{17} < 151,000 \leq 2^{18}, \qquad\qquad m_1 = 18$$

$$(5.8 - 4.1) \times 10,000 = 17,000$$

$$2^{14} < 17,000 \leq 2^{15}, \qquad\qquad m_2 = 15$$

$$m = m_1 + m_2 = 18 + 15 = 33$$

The total length of a chromosome is 33 bits which can be represented as follows:

$$\longleftarrow \text{33 bits} \longrightarrow$$

v_j 000001010100101001 101111011111110

$$\longleftarrow \text{18 bits} \longrightarrow \longleftarrow \text{15 bits} \longrightarrow$$

The corresponding values for variables x_1 and x_2 are given below:

	Binary Number	Decimal Number
x_1	000001010100101001	5417
x_2	101111011111110	24318

$$x_1 = -3.0 + 5417 \times \frac{12.1 - (-3.0)}{2^{18} - 1} = -2.687969$$

$$x_2 = 4.1 + 24318 \times \frac{5.8 - 4.1}{2^{15} - 1} = 5.361653$$

Initial Population. Initial population is randomly generated as follows:

$$v_1 = [000001010100101001101111011111110]$$
$$v_2 = [001110101110011000000010101001000]$$
$$v_3 = [111000111000001000010101001000110]$$
$$v_4 = [100110110100101101000000010111001]$$
$$v_5 = [000010111101100010001110001101000]$$
$$v_6 = [111110101011011000000010110011001]$$
$$v_7 = [110100010011111000100110011101101]$$
$$v_8 = [001011010100001100010110011001100]$$
$$v_9 = [111110001011101100011101000111101]$$
$$v_{10} = [111101001110101010000010101101010]$$

The corresponding decimal values are

$$v_1 = [x_1, x_2] = [-2.687969, 5.361653]$$
$$v_2 = [x_1, x_2] = [\ 0.474101, 4.170144]$$
$$v_3 = [x_1, x_2] = [10.419457, 4.661461]$$
$$v_4 = [x_1, x_2] = [\ 6.159951, 4.109598]$$
$$v_5 = [x_1, x_2] = [-2.301286, 4.477282]$$
$$v_6 = [x_1, x_2] = [11.788084, 4.174346]$$
$$v_7 = [x_1, x_2] = [\ 9.342067, 5.121702]$$
$$v_8 = [x_1, x_2] = [-0.330256, 4.694977]$$
$$v_9 = [x_1, x_2] = [11.671267, 4.873501]$$
$$v_{10} = [x_1, x_2] = [11.446273, 4.171908]$$

Evaluation. The process of evaluating the fitness of a chromosome consists of the following three steps:

Procedure: Evaluation

Step 1. Convert the chromosome's genotype to its phenotype. Here, this means converting binary string into relative real values $x^k = (x_1^k, x_2^k), k = 1, 2, \ldots, pop\_size$.

Step 2. Evaluate the objective function $f(x^k)$.

Step 3. Convert the value of objective function into fitness. For the maximization problem, the fitness is simply equal to the value of objective function $eval(v_k) = f(x^k), k = 1, 2, \ldots, pop\_size$.

An evaluation function plays the role of the environment, and it rates chromosomes in terms of their fitness.

The fitness function values of above chromosomes are as follows:

$$eval(v_1) = f(-2.687969, 5.361653) = 19.805119$$
$$eval(v_2) = f(\ 0.474101, 4.170144) = 17.370896$$
$$eval(v_3) = f(10.419457, 4.661461) = 9.590546$$
$$eval(v_4) = f(\ 6.159951, 4.109598) = 29.406122$$
$$eval(v_5) = f(-2.301286, 4.477282) = 15.686091$$
$$eval(v_6) = f(11.788084, 4.174346) = 11.900541$$
$$eval(v_7) = f(\ 9.342067, 5.121702) = 17.958717$$
$$eval(v_8) = f(-0.330256, 4.694977) = 19.763190$$
$$eval(v_9) = f(11.671267, 4.873501) = 26.401669$$
$$eval(v_{10}) = f(11.446273, 4.171908) = 10.252480$$

It is clear that chromosome v_4 is the strongest one and that chromosome v_3 is the weakest one.

Selection. In most practices, a *roulette wheel* approach is adopted as the selection procedure; it belongs to the fitness-proportional selection and can select a new population with respect to the probability distribution based on fitness values. The roulette wheel can be constructed as follows:

1. Calculate the fitness value $eval(v_k)$ for each chromosome v_k:

$$eval(v_k) = f(x), \qquad k = 1, 2, \ldots, pop\_size$$

2. Calculate the total fitness for the population:

$$F = \sum_{k=1}^{pop\_size} eval(v_k)$$

3. Calculate selection probability p_k for each chromosome v_k:

$$p_k = \frac{eval(v_k)}{F}, \qquad k = 1, 2, \ldots, pop\_size$$

4. Calculate cumulative probability q_k for each chromosome v_k:

$$q_k = \sum_{j=1}^{k} p_j, \qquad k = 1, 2, \ldots, pop\_size$$

The selection process begins by spinning the roulette wheel *pop_size* times; each time, a single chromosome is selected for a new population in the following way:

Procedure: Selection

Step 1. Generate a random number r from the range $[0, 1]$.
Step 2. If $r \leq q_1$, then select the first chromosome v_1; otherwise, select the kth chromosome $v_k (2 \leq k \leq pop\_size)$ such that $q_{k-1} < r \leq q_k$.

The total fitness F of the population is

$$F = \sum_{k=1}^{10} eval(v_k) = 178.135372$$

The probability of a selection p_k for each chromosome $\boldsymbol{v}_k (k = 1, \ldots, 10)$ is as follows:

$$p_1 = 0.111180, \qquad p_2 = 0.097515, \qquad p_3 = 0.053839$$
$$p_4 = 0.165077, \qquad p_5 = 0.088057, \qquad p_6 = 0.066806$$
$$p_7 = 0.100815, \qquad p_8 = 0.110945, \qquad p_9 = 0.148211$$
$$p_{10} = 0.057554$$

The cumulative probabilities q_k for each chromosome $\boldsymbol{v}_k (k = 1, \ldots, 10)$ is as follows:

$$q_1 = 0.111180, \qquad q_2 = 0.208695, \qquad q_3 = 0.262534$$
$$q_4 = 0.427611, \qquad q_5 = 0.515668, \qquad q_6 = 0.582475$$
$$q_7 = 0.683290, \qquad q_8 = 0.794234, \qquad q_9 = 0.942446$$
$$q_{10} = 1.000000$$

Now we are ready to spin the roulette wheel 10 times, and each time we select a single chromosome for a new population. Let us assume that a random sequence of 10 numbers from the range [0, 1] is as follows:

$$0.301431 \quad 0.322062 \quad 0.766503 \quad 0.881893$$
$$0.350871 \quad 0.583392 \quad 0.177618 \quad 0.343242$$
$$0.032685 \quad 0.197577$$

The first number $r_1 = 0.301431$ is greater than q_3 and smaller than q_4, meaning that the chromosome $\boldsymbol{v}_4$ is selected for the new population; the second number $r_2 = 0.322062$ is greater than q_3 and smaller than q_4, meaning that the chromosome $\boldsymbol{v}_4$ is again selected for the new population; and so on. Finally, the new population consists of the following chromosomes:

$$\boldsymbol{v}_1' = [1001101101001011010000000010111001] \quad (\boldsymbol{v}_4)$$
$$\boldsymbol{v}_2' = [1001101101001011010000000010111001] \quad (\boldsymbol{v}_4)$$
$$\boldsymbol{v}_3' = [0010110101000011000101100110011001100] \quad (\boldsymbol{v}_8)$$
$$\boldsymbol{v}_4' = [1111100010111011000111010000111101] \quad (\boldsymbol{v}_9)$$
$$\boldsymbol{v}_5' = [1001101101001011010000000010111001] \quad (\boldsymbol{v}_4)$$
$$\boldsymbol{v}_6' = [1101000100111110001001100111011101] \quad (\boldsymbol{v}_7)$$
$$\boldsymbol{v}_7' = [0011101011100110000000010101001000] \quad (\boldsymbol{v}_2)$$
$$\boldsymbol{v}_8' = [1001101101001011010000000010111001] \quad (\boldsymbol{v}_4)$$
$$\boldsymbol{v}_9' = [0000010101001010011011110111111110] \quad (\boldsymbol{v}_1)$$
$$\boldsymbol{v}_{10}' = [0011101011100110000000010101001000] \quad (\boldsymbol{v}_2)$$

Crossover. Crossover used here is one-cut-point method, which randomly selects one cut-point and exchanges the right parts of two parents to generate offspring. Consider two chromosomes as follows, and the cut-point is randomly selected after the 17th gene:

$$\downarrow$$
$$v_1 = [1001101101001011 01000000010111001]$$
$$v_2 = [0010110101000011000101 10011001100]$$

The resulting offspring by exchanging the right parts of their parents would be as follows:

$$v_1' = [10011011010010110\ 0010110011001100]$$
$$v_2' = [00101101010000110\ 1000000010111001]$$

The probability of crossover is set as $p_c = 0.25$, so we expect that, on average, 25% of chromosomes undergo crossover. Crossover is performed in the following way:

Procedure: Crossover

begin
  $k \leftarrow 0$;
  while $(k \leq 10)$ **do**
    $r_k \leftarrow$ random number from [0, 1];
    if $(r_k < 0.25)$ **then**
      select v_k as one parent for crossover;
    end
    $k \leftarrow k + 1$;
  end
end

Assume that the sequence of random numbers is

    0.625721   0.266823   0.288644   0.295114
    0.163274   0.567461   0.085940   0.392865
    0.770714   0.548656

This means that the chromosomes v_5' and v_7' were selected for crossover. We generate a random integer number *pos* from the range [1, 32] (because 33 is the total length of a chromosome) as cutting point or in other words, the position of the crossover point. Assume that the generated number *pos* equals 1, the two chromosomes are cut after the first bit, and offspring are generated by exchanging the right parts of them as follows:

$$v_5' = [1001101101001011010000000010111001]$$
$$v_7' = [0001101101001011010000000010111001]$$
$$\downarrow$$
$$v_5' = [1001101101001011010000000010111001]$$
$$v_7' = [0001101101001011010000000010111001]$$

Mutation. Mutation alters one or more genes with a probability equal to the mutation rate. Assume that the 18th gene of the chromosome v_1 is selected for a mutation. Since the gene is 1, it would be flipped into 0. Thus the chromosome after mutation would be

$$v_1 = [1001101101001011010100000010111001]$$
$$\downarrow$$
$$v_1' = [1001101101001011000010000010111001]$$

The probability of mutation is set as $p_m = 0.01$, so we expect that, on average, 1% of total bit of population would undergo mutation. There are $m \times pop\_size = 33 \times 10 = 330$ bits in the whole population; we expect 3.3 mutations per generation. Every bit has an equal chance to be mutated. Thus we need to generate a sequence of random numbers $r_k(k = 1, \ldots, 330)$ from the range [0, 1]. Suppose that the following genes will go through mutation:

bit_pos	chrom_num	bit_no	random_num
105	4	6	0.009857
164	5	32	0.003113
199	7	1	0.000946
329	10	32	0.001282

After mutation, we get the final population as follows:

$$v_1' = [1001101101001011010000000010111001]$$
$$v_2' = [1001101101001011010000000010111001]$$
$$v_3' = [0010110101000011000101100110011001100]$$
$$v_4' = [1111110010111011000111010001111101]$$
$$v_5' = [1011101011100110000000010101001010]$$
$$v_6' = [1101000100111110001001100111011101]$$
$$v_7' = [1001101101001011010000000010111001]$$
$$v_8' = [1001101101001011010000000010111001]$$

$$v_9' = [00000101010010100110111101111110]$$
$$v_{10}' = [00111010111001100000010101001010]$$

The corresponding decimal values of variables $[x_1, x_2]$ and fitness are as follows:

$$f(\ 6.159951, 4.109598) = 29.406122$$
$$f(\ 6.159951, 4.109598) = 29.406122$$
$$f(-0.330256, 4.694977) = 19.763190$$
$$f(11.907206, 4.873501) = \ 5.702781$$
$$f(\ 8.024130, 4.170248) = 19.91025$$
$$f(\ 9.342067, 5.121702) = 17.958717$$
$$f(\ 6.159951, 4.109598) = 29.406122$$
$$f(\ 6.159951, 4.109598) = 29.406122$$
$$f(-2.687969, 5.361653) = 19.805119$$
$$f(\ 0.474101, 4.170248) = 17.370896$$

Now we just completed one iteration of genetic algorithm. The test run is terminated after 1000 generations. We have obtained the best chromosome in the 419th generation as follows:

$$v^* = (11111000000001110001111101001010110)$$
$$eval(v^*) = f(11.631407, 5.724824) = 38.818208$$
$$x_1^* = 11.631407$$
$$x_2^* = 5.724824$$
$$f(x_1^*, x_2^*) = 38.818208$$

1.2.2 Word-Matching Problem

This example was given by Freeman [143]. It is a nice example to show the power of genetic algorithms. The *word-matching problem* tries to evolve an expression of "to be or not to be" from the randomly-generated lists of letters with genetic algorithm. Since there are 26 possible letters for each of 13 locations in the list, the probability that we get the correct phrase in a pure random way is $(1/26)^{13} = 4.03038 \times 10^{-19}$, which is about two chances out of a billion.

We use a list of ASCII integers to encode the string of letters. The lowercase letters in ASCII are represented by numbers in the range [97, 122] in the decimal number system. For example, the string of letters tobeornottobe is converted into the following chromosome represented with ASCII integers:

[116, 111, 98, 101, 111, 114, 110, 111, 116, 116, 111, 98, 101]

Generate an initial population of 10 random phrases as follows:

[114, 122, 102, 113, 100, 104, 117, 106, 97, 114, 100, 98, 101]
[110, 105, 101, 100, 119, 118, 121, 118, 106, 97, 104, 102, 106]
[115, 99, 121, 117, 101, 105, 115, 111, 115, 113, 118, 99, 98]
[102, 98, 102, 118, 114, 97, 109, 116, 101, 107, 117, 118, 115]
[107, 98, 117, 113, 114, 116, 106, 116, 106, 101, 110, 115, 98]
[102, 119, 121, 113, 121, 107, 107, 116, 122, 121, 111, 106, 104]
[116, 98, 120, 98, 108, 115, 111, 105, 122, 103, 103, 119, 109]
[101, 111, 111, 117, 114, 104, 100, 120, 98, 118, 116, 120, 97]
[100, 116, 114, 105, 117, 111, 115, 114, 103, 107, 109, 98, 103]
[106, 118, 112, 98, 103, 101, 109, 116, 112, 106, 97, 108, 113]

Now, we convert this population to string to see what they look like:

rzfqdhujardbe
niedwvyvjahfj
scyueisosqvcb
fbfvramtekuvs
kbuqrtjtjensb
fwyqykktzyojh
tbxblsoizggwm
dtriuosrgkmbg
jvpbgemtpjalq

Fitness is calculated as the number of matched letters. For example, the fitness for string "rzfqdhujardbe" is 2. Only mutation is used which results in a change to a given letter with a given probability. Now, we run our genetic algorithm with 30 generations to see how well it works. The best one of each generation is listed in Table 1.2.

After only 23 generations, the population produced the desired phrase. The total examined chromosomes are 230. If we use pure random method to produce 230 random phrases, could we have a match?

1.3 ENCODING PROBLEM

How to encode a solution of the problem into a chromosome is a key issue for genetic algorithms. In Holland's work, encoding is carried out using binary

Table 1.2   The Best String for Each Generation

Generation	String	Fitness	Generation	String	Fitness
1	rzfqdhujardbe	2	16	rzbwornottobe	10
2	rzfqdhuoardbe	3	17	rzbwornottobe	10
3	rzfqghuoatdbe	4	18	rzbwornottobe	10
4	rzfqghuoztobe	5	19	rzbwornottobe	10
5	rzfqghhottobe	6	20	robwornottobe	11
6	rzfqohhottobe	7	21	tobwornottobe	12
7	rzfqohnottobe	8	22	tobwornottobe	12
8	rzfqohnottobe	8	23	tobeornottobe	13
9	rzfqohnottobe	8	24	tobeornottobe	13
10	rzfqohnottobe	8	25	tobeornottobe	13
11	rzfqornottobe	9	26	tobeornottobe	13
12	rzfqornottobe	9	27	tobeornottobe	13
13	rwfwornottobe	9	28	tobeornottobe	13
14	rwcwornottobe	9	29	tobeornottobe	13
15	rzcwornottobe	9	30	tobeornottobe	13

strings. For many GA applications, especially for the problems from industrial engineering world, the simple GA was difficult to apply directly because the binary string is not a natural coding. During the past 10 years, various nonstring encoding techniques have been created for particular problems—for example, *real number coding* for constrained optimization problems and *integer coding* for combinatorial optimization problems. Choosing an appropriate representation of candidate solutions to the problem at hand is the foundation for applying genetic algorithms to solve real world problems, which conditions all the subsequent steps of genetic algorithms. For any application case, it is necessary to perform analysis carefully to ensure an appropriate representation of solutions together with meaningful and problem-specific genetic operators.

One of the basic features of genetic algorithms is that they work on *coding space* and *solution space* alternatively: genetic operations work on coding space (chromosomes), while evaluation and selection work on solution space as shown in Figure 1.4. Natural selection is the link between chromosomes and the performance of their decoded solutions. For the nonstring coding approach, three critical issues emerged concerning with the encoding and decoding between chromosomes and solutions (or the mapping between phenotype and genotype):

- The feasibility of a chromosome
- The legality of a chromosome
- The uniqueness of mapping

Feasibility refers to the phenomenon of whether a solution decoded from a chromosome lies in the feasible region of a given problem. *Legality* refers to

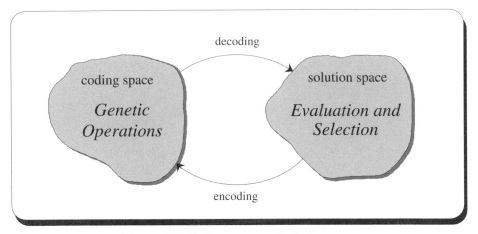

Figure 1.4. Coding space and solution space.

the phenomenon of whether a chromosome represents a solution to a given problem as shown in Figure 1.5.

The *infeasibility* of chromosomes originates from the nature of the constrained optimization problem. All methods, conventional ones or genetic algorithms, must handle the constraints. For many optimization problems, the feasible region can be represented as a system of equalities or inequalities (linear or nonlinear). For such cases, many efficient penalty methods have been proposed to handle infeasible chromosomes [152, 289, 382]. In constrained optimization problems, the optimum typically occurs at the boundary between feasible and infeasible areas. The penalty approach will force the genetic search to approach the optimum from both feasible and infeasible regions.

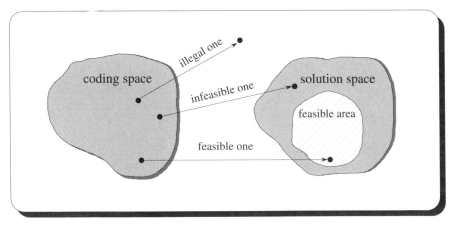

Figure 1.5. Feasibility and legality.

The *illegality* of chromosomes originates from the nature of encoding techniques. For many combinatorial optimization problems, problem-specific encodings are used and such encodings usually yield to illegal offspring by a simple one-cut-point crossover operation. Because an illegal chromosome cannot be decoded to a solution, it means that such chromosomes cannot be evaluated; thus the penalty approach is inapplicable to this situation. Repairing techniques are usually adopted to convert an illegal chromosome to a legal one. For example, the well-known PMX operator (which will be introduced in Chapter 3) is essentially a kind of two-cut-point crossover for permutation representation together with a repairing procedure to resolve the illegitimacy caused by the simple two-cut-point crossover. Orvosh and Davis [322] have shown that for many combinatorial optimization problems, it is relatively easy to repair an infeasible or illegal chromosome; the repair strategy does indeed surpass other strategies such as rejecting strategy or penalizing strategy.

The *mapping* from chromosomes to solutions (decoding) may belong to one of the following three cases:

- 1-to-1 mapping
- *n*-to-1 mapping
- 1-to-*n* mapping

as shown in Figure 1.6. The 1-to-1 mapping is the best one among three cases and 1-to-*n* mapping is the most undesired one. We need to consider these problems carefully when designing a new non-binary-string coding so as to build an effective genetic algorithm.

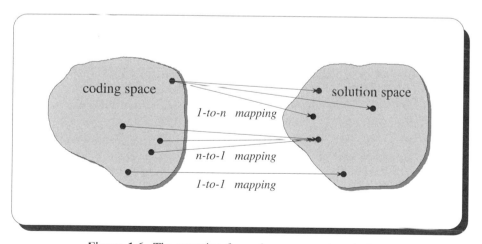

Figure 1.6. The mapping from chromosomes to solutions.

1.4 SELECTION

The principle behind genetic algorithms is essentially Darwinian natural selection. Selection provides the driving force in a genetic algorithm, and the selection pressure is critical in it. At one extreme, the search will terminate prematurely; while at the other extreme progress will be slower than necessary. Typically, low selection pressure is indicated at the start of the genetic algorithm search in favor of a wide exploration of the search space, while high selection pressure is recommended at the end in order to exploit the most promising regions of the search space. The selection directs a genetic algorithm search toward promising regions in the search space. During the past few years, many selection methods have been proposed, examined, and compared.

There are three basic issues involved in selection phase:

- Sampling space
- Sampling mechanism
- Selection probability

Each of these gives a significant influence on selective pressure and thereby genetic algorithm behavior. We will present a detailed discussion on each of these.

1.4.1 Sampling Space

Selection procedure may create a new population for the next generation based on either all parents and offspring or part of them. This leads to the problem of sampling space. A sample space is characterized by two factors: *size* and *ingredient* (parent or offspring). Let *pop_size* denote the size of population and let *off_size* denote the size of offspring produced at each generation. The *regular sampling* space has the size of *pop_size* and contains all offspring but just part of parents. The *enlarged sampling space* has the size of *pop_size* + *off_size* and contains whole of parents and offspring.

Regular Sampling Space. In Holland's original genetic algorithm, parents are replaced by their offspring soon after they give birth. This is called *generational replacement*. Because genetic operations are blind in nature, offspring may be worse than their parents. With the strategy of replacing each parent with his offspring directly, some fitter chromosomes will be lost from the evolutionary process. To overcome this problem, several *replacement strategies* have been examined. Holland suggested that when each offspring is born, it replaces a randomly chosen chromosome of the current population [220]. De Jong proposed a *crowding strategy* [104]. In the crowding model, when an offspring was born, one parent was selected to die. The dying parent was chosen as that parent who most closely resembled the new offspring, using a simple bit-by-bit similarity count to measure resemblance. Note that in Holland's works, *selection* refers to choosing parents for recombination; a new population was formed by replacing parents with their offspring. This was called a *reproductive plan*. Since Grefenstette and Baker's work [192], selection is used to form the next generation, usually with a probabilistic mechanism. Michalewicz gave a detailed description on simple genetic algorithms where offspring in each generation replaced their parents soon after they were born; the next generation was formed by *roulette wheel selection* [287]. Figure 1.7 illustrates selection based on regular sampling space.

Enlarged Sampling Space. When selection performs on enlarged sampling space, both parents and offspring have the same chance of competing for survival. A typical case is $(\mu + \lambda)$ selection [130]. This strategy was originally used in evolution strategies [373]. Bäck and Hoffmeister introduced it in genetic algorithms [10, 13]. With this strategy, μ parents and λ offspring compete for survival and the μ best out of offspring and old parents are selected as parents of the next generation. Another case from evolution strategies is (μ, λ) selection, which selects the μ best offspring as parents of the next generation $(\mu < \lambda)$. Both methods are completely deterministic and can be transferred to a probabilistic method. Although most of the selection methods are based on regular sampling space, it is easy to implement them on enlarged sampling space. Figure 1.8 illustrates the selection based on enlarged sampling space. An evident advantage of this approach is that we can improve genetic algorithm performance by increasing the crossover and mutation rates. We do not worry that the high rate will introduce too much random perturbation if selection is performed on enlarged sampling space.

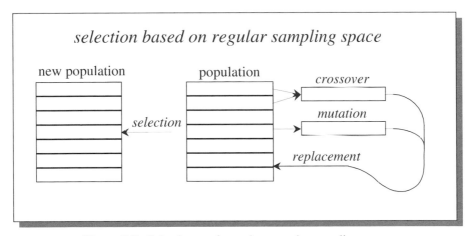

Figure 1.7. Selection performed on regular sampling space.

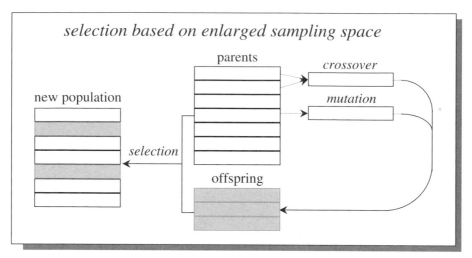

Figure 1.8. Selection performed on enlarged sampling space.

1.4.2   Sampling Mechanism

Sampling mechanism concerns the problem of how to select chromosomes from sampling space. Three basic approaches have been used to sampling chromosomes:

- Stochastic sampling
- Deterministic sampling
- Mixed sampling

Stochastic Sampling. Major early works have concentrated on this approach. A common feature of methods in this class as pointed by Baker is that the selection phase determines the actual number of copies that each chromosome will receive based on its survival probability [20]. Thus the selection phase is composed of two parts:

1. Determine the chromosome's expected value.
2. Convert the expected values to the number of offspring.

A chromosome's expected value is a real number indicating the average number of offspring that a chromosome should receive. The sampling procedure is used to convert the real expected value to the number of offspring.

   The best known of these class methods is Holland's *proportionate selection* or *roulette wheel selection*. The basic idea is to determine *selection probability* (also called *survival probability*) for each chromosome proportional to the fitness value. For chromosome k with fitness f_k, its selection probability p_k is calculated as follows:

$$p_k = f_k \bigg/ \sum_{j=1}^{pop\_size} f_j$$

Then we can make a wheel according to these probabilities. The selection process is based on spinning the roulette wheel *pop_size* times; each time, we select a single chromosome for the new population as described in the last section.

Baker proposed *stochastic universal sampling* [20], which uses a single wheel spin. The wheel is constructed as a roulette wheel and is spun with a number of equally spaced markers equal to the population size. The expected value e_k for chromosome k is calculated as $e_k = pop\_size \times p_k$. The procedure of stochastic universal sampling can be described as follows:

Procedure: Stochastic Universal Sampling

begin
  sum ← 0;
  ptr ← *rand*( );
  for k ← 1 **to** *pop_*size **do**
    sum ← *sum* + e_k;
    while (*sum* > *ptr*) **do**
      select chromosome k;
      ptr ← *ptr* + 1;
    end
  end
end

where *rand*( ) returns a random real number uniformly distributed within the range [0, 1).

The basic consideration of this approach is to keep the expected number of copies of each chromosome in the next generation. An interesting concern is the *prohibition of duplicate chromosome* in population. There are two reasons for us using this strategy:

1. Prevent *super chromosomes* from dominating population by maintaining too many copies in population. This is one cause of rapid convergence to local optimum (premature).

2. Keep the diversity of population so that the constant generation pool can contain much more information for genetic search.

A related problem is that when duplicate chromosomes are discarded, the size of the population formed with remained parents and offspring may be less than the predetermined size *pop_size*. In this case, the initial procedure is usually used to fill the vacant space of the population pool.

Deterministic Sampling. This approach usually selects the best *pop_size* chromosomes from the sampling space. Both $(\mu + \lambda)$ selection and (μ, λ) selection belong to this method. Bäck has discussed how to transfer them to probability methods [10]. Note that both approaches prohibit duplicate chromosomes from entering the population during selection, so most researchers prefer to use this method to deal with combinatorial optimization problems.

Truncation selection and *block selection* also belong to this method, which rank all chromosomes according to their fitness and selects the best ones as parents [401]. In truncation selection, a threshold T is defined such that the $T\%$ best chromosomes are selected and each one receives nearly $100/T$ copies. Block selection is equivalent to truncation selection since for a given population size *pop_size*, one simply gives s copies to the *pop_size*/s best chromosomes. Both implementations are identical when $s = pop\_size/T$.

Elitist selection ensures that the best chromosome is passed onto the new generation if it is not selected through another process of selection.

Brindle's deterministic sampling is based on the concept of *expected number* [48]. The selection probability for each chromosome is calculated as usual, $p_k = f_k/\Sigma f_j$. The expected number of each chromosome is calculated as $e_k = p_k \times$ *pop_size*. Each chromosome is allocated samples according to the integer part of the expected number, and then population is sorted according to the fractional parts of the expected number. The remainder of chromosomes needed to fill are drawn from the top of the sorted list.

The generational replacement (replacing entire set of parents by their offspring) can be viewed as another version of the deterministic approach. A modification to the method is to replace n worst old chromosomes with offspring (n is the number of offspring). It was first given by Whitely [414] (Whitley called it *GENITOR*). It also was described by Syswerda [391] (Syswerda called it *steady-state reproduction*). Michalewicz gave a stochastic version of this approach called modGA (modified genetic algorithms) [287]. The chromosomes to be replaced by offspring are selected according to their survival probabilities. Chromosomes which are worse in performance than the average have higher chances to be selected to die.

Mixed Sampling. This approach contains both random and deterministic features simultaneously. A typical example is *tournament selection* given by Goldberg et al. [173]. This method randomly chooses a set of chromosomes and picks out the best one from the set for reproduction. The number of chromosomes in the set is called *tournament size*. A common tournament size is 2. This is called *binary tournament*.

Stochastic tournament selection was suggested by Wetzel [413]. In this method, selection probabilities are calculated normally and successive pairs of chromosomes are drawn using roulette wheel selection. After drawing a pair, the chromosome with higher fitness is inserted in the new population. The process continues until the population is full.

Remainder stochastic sampling proposed by Brindle is a modified version of his deterministic sampling [48]. In this method, each chromosome is allocated samples according to the integer part of the expected number, and then chromosomes compete according to the fractional parts of the expected number for the remaining places in the population.

1.4.3   Selection Probability

This issue concerns how to determine selection probability for each chromosome. In proportional selection procedure, the selection probability of a chromosome is proportional to its fitness. This simple scheme exhibits some undesirable properties. For example, in early generations, there is a tendency for a few super chromosomes to dominate the selection process; in later generations, when population is largely converged, competition among chromosomes is less strong and a random search behavior will emerge.

Scaling and ranking mechanisms are proposed to mitigate these problems. Scaling method maps raw objective function values to some positive real values, and the survival probability for each chromosome is determined according to these values. Ranking method ignores the actual objective function values and uses a ranking of chromosomes instead to determine survival probability. Fitness scaling has a twofold intention:

1. To maintain a reasonable differential between relative fitness ratings of chromosomes
2. To prevent a too-rapid takeover by some super chromosomes in order to meet the requirement to limit competition early on, but to stimulate it later

For most scaling methods, scaling parameters are problem-dependent. Fitness ranking has a similar effect as the fitness scaling, but avoids the need for extra scaling parameters [348].

Since De Jong's works, scaling of objective function values has become a widely accepted practice and several scaling mechanisms have been proposed. Goldberg [171] and Michalewicz [287] have summarized them well. In general, the scaled fitness f'_k from the raw fitness (e.g., objective function value) f_k for chromosome k can be expressed as follows [192]:

$$f'_k = g(f_k)$$

where function $g(\cdot)$ transforms the raw fitness into scaled fitness. The function $g(\cdot)$ may take different forms to yield different scaling methods, such as linear scaling sigma truncation, power law scaling, logarithmic scaling, and so on.

These methods can be roughly classified into two categories:

- Static scaling
- Dynamic scaling

The mapping relation between the scaled fitness and raw fitness can be constant to yield static scaling methods, or it can vary according to some factors to yield dynamic scaling methods. The dynamic scaling methods are further divided into two cases:

1. Scaling parameters are adaptively adjusted according to the scatter situation of fitness values in each generation in order to keep constant selective pressure.
2. Scaling parameters are dynamically changed along with the increase of the number of generations in order to increase the selective pressure accordingly.

Note that our definitions of static and dynamic are different from those given by Bäck and Hoffmeister [13]. They examined the mapping relation between selection probability and raw fitness but not between the scaled fitness and raw fitness.

Linear Scaling. Linear scaling adjusts the fitness values of all chromosomes such that the best chromosome gets a fixed number of expected offspring and thus prevent it from reproducing too many.

When function $g(\cdot)$ takes the form of linear transformation, we have the following linear scaling method:

$$f'_k = a \times f_k + b$$

The parameters a and b are normally selected so that average chromosomes receive one offspring copy on average, and the best receives the specified number of copies (usually two). This method may give negative fitness values that must be dealt with.

Dynamic Linear Scaling. When parameter b varies with generation, we have the following dynamic linear scaling method [192]:

$$f'_k = a \times f_k + b_t$$

One possible definition for b_t is to take the minimal raw fitness value of current population $b_t = -f_{min}$.

Sigma Truncation. This method was suggested by Forrest to improve linear scaling both to deal with the negative values and to incorporate the problem-dependent information into the mapping [134]. The reformulated form by Goldberg [171] is:

$$f'_k = f_k - (\bar{f} - c \times \sigma)$$

where c is a small integer, σ is the population standard deviation, and $\bar{f}$ is the average raw fitness. Possible negative scaled fitness f'_k is set to zero.

The selection pressure is related to the scatter of the fitness values in the population. Sigma scaling exploits the observation and sets the scaling factor as the difference between the mean and standard deviation. Chromosomes below this score are assigned a fitness of zero. This method helps to overcome a potential problem with particularly poor chromosomes which would make the scatter of the fitness very large, thus reducing selection pressure [206].

Power Law Scaling. This method was proposed by Gillies [169], whose function $g(\cdot)$ takes the form of some specified power of the raw fitness as follows:

$$f'_k = f_k^\alpha$$

In general, the value α is problem-dependent. Gillies reported an α value of 1.005.

The gap of scaled fitness between the best and the worst chromosomes increases with the value of α. When α approaches to zero, the gap approaches zero and sampling becomes random search; while when $\alpha > 1$, the gap is enlarged and sampling will be allocated to fitter chromosomes. In a run, the value of α may require dynamic adjustment to stretch or shrink the range as needed. Michalewicz proposed a method to dynamically change the value of α [287].

Another possible definition of power law scaling is given as follows [192]:

$$f'_k = (a \times f_k + b)^\alpha$$

Logarithmic Scaling. This method was examined by Fitzpatrick and Grefenstette for mapping the objective function of minimization problem [192], where function $g(\cdot)$ takes the following logarithmic form:

$$f'_k = b - \log(f_k)$$

where b is chosen to be larger than any value of $\log(f_k)$.

Windowing. Consider the following situation discussed by Hancock [206]. Given two chromosomes, with fitness 2 and 1 respectively, the first will get twice as many offspring as the second. If the underlying function is changed simply by adding 1 to all the values, the two chromosomes will score 3 and 2, a ratio of only 1.5. It shows that a mere modification of the *baseline* of the fitness function will have a significant effect on the rate of progress.

The windowing technique introduces a *moving baseline* technique into fitness-proportional selection to maintain much more constant selection pressure. This method can be viewed as one kind of dynamic linear scaling shown as follows:

$$f'_k = f_k - f_w$$

where w is known as the window size and is typically of the order of 2 ~ 10, and f_w is the worst value observed in the w most recent generations.

Normalizing. Normalizing technique is also one kind of dynamic scaling given by Cheng and Gen [62]. For the maximization problem, it takes the following form:

$$f'_k = \frac{f_k - f_{\min} + \gamma}{f_{\max} - f_{\min} + \gamma}$$

where $f_{\max}$ and $f_{\min}$ are the best and the worst raw fitnesses in current population, respectively. It also can be viewed as a normalized special windowing method where window size w equals 1. γ is a small positive real number which is usually restricted within the open interval $(0, 1)$. The purpose of using such transformation is twofold:

1. To prevent the equation from zero division
2. To make it possible to adjust the selection behavior from fitness-proportional selection to pure random selection.

For the minimization problem, we have

$$f'_k = \frac{f_{\max} - f_k + \gamma}{f_{\max} - f_{\min} + \gamma}$$

Boltzmann Selection. Boltzmann selection is a misleading name for yet another scaling method for proportional selection, using the following scaling function [10, 287].

$$f'_k = e^{f_k/T}$$

Selection pressure is low when the control parameter T is high.

Ranking. Baker introduced the notion of *ranking selection* to genetic algorithms to overcome the scaling problems of the direct fitness-based approach [19, 172, 206]. The idea is straightforward: Sort the population from the best to the worst and assign the selection probability of each chromosome according to the ranking but not its raw fitness. Two methods are in common use: *linear ranking* and *exponential ranking.*

Let p_k be the selection probability for the kth chromosome in the ranking of population; the linear ranking takes the following form:

$$p_k = q - (k - 1) \times r$$

where parameter q is the probability for the best chromosome. Let q_0 is the probability for the worst chromosome; the parameter r can be determined as follows:

$$r = \frac{q - q_0}{pop\_size - 1}$$

Intermediate chromosomes' fitness values are decreased from q to q_0 proportional to their rank. When set q_0 be 0, it provides the maximum selective pressure.

Michalewicz proposed the following *exponential ranking* method [287]:

$$p_k = q(1 - q)^{k-1}$$

A larger value of q implies stronger selective pressure. Hancock proposed the following *exponential ranking* method [206]:

$$p_k = q^{k-1}$$

where q is typically about 0.99. The best chromosome has a fitness of 1, and the last one receives $q^{pop\_size - 1}$.

1.4.4   Selective Pressures

According to the viewpoint of *Neo-Darwinism*, the process of evolution can be divided into three categories [281]:

- Stabilizing selection
- Directional selection
- Disruptive selection

Stabilizing selection is also called *normalizing selection* because it tends to eliminate chromosomes with extreme values. Directional selection has the effect of either increasing or decreasing the mean value of the population. Disruptive selection tends to eliminate chromosomes with moderate values. Most of selection methods are based on the directional selection.

Kuo and Hwang proposed a method based on disruptive selection [257]. The method adopted a nonmonotonic fitness function that is quite different from traditional monotonic fitness functions. They gave a concept of *normalized-by-mean-fitness* function as follows:

$$f'_k = |f_k - \bar{f}|$$

It is clear that this function is a kind of nonmonotonic function. A selection procedure using the normalized-by-mean-fitness function is called the *disruptive selection*. Such a method gives a higher probability to both the best and the worst chromosomes. Their experimental results showed that genetic algorithms using this method can easily find the optimum of the *needle-in-a-haystack* function.

An example of stabilizing selection method was given by Gen, Liu, and Ida [163]. In their method, chromosomes are ranked from the best to the worst according to their raw fitness values. And then three preference parameters p_1, p_0, and $p_2 (0 < p_1 < p_0 < p_2 < 1)$ are defined, which are used to determine three critical chromosomes with ranking u_1, u_0, and u_2, respectively, such that $u_1 = \lfloor p_1 \cdot pop\_size \rfloor, u_0 = \lfloor p_0 \cdot pop\_size \rfloor$, and $u_2 = \lfloor p_2 \cdot pop\_size \rfloor$. Lastly, the chromosome with the ranking u_1 is assigned a fitness value of $e^{-1} \approx 0.37$, the one with ranking u_0 is assigned a fitness value of 1, and the one with ranking u_2 is assigned a fitness value of $2 - e^{-1} \approx 1.63$. For an chromosome k with ranking u, the relation between the ranking u and the exponential fitness f_k is given as follows:

$$f_k = \begin{cases} \exp\left[-\dfrac{u - u_0}{u_1 - u_0}\right], & u < u_0 \\[2em] 2 - \exp\left[-\dfrac{u - u_0}{u_2 - u_0}\right], & u \geq u_0 \end{cases} \tag{1.1}$$

which is shown in Figure 1.9. Obviously, the preference parameters p_1 and p_2 are used to designate the chromosomes with extreme values to be eliminated.

Bäck and Hoffmeister have introduced the concepts of *extinctive* and *preservative* selection [13]. The preservative selection requires nonzero selection probability for each chromosome, whereas extinctive selection does not. Extinctive selections are further divided into *left* and *right* selections: In left extinctive selection the best chromosomes are prevented from reproduction in order to avoid premature convergence due to super chromosomes; while in right extinc-

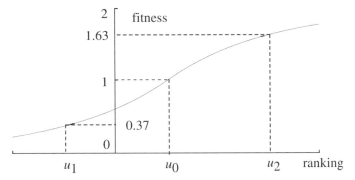

Figure 1.9. Stabilizing scaling.

tive selection the worst chromosomes are prevented from reproduction in order to reduce selective pressure due to poor chromosomes which would put the scatter of the fitness very large. The stabilizing scaling method introduced by Gen, Liu, and Ida [163] can be viewed as a combination form of the left and right extinctive selections, which prevents both super and poor chromosomes from reproduction.

1.5   HYBRID GENETIC ALGORITHMS

Genetic algorithms have proved to be a versatile and effective approach for solving optimization problems. Nevertheless, there are many situations in which the simple genetic algorithm does not perform particularly well, and various methods of *hybridization* have been proposed. For example, Goldberg used the idea of G-bit improvement for binary strings [171]. Davis advocated hybridization of genetic algorithms with domain-specific techniques for real-world optimization, by incorporating extra move operators [101]. Ackley recommended genetic hillclimbing, in which crossover plays a rather less dominant role [2]. Reeves explored an alternative perspective which views genetic algorithms as a generalization of the neighborhood search method [349]. Mühlenbein [301] argued theoretically and Gorges-Schleuter [180] provided empirical demonstrations that local search can play a key role. Miller et al. included local improvement operators in pure genetic algorithm for NP-hard optimization problems [295].

One of the most common forms of hybrid genetic algorithms is to incorporate local optimization as an add-on extra to the simple genetic algorithm loop of recombination and selection. With the hybrid approach, local optimization is applied to each newly generated offspring to move it to a local optimum before injecting it into the population. Genetic algorithms are used to perform global exploration among a population, while heuristic methods are used to perform local exploitation around chromosomes. Because of the complementary properties of genetic algorithms and conventional heuristics, the hybrid approach often outperforms either method operating alone. Some studies have been performed to reveal the nature mechanism behind such a hybrid approach. Among them are Lamarckian evolution and memetic algorithms.

1.5.1 Lamarckian Evolution

Perhaps the early studies which linked genetic algorithm and Lamarckian evolution theory were due to the following individuals: Grefenstette, who introduced Lamarckian operators into genetic algorithms [190]; Davidor, who defined Lamarckian probability for mutations in order to enable mutation operators to be more controlled and to introduce some qualities of a local hill-climbing operator [93], and Shaefer, who added an intermediate mapping between the chromosome space and solution space into standard genetic algorithms, which is Lamarckian in nature [377].

Kennedy gave an explanation for hybrid genetic algorithms with Lamarckian evolution theory [246, 445]. The simple genetic algorithm of Holland was inspired by Darwin's theory of natural selection. In the nineteenth century, Darwin's theory was challenged by Lamarck, who proposed that environmental changes throughout an organism's life cause structural changes that are transmitted to offspring. This theory lets organisms pass along the knowledge and experience they acquire in their lifetime. While no biologist today believes that traits acquired in the natural world can be inherited, the power of Lamarckian theory is illustrated by the evolution of our society. Ideas and knowledge are passed from generation to generation through structured language and culture. Genetic algorithms, the artificial organisms, can benefit from the advantages of Lamarckian theory. By letting some of the individuals' "experiences" be passed along to future individuals, we can improve the genetic algorithms' ability to focus on the most promising areas. Following a more Lamarckian approach, first a traditional hill-climbing routine could use the offspring as a starting point and perform quick and localized optimization. After it has *learned* to climb the local landscape, we can put the offspring through the evaluation and selection phases. The offspring has a chance to pass its experience to future offspring through common crossover.

Let $P(t)$ and $C(t)$ be parents and offspring in current generation t. The general structure of hybrid genetic algorithms is described as follows (see Figure 1.10):

Procedure: Hybrid Genetic Algorithms

begin
    $t \leftarrow 0$;
    initialize $P(t)$;
    evaluate $P(t)$;
    while (not termination condition) **do**
      recombine $P(t)$ to yield $C(t)$;

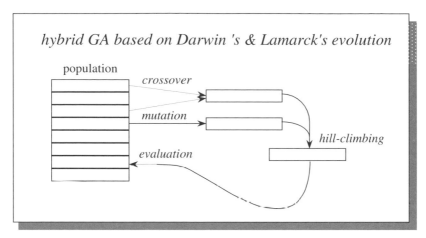

Figure 1.10. General structure of hybrid genetic algorithms.

```
    locally climb C(t);
    evaluate C(t);
    select P(t + 1) from P(t) and C(t);
    t ← t + 1;
  end
end
```

In the hybrid approach, artificial organisms first pass through Darwin's biological evolution and then pass through Lamarckian's intelligent evolution. A traditional hill-climbing routine is used as Lamarckian's evolution to try to inject some "smarts" into the offspring organism before returning it to be evaluated.

1.5.2 Memetic Algorithms

Moscato and Norman have introduced the term *memetic algorithm* to describe genetic algorithms in which local search plays a significant part [300]. The term is motivated by Dawkins's notion of a *meme* as a unit of information that reproduces itself as people exchange ideas [103]. A key difference exists between genes and memes: Before a meme is passed on, it is typically adapted by the person who transmits it as that person thinks, understands, and processes the meme, whereas genes get passed on whole. Moscato and Norman link this thinking to local refinement, and therefore they promoted the term *memetic algorithm* to describe genetic algorithms that use local search heavily.

Radcliffe and Surry gave a formal description of *memetic algorithms* [344], which provided a homogeneous formal framework for considering memetic and genetic algorithms. According to Radcliffe and Surry, if a local optimizer is added to a genetic algorithm and applied to every child before it is inserted

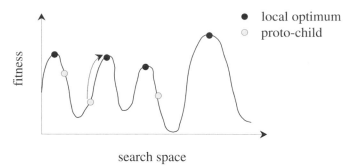

Figure 1.11. Memetic algorithm and local optimizer. (Adapted from Radcliffe and Surry [344].)

into the population, then a memetic algorithm can be thought of simply as a special kind of genetic search over the subspace of local optima. Recombination and mutation will usually produce solutions that are outside this space of local optima, but a local optimizer can then repair such solutions to produce final children that lie within this subspace, yielding a memetic algorithm as shown in Figure 1.11.

The role of local search in the context of genetic algorithms has been receiving serious consideration and many successful applications are strongly in favor of such hybrid approach. Both memetic algorithms and Lamarckian evolution try to give a reasonable explanation on the hybrid approach based on different nature phenomena.

1.6   IMPORTANT EVENTS IN THE GENETIC ALGORITHM COMMUNITY

1.6.1   Books on Genetic Algorithms

Probably the earliest predecessor of genetic algorithms emerged in the work of Fraser, a biologist who wanted to simulate evolution with special emphasis on the interaction of epistasis with selection [139–142]. The term *epistasis* is used to denote the impact of one gene—the epistatic one—on the expression of another gene. In the field of genetic algorithms, the term is used to denote the effect on chromosome fitness of a combination of alleles which is not merely a linear function of the effects of the individual alleles [351].

Even to attentive scientists, genetic algorithms did not become apparent before 1975 when the first book *Adaptation in Natural and Artificial Systems* of Holland and the dissertation *An Analysis of the Behavior of a Class of Genetic Adaptive Systems* of De Jong [104] were published [373]. Davis said: Holland created the genetic algorithms field. The unique features of genetic algorithms have been shaped by the careful and insightful work of Holland and his students [101]. In Holland's works, the motivation was the design and implementation of robust adaptive systems, capable of dealing with an uncertain and changing environment. His view emphasized the need for systems which self-adapt over time as a function of feedback obtained from interacting with the environment in which they operate. This led to an initial family of reproductive plans which formed the basis for what we call *simple genetic algorithms* today [106].

Since then, genetic algorithm has been covering three major fields: research into basic genetic algorithm, optimization using genetic algorithm, and machine learning with classifier systems [101]. This research thrust was well described in Goldberg's book of *Genetic Algorithms in Search, Optimization and Machine Learning*.

Over the past 10 years, application of genetic algorithms to real-world problems has increased greatly. Many researchers started to adapt the algorithm to the natural representation of the search space for a given problem, thus developing new genetic operators that are suited to the special data structures. As a consequence, such extensions and modifications of genetic algorithms led to a new promising direction for problem solving in general, especially for the difficult-to-solve optimization problems from the industrial engineering world. This problem-oriented approach demonstrates an enormous difference with respect to basic genetic algorithms, such that the boundaries to other evolutionary algorithms become blurred [11]. The major progress of the research was excellently discussed in Michalewicz's book entitled *Genetic Algorithms + Data Structures = Evolution Programs*.

Several books on genetic algorithms have been published; these are listed in Table 1.3. Also, other related edited books and proceedings can be found in references [436, 440, 447, 450].

1.6.2 Conferences and Workshops

Since 1985, several conferences and workshops have been held to provide an international forum for exchanging new ideas, progress, or experience on genetic algorithms and to promote better understanding and collaborations between the theorists and practitioners in this field. The major meetings are listed in Table 1.4.

In 1985, a series of biannual International Conferences on Genetic Algorithms (ICGA) were started to bring together those who are interested in the theory or application of genetic algorithms (see Table 1.5).

The biannual international conference on Parallel Problem Solving from Nature (PPSN) is the European sister conference to ICGA held in even-numbered years. The first meeting was held in Germany, 1990. The unifying theme of the PPSN conferences is natural computation, i.e., the design, the theoretical and empirical understanding, and the comparison of algorithms gleaned from nature together with their application to real-world problems in science and technology.

The workshops on Foundation of Genetic Algorithms (FOGA) have been

Table 1.3   Books on Genetic Algorithms

Year	Authors	Book Title
1975	Holland [220]	*Adaptation in Natural and Artificial Systems*
1987	Davis [102]	*Genetic Algorithms and Simulated Annealing*
1989	Goldberg [171]	*Genetic Algorithms in Search, Optimization and Machine Learning*
1991	Davis [101]	*Handbook of Genetic Algorithms*
	Davidor [94]	*Genetic Algorithms and Robotics*
1992	Koza [253]	*Genetic Programming*
	Michalewicz [287]	*Genetic Algorithms + Data Structures = Evolution Programs* (2nd edition 1994, 3rd edition 1996)
1994	Bauer [25]	*Genetic Algorithms and Investment Strategies*
	Grefenstette [191]	*Genetic Algorithms for Machine Learning*
	Bhanu and Lee [33]	*Genetic Learning for Adaptive Image Segmentation*
	Koza [254]	*Genetic Programming II*
1995	Chambers [54]	*Practical Handbook of Genetic Algorithms, vols. 1 and 2*
	Schwefel [373]	*Evolution and Optimum Seeking*
	Biethahn and Nissen [34]	*Evolutionary Algorithms in Management Application*
	Fogel [131]	*Evolutionary Computation*
1996	Mitchell [296]	*An Introduction to Genetic Algorithms*
	Bäck [11]	*Evolutionary Algorithms in Theory and Practice*
	Lawton [204]	*A Practical Guide to Genetic Algorithms in C++*
	Winter et al. [264]	*Genetic Algorithms in Engineering and Computer Science*
	Herrera and Verdegay [448]	*Genetic Algorithms and Soft Computing*
1997	Bäck et al. [12]	*Handbook of Evolutionary Computation*
	Gen and Tsujimura [442]	*Evolutionary Computations and Intelligent Systems*

Table 1.4   Conferences on Genetic Algorithms

Abbreviation	Conference Name
ICGA	International Conference on Genetic Algorithms
PPSN	International Conference on Parallel Prolem Solving from Nature
ICEC	IEEE International Conference on Evolutionary Computations
ANN&GA	International Conference on Artificial Neural Nets & Genetic Algorithms
EP	Annual Conference on Evolutionary Programming
FOGA	Workshop on Foundation of Genetic Algorithms
COGANN	International Workshop on Combinations of Genetic Algorithms and Neural Networks
EC	AISB Workshop on Evolutionary Computing
GP	Genetic Programming Conference
SEAL	The Asia-Pacific Conference on Simulated Evolution And Learning

Table 1.5   International Conference on Genetic Algorithms

Year	Proceeding Editors	Meeting Place
1985	Grefenstelle [186]	Pittsburgh, USA
1987	Grefenstelle [189]	Cambridge, USA
1989	Schaffer [370]	George Mason University, USA
1991	Belew and Booker [30]	San Diego, USA
1993	Forrest [137]	Urbana-Champaign, USA
1995	Eshelman [120]	Pittsburgh, USA
1997	—	Ann Arbor, USA

held biannually, starting in 1990 (see Table 1.7). FOGA alternates with the ICGA. The ICGA conferences have been held in the odd-numbered years, while the FOGA conferences took place in the even-numbered years. Both events are sponsored and organized under the auspices of the International Society for Genetic Algorithms. This series meetings provide forums specifically targeting theoretical publications on genetic algorithms.

In 1992, the first annual conference on evolutionary programming was held in San Diego. The series annual meetings were sponsored by the Evolutionary Programming Society (see Table 1.8). Participants enjoyed a broad range of exposure to the field of evolutionary computation in general, and evolutionary programming, evolution strategies, genetic algorithms, genetic programming, and cultural algorithms in particular.

In 1993, another series of biannual meetings were held in Austria, called the International Conference on Artificial Neural Nets and Genetic Algorithms (ANN&GA) (see Table 1.9). The series meetings are devoted to the topics of ANN and GA as well as to the interactions between them.

Another important annual conference is the IEEE International Conference on Evolutionary Computation (ICEC) (see Table 1.10). The first ICEC meeting was held in Orlando in 1994. The ICEC conferences are sponsored by the IEEE Neural Network Council and encompass all the various flavors of this technology including evolution strategies, evolutionary programming, genetic algorithms, and genetic programming.

The AISB workshop on evolutionary computing [128] was held at the University of Leeds, United Kingdom, in 1994. The workshop was sponsored by the Society for the Study of Artificial Intelligence and Simulation of Behavior and brought together most of the people doing research on evolutionary com-

Table 1.6   International Conference on Parallel Problem Solving from Nature

Year	Proceeding Editors	Meeting Place
1990	Schwefel and Männer [374]	Dortmund, Germany
1992	Männer and Manderick [285]	Brussels, Belgium
1994	Davidor, Schwefel and Männer [95]	Jerusalem, Israel
1996	Ebeling and Voigt [438]	Dortmund, Germany

Table 1.7   Workshop on Foundation of Genetic Algorithms

Year	Proceeding Editors	Meeting Place
1990	Rawlins [347]	Bloomington, USA
1992	Whitley [416]	Vail, USA
1994	Whitley and Vose [417]	Estes Park, USA

Table 1.8   Annual Conference on Evolutionary Programming

Year	Proceeding Editors	Meeting Place
1992	Fogel and Atmar [133]	San Diego, USA
1993	Fogel and Atmar [134]	La Jolla, USA
1994	Sebald and Fogel [375]	San Diego, USA
1995	McDonnell, Reynolds, and Fogel [284]	San Diego, USA
1996	Angeline and Bäck [435]	San Diego, USA

Table 1.9   International Conference on Artificial Neural Nets and Genetic Algorithms

Year	Proceeding Editors	Meeting Place
1993	Albrech, Reeves, and Steele [5]	Innsbruck, Austria
1995	Pearson, Steels, and Albrecht [335]	Ales, France

Table 1.10   IEEE International Conference on Evolutionary Computations

Year	Proceeding Editors	Meeting Place
1994	Fogel [132]	Orlando, USA
1995	deSilva [108]	Perth, Australia
1996	Fogel [129]	Nagoya, Japan

puting in the United Kingdom. Another European conference on artificial evolution was held in Brest in 1995 [434].

Two workshops, *AI'93* and *AI'94 Workshops on Evolutionary Computation*, were held in Melbourne and Armidale, Australia, respectively. Some selected papers from the workshops were included in an edited volume *Progress in Evolutionary Computation* by X. Yao [462, 463]. Another international workshop on genetic algorithms and engineering design was held in Ashikaga, Japan, May 1996. All papers of this workshop were included in the *Proceedings of Mini-Symposium on Genetic Algorithms and Engineerng Design*, edited by M. Gen and K. Ida [441].

The first Asia-Pacific conference on simulated evolution and learning (SEAL'96) was held at KAIST, Korea, November 1996, in conjunction with Micro-Robot World Cup Soccer Tournament (MIROSOT'96) [464]. The first genetic programming (GP) conference was held at Stanford University, July 1996, in cooperation with the Association for Computing Machinery, SIGART, the American Association for Artificial Intelligence, and the IEEE Neural Networks Council [449]. About genetic programming in general, the following WWW server is available:

http://www-cs-faculty.stanford.edu/koza/

The first international conference on evolvable systems was held at Tsukuba, Japan, October 1996 [471].

1.6.3 Journals and Special Issues on Genetic Algorithms

The journal of *Evolutionary Computation* (De Jong, editor-in-chief, MIT Press, started in 1993) provides a forum specifically targeting theoretical publication on genetic algorithms. MIT Press has a WWW (World Wide Web) server on line:

http://www-mitpress.mit.edu

and is maintaining a section including subscription information and the titles and abstracts of all published issues.

A new international journal *IEEE Transactions on Evolutionary Computation*, will be started from May 1997. The journal particularly emphasizes the practical application of evolutionary computation and related techniques to solving real problems. The Editor-in-Chief is David B. Fogel (dfogel@natural-selection.com). Another new one *Evolutionary Optimization* will also be started in 1997. The journal is primarily concerned with evolutionary optimization. The Editor-in-Chief is A. Osyczka (osyczka@control.prec.metro-u.ac.jp).

In recent years, several journals have published *special issues* on genetic algorithms and related topics, which are listed in the following:

1. D. B. Fogel and L. J. Fogel, guest editors, special issue on *Evolutionary Computation, IEEE Transactions on Neural Networks*, vol. 5, no. 1, 1994.

2. Z. Michalewicz, guest editor, special issue on *Evolutionary Computation, Statistics and Computing*, vol. 4, no. 2, 1994.

3. S. J. Raff, guest editor, special issue on *Genetic Algorithms, International Journal of Computers and Operations Research*, vol. 22, no. 1, 1995.

4. M. Gen, G. S. Wasserman, and A. E. Smith, guest editors, special issue on *Genetic Algorithms and Industrial Engineering, International Journal of Computers and Industrial Engineering*, vol. 30, no. 4, 1996.

5. M. Gen and A. E. Smith, guest editors, special issue on *Intelligent Engineering Design, International Journal of Engineering Design and Automation*, vol. 3, no. 3, 1997.

6. T. Ibaraki, guest editor, special issue on *Combinatorial Optimization, The Transactions of the Institute of Electrical Engineers of Japan*, vol. 114-C, no. 4, 1994 (in Japanese).

7. J. Watada, editor-in-chief, special issue on *Fuzzy and Genetic Algorithms, Journal of Japan Society for Fuzzy Theory and Systems*, vol. 7, no. 5, 1995 (in Japanese).

8. H. Hirata and J. Miyamichi, guest editors, special issue on *New Approach to Combinatorial Problems and Scheduling Problems, Transactions of the Society of Instrument and Control Engineers of Japan*, vol. 31, no. 5, 1995 (in Japanese).

1.6.4 Public-Accessible Internet Service for Genetic Algorithm Information

Because of the popularity of the Internet, there are many sites where various genetic algorithm information is open to the public through anonymous-FTP or WWW server. Among them, *GA Archives* is the most famous one, which is available by access to the following site:

http://www.aic.nrl.navy.mil/galist/

The *GA Archives* is a repository for information related to research in genetic algorithms. Available from this site are past issues of the GA-List digest, source codes for many genetic algorithm implementations, and announcements about genetic-algorithm-related conferences. Also, links are given to many interesting sites around the world with material related to evolutionary computation. This archive is maintained by the Navy Center for Applied Research in Artificial Intelligence.

The major contents are as follows:

- Calendar of EC-related events
- GA-List archives: back issues, source code, information

- Links to other genetic-algorithm-related information
- Links to EC research groups

Another important one in the Internet is the ENCORE: the EvolutioNary COmputation REpository network. ENCORE contains a documented collection (i.e., compilation) of electronically available resources related to the field of evolutionary computation. It is divided into several categories that reflect the current main research paradigms: genetic algorithms, evolution strategies, evolutionary programming, genetic programming, and classifier systems.

ENCORE is not located at a single location but is instead a network of servers carrying the same information. You thus can choose your nearest EClair (a single node in the ENCORE network):

The EClair at EUnet Deutschland GmbH:
   ftp://ftp.Germany.EU.net/pub/research/softcomp/EC/
The EClair at the Santa Fe Institutte:
   ftp://alife.santafe.edu/pub/USER-AREA/EC/
The EClair at the Chinese University of Hong Kong:
   ftp://ftp.dcs.warwick.ac.uk/pub/mirrors/EC/
The EClair at the California Institute of Technology:
   ftp://ftp.krl.caltech.edu/pub/EC/
The EClair at Wayne State University, Detroit:
   ftp://ftp.cs.wayne.edu/pub/EC/
The EClair at the Michigan State University:
   ftp://ftp.egr.msu.edu/pub/EC/
The EClair at the University of Capetown, South Africa:
   ftp://ftp.uct.ac.za/pub/mirrors/EC/
The EClair at Technical University of Berlin, Germany:
   ftp://ftp-bionik,fb10.tu-berlin.de/pub/EC/
The EClair at Purdue University, West Lafayette, USA:
   ftp://coast.cs.purdue.edu/pub/EC/
The EClair at the University of Oviedo, Spain:
   ftp://zeus.etsimo.uniovi.es/pub/EC/
The EClair at CEFET-PR, Curitiba, Brazil:
   ftp://ftp.cefetpr.br/pub/EC/

2

CONSTRAINED OPTIMIZATION PROBLEMS

Optimization plays a central role in operations research/management science and engineering design problems. Optimization deals with problems of minimizing or maximizing a function with several variables usually subject to equality and/or inequality constraints. Optimization techniques have had an increasingly great impact on our society. Both the number and variety of their applications continue to grow rapidly, and no slowdown is in sight.

However, many engineering design problems are very complex in nature and difficult to solve with conventional optimization techniques. In recent years, genetic algorithms have received considerable attention regarding their potential as a novel optimization technique. In this chapter, we will discuss the applications of genetic algorithms to unconstrained optimization, nonlinear programming, stochastic programming, goal programming, and interval programming.

2.1 UNCONSTRAINED OPTIMIZATION

Unconstrained optimization deals with the problem of minimizing or maximizing a function in the absence of any restrictions [28]. In general, an unconstrained optimization problem can be mathematically represented as follows [279]:

$$\min \quad f(x)$$
$$\text{s.t.} \quad x \in \Omega$$

where f is a real-valued function and Ω, the feasible set, is a subset of E^n. When attention is restricted to the case where $\Omega = E^n$, it corresponds to the completely unconstrained case. In this section we consider the case where Ω is some particularly simple subset of E^n, and then we can direct our attention to genetic algorithms techniques.

A point $x* \in \Omega$ is said to be a *local minima* of f over Ω if there is an $\varepsilon > 0$ such that $f(x) \geq f(x*)$ for all $x \in \Omega$ within a distance ε of $x*$. A point $x* \in \Omega$ is said to be a *global minima* of f over Ω if $f(x) \geq f(x*)$ for all $x \in \Omega$. The necessary conditions for local minima are based on the differential calculus of f, that is, the *gradient* of f defined as follows:

$$\nabla f(x) = \left[\frac{\partial f(x)}{\partial x_1}, \ \frac{\partial f(x)}{\partial x_2}, \ \ldots, \ \frac{\partial f(x)}{\partial x_n} \right]$$

and the *Hessian* of f at x denoted as $\nabla^2 f(x)$ or $\mathbf{F}(x)$ is defined as

$$\mathbf{F}(x) = \left[\frac{\partial^2 f(x)}{\partial x_i \partial x_j} \right]$$

Even though most practical optimization problems have side restrictions that must be satisfied, the study of techniques for unconstrained optimization provides a basis for the further studies. In this section, we will discuss how to solve the unconstrained optimization problem with genetic algorithms. In subsequent sections we can see that the method used here can be extended in a natural way to provide and motivate the solution procedures for constrained problems.

2.1.1 Ackley's Function

Ackley's function is a continuous and multimodal test function obtained by modulating an exponential function with a cosine wave of moderate amplitude [2, 11]. Its topology, as shown in Figure 2.1, is characterized by an almost flat outer region and a central hole or peak where modulations by the cosine wave become more and more influential. Ackley's function is as follows:

$$\min f(x_1, x_2) = -c_1 \cdot \exp\left(-c_2 \sqrt{\frac{1}{2} \sum_{j=1}^{2} x_j^2} \right) - \exp\left(\frac{1}{2} \sum_{j=1}^{2} \cos(c_3 \cdot x_j) \right)$$

$$+ c_1 + e \qquad -5 \leq x_j \leq 5, \qquad j = 1, 2$$

where $c_1 = 20, c_2 = 0.2, c_3 = 2\pi$, and $e = 2.71282$. The known optimal solution is $(x_1^*, x_2^*) = (0, f(x_1^*, x_2^*) = 0$.

As Ackley pointed out, this function causes moderate complications to the search, since though a strictly local optimization algorithm that performs hill-

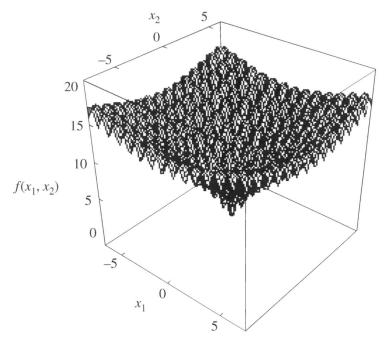

Figure 2.1. Ackley's function.

climbing would surely get trapped in a local optimum, a search strategy that scans a slightly bigger neighborhood would be able to cross intervening valleys toward increasingly better optima. Therefore, Ackley's function provides one of the reasonable test cases for genetic search.

2.1.2 Genetic Algorithm Approach for Minimization of Ackley's Function

To minimize Ackley's function, we simply use the following implementation of the genetic algorithm:

1. Real number encoding
2. Arithmetic crossover
3. Nonuniform mutation
4. Top *pop_size* selection

The arithmetic crossover is defined as the combination of two chromosomes v_1 and v_2 as follows [287]:

$$v'_1 = v_1 + (1 - \lambda)v_2$$
$$v'_2 = v_2 + (1 - \lambda)v_1$$

where $\lambda \in (0, 1)$.

The nonuniform mutation is given as follows [287]: For a given parent v, if the element x_k of it is selected for mutation, the resulting offspring is $v' = [x_1, \ldots, x'_k, \ldots, x_n]$, where x'_k is randomly selected from two possible choices:

$$x'_k = x_k + \Delta(t, x_k^U \quad x_k) \text{ or}$$
$$x'_k = x_k - \Delta(t, x_k - x_k^L)$$

where x_k^U and x_k^L are the upper and lower bounds for x_k. The function $\Delta(t, y)$ returns a value in the range $[0, y]$ such that the value of $\Delta(t, y)$ approaches to 0 as t increases (t is the generation number) as follows:

$$\Delta(t, y) = y \cdot r \cdot \left(1 - \frac{t}{T}\right)^b$$

where r is a random number from $[0, 1]$, T is the maximal generation number, and b is a parameter determining the degree of nonuniformity.

Top *pop_size* selection produces the next generation by selecting the best *pop_size* chromosomes from parents and offspring. For this case, we can simply use the values of objective as fitness values and sort chromosomes according to these values.

The parameters of the genetic algorithm are set as follows:

pop_size	10
maxgen	1000
p_m	0.1
p_c	0.3

The initial population is randomly created within the region $[-5, 5]$ as follows:

$$
\begin{array}{ccc}
 & x_1 & x_2 \\
v_1 = [& 4.954222, & 0.169225] \\
v_2 = [& -4.806207, & -1.630757] \\
v_3 = [& 4.672536, & -1.867275] \\
v_4 = [& 1.897794, & -0.196387] \\
v_5 = [& -2.127598, & 0.750603] \\
v_6 = [& -3.832667, & -0.959655] \\
v_7 = [& -3.792383, & 4.064608] \\
v_8 = [& 1.182745, & -4.712821] \\
v_9 = [& 3.812220, & -3.441115] \\
v_{10} = [& -4.515976, & 4.539171]
\end{array}
$$

The corresponding fitness function values are

$$
\begin{aligned}
eval(v_1) &= f(4.954222, 0.169225) = 10.731945 \\
eval(v_2) &= f(-4.806207, -1.630757) = 12.110259 \\
eval(v_3) &= f(4.672536, -1.867275) = 11.788221 \\
eval(v_4) &= f(1.897794, -0.196387) = 5.681900 \\
eval(v_5) &= f(-2.127598, 0.750603) = 6.757691 \\
eval(v_6) &= f(-3.832667, -0.959655) = 9.194728 \\
eval(v_7) &= f(-3.792383, 4.064608) = 11.795402 \\
eval(v_8) &= f(1.182745, -4.712821) = 11.559363 \\
eval(v_9) &= f(3.812220, -3.441115) = 12.279653 \\
eval(v_{10}) &= f(-4.515976, 4.539171) = 14.251764
\end{aligned}
$$

Now we are ready to apply the crossover operation to the chromosome. Let the sequence of random numbers be

$$
\begin{array}{ccccc}
0.828211 & 0.199683 & 0.639149 & 0.629170 & 0.957427 \\
0.149358 & 0.304788 & 0.058504 & 0.149693 & 0.326670
\end{array}
$$

This means that the chromosomes v_2, v_6, v_8, and v_9 were selected for crossover. The offspring were generated as follows:

$$v_2 = [-4.806207, \ -1.630757]$$
$$v_6 = [-3.832667, \ -0.959655]$$
$$\downarrow$$
$$v_1' = [-4.444387, \ -1.383817]$$
$$v_2' = [-4.194488, \ -1.206594]$$

and

$$v_8 = [\ \ 1.182745, \ -4.712821]$$
$$v_9 = [\ \ 3.812220, \ -3.441115]$$
$$\downarrow$$
$$v_3' = [\ \ 3.683262, \ -4.521950]$$
$$v_4' = [\ \ 1.311703, \ -3.631985]$$

The mutation then is performed. Because there are a total $2 \times 10 = 20$ genes in whole population, we generate a sequence of random numbers r_k ($k = 1, \ldots, 20$) from the range $[0, 1]$. The corresponding gene to be mutated is

bit_pos	chrom_num	variable	random_num
11	6	x_1	0.081393

and the resulting offspring is as follows:

$$v_5' = [-4.068506, \ -0.959655]$$

The fitness value for each offspring is as follows:

$$eval(v_1') = f(-4.444387, \ -1.383817) = 11.927451$$
$$eval(v_2') = f(-4.194488, \ -1.206594) = 10.566867$$
$$eval(v_3') = f(\ \ 3.683262, \ -4.521950) = 13.449167$$
$$eval(v_4') = f(\ \ 1.311703, \ -3.631985) = 10.538330$$
$$eval(v_5') = f(-4.068506, \ -0.959655) = \ \ 9.083240$$

The best 10 chromosomes among parents and offspring form a new population as follows:

$$
\begin{array}{lrr}
 & x_1 & x_2 \\
\boldsymbol{v}_1 = [& 4.954222, & 0.169225] \\
\boldsymbol{v}_2 = [& 1.311703, & -3.631985] \\
\boldsymbol{v}_3 = [& 4.672536, & -1.867275] \\
\boldsymbol{v}_4 = [& 1.897794, & -0.196387] \\
\boldsymbol{v}_5 = [& -2.127598, & 0.750603] \\
\boldsymbol{v}_6 = [& -3.832667, & -0.959655] \\
\boldsymbol{v}_7 = [& -3.792383, & 4.064608] \\
\boldsymbol{v}_8 = [& 1.182745, & -4.712821] \\
\boldsymbol{v}_9 = [& -4.194488, & -1.206594] \\
\boldsymbol{v}_{10} = [& -4.068506, & -0.959655]
\end{array}
$$

The corresponding fitness values of variables $[x_1, x_2]$ are as follows:

$$
\begin{aligned}
eval(\boldsymbol{v}_1) &= f(\ 4.954222, \ \ 0.169225) = 10.731945 \\
eval(\boldsymbol{v}_2) &= f(\ 1.311703, \ -3.631985) = 10.538330 \\
eval(\boldsymbol{v}_3) &= f(\ 4.672536, \ -1.867275) = 11.788221 \\
eval(\boldsymbol{v}_4) &= f(\ 1.897794, \ -0.196387) = \ 5.681900 \\
eval(\boldsymbol{v}_5) &= f(-2.127598, \ \ 0.750603) = \ 6.757691 \\
eval(\boldsymbol{v}_6) &= f(-3.832667, \ -0.959655) = \ 9.194728 \\
eval(\boldsymbol{v}_7) &= f(-3.792383, \ \ 4.064608) = 11.795402 \\
eval(\boldsymbol{v}_8) &= f(\ 1.182745, \ -4.712821) = 11.559363 \\
eval(\boldsymbol{v}_9) &= f(-4.194488, \ -1.206594) = 10.566867 \\
eval(\boldsymbol{v}_{10}) &= f(-4.068506, \ -0.959655) = \ 9.083240
\end{aligned}
$$

Now we just completed one iteration of the genetic procedure (one generation). At the 1000th generation, we have the following chromosomes:

$$
\begin{array}{lrr}
 & x_1 & x_2 \\
\boldsymbol{v}_1 = [& -0.000002, & -0.000000] \\
\boldsymbol{v}_2 = [& -0.000002, & -0.000000] \\
\boldsymbol{v}_3 = [& -0.000002, & -0.000000] \\
\boldsymbol{v}_4 = [& -0.000002, & -0.000000] \\
\boldsymbol{v}_5 = [& -0.000002, & -0.000000]
\end{array}
\qquad
\begin{array}{lrr}
 & x_1 & x_2 \\
\boldsymbol{v}_6 = [& -0.000002, & -0.000000] \\
\boldsymbol{v}_7 = [& -0.000002, & 0.000000] \\
\boldsymbol{v}_8 = [& -0.000002, & -0.000000] \\
\boldsymbol{v}_9 = [& -0.000002, & -0.000000] \\
\boldsymbol{v}_{10} = [& -0.000002, & -0.000000]
\end{array}
$$

The fitness value is $f(x_1^*, x_2^*) = -0.005456$.

2.2   NONLINEAR PROGRAMMING

Nonlinear programming (or constrainted optimization) deals with the problem of optimizing an objective function in the presence of equality and/or inequality constraints [452, 454, 465]. Nonlinear programming is an extremely important tool used in almost every area of engineering, operations research, and mathematics because many practical problems cannot be successfully modeled as a linear program. The general nonlinear programming may be written as follows:

$$\max \quad f(\boldsymbol{x}) \tag{2.1}$$

$$\text{s.t.} \quad g_i(\boldsymbol{x}) \leq 0, \qquad i = 1, 2, \ldots, m_1 \tag{2.2}$$

$$h_i(\boldsymbol{x}) = 0, \qquad i - m_1 + 1, \ldots, m(= m_1 + m_2) \tag{2.3}$$

$$\boldsymbol{x} \in X \tag{2.4}$$

where $f, g_1, g_2, \ldots, g_{m_1}, h_{m_1+1}, h_{m_1+2}, \ldots, h_m$ are real valued functions defined on E^n, X is a subset of E^n, and $\boldsymbol{x}$ is an n-dimensional real vector with components $x_1, x_2, \ldots, x_n$. The above problem must be solved for the values of the variables $x_1, x_2, \ldots, x_n$ that satisfy the restrictions and meanwhile minimize the function f. The function f is usually called the *objective function* or *criterion function*. Each of the constraints $g_i(\boldsymbol{x}) \leq 0$ is called an *inequality constraint*, and each of the constraints $h_i(\boldsymbol{x}) = 0$ is called an *equality constraint*. The set X might typically include lower and upper bounds on the variables, which is usually called *domain constraint*. A vector $\boldsymbol{x} \in X$ satisfying all the constraints is called a *feasible solution* to the problem. The collection of all such solutions forms the *feasible region*. The nonlinear programming problem then is to find a feasible point $\bar{x}$ such that $f(x) \leq f(\bar{x})$ for each feasible point $\boldsymbol{x}$. Such a point is called an *optimal solution*. Unlike linear programming problems, the conventional solution methods for nonlinear programming are very complex and not very efficient. In the past few years, there has been a growing effort to apply genetic algorithms to the nonlinear programming problem [354]. This section shows how to solve the nonlinear programming problem with genetic algorithms in general.

2.2.1   Handling Constraints

The central problem for applying genetic algorithms to the constrained optimization is how to handle constraints because genetic operators used to manipulate the chromosomes often yield infeasible offspring. Recently, several techniques have been proposed to handle constraints with genetic algorithms [248, 288, 308, 322]. Michalewicz has published a very good survey on this problem [289, 291]. The existing techniques can be roughly classified as follows:

- Rejecting strategy
- Repairing strategy

- Modifying genetic operators strategy
- Penalizing strategy

Each of these strategies have advantages and disadvantages.

Rejecting Strategy. Rejecting strategy discards all infeasible chromosomes created throughout the evolutionary process. This is a popular option in many genetic algorithms. The method may work reasonably well when the feasible search space is convex, and it constitutes a reasonable part of the whole search space. However, such an approach has serious limitations. For example, for many constrained optimization problems where the initial population consists of infeasible chromosomes only, it might be essential to improve them. Moreover, quite often the system can reach the optimum more easily if it is possible to "cross" an infeasible region (especially in nonconvex feasible search spaces).

Repairing Strategy. Repairing a chromosome involves taking an infeasible chromosome and generating a feasible one through some repairing procedure. For many combinatorial optimization problems, it is relatively easy to create a repairing procedure. Liepins and collaborators have shown, through an empirical test of genetic algorithms performance on a diverse set of constrained combinatorial optimization problems, that the repair strategy did indeed surpass other strategies in both speed and performance [274, 275].

Repairing strategy depends on the existence of a deterministic repair procedure to convert an infeasible offspring into a feasible one. The weakness of the method is in its problem dependence. For each particular problem, a specific repair algorithm should be designed. Also, for some problems, the process of repairing infeasible chromosomes might be as complex as solving the original problem.

The repaired chromosome can be used either for evaluation only, or it can also replace the original one in population. Liepins and coworkers [274, 275] took the *never replacing* approach (that is, the repaired version is never returned to the population), while Nakano and Yamada [313] took *always replacing* approach. Recently, Orvosh and Davis reported a so-called 5% rule: This heuristic rule states that in many combinatorial optimization problems, genetic algorithms with a repairing procedure provide the best result when 5% of repaired chromosomes replace their infeasible originals [322]. Michalewicz et al. reported that 15% replacement rule is a clear winner for numerical optimization problems with nonlinear constraints [291].

Modifying Genetic Operator Strategy. One reasonable approach for dealing with the issue of feasibility is to invent problem-specific representation and specialized genetic operators to maintain the feasibility of chromosomes. Michalewicz et al. have pointed out that often such systems are much more reliable than any other genetic algorithms based on the penalty approach. This is a quite popular trend: Many practitioners use problem-specific representation and specialized operators in building very successful genetic algorithms in many areas [291]. However, the genetic search of this approach is confined within the feasible region.

Penalty Strategy. These strategies above have the advantage that they never generate infeasible solutions but have the disadvantage that they consider no points outside the feasible regions. For highly constrained problem, infeasible solutions may take a relatively big portion of the population. In such case, feasible solutions may be difficult to be found if we just confine genetic search within feasible regions. Glover and Greenberg have suggested that constraint management techniques allowing movement through infeasible regions of the search space tend to yield more rapid optimization and produce better final solutions than do approaches limiting search trajectories only to feasible regions of the search space [170]. The *penalizing strategy* is such kind of techniques proposed to consider infeasible solutions in genetic search.

2.2.2   Penalty Function

The penalty technique is perhaps the most common technique used to handle infeasible solutions in the genetic algorithms for constrained optimization problems. In essence, this technique transforms the constrained problem into an unconstrained problem by penalizing infeasible solutions, in which a penalty term is added to the objective function for any violation of the constraints.

The basic idea of the penalty technique is borrowed from conventional optimization. It is a nature question: is there any difference when we use the penalty method in conventional optimization and in genetic algorithms? In conventional optimization, the penalty technique is used to generate a sequence of infeasible points whose limit is an optimal solution to the original problem. The major concern is how to choose a proper value of penalty so as to hasten convergence and avoid premature termination. In genetic algorithms, the penalty technique is used to keep a certain amount of infeasible solutions in each generation so as to enforce genetic search towards an optimal solution from both sides of feasible and infeasible regions. We do not simply reject the infeasible solutions in each generation because some may provide much more useful information about optimal solution than some feasible solutions. The major concern is how to determine the penalty term so as to strike a balance between the information preservation (keeping some infeasible solutions) and the selective pressure (rejecting some infeasible solutions), and void both under-penalty and over-penalty.

In general, solution space contains two parts: feasible area and infeasible area. We do not make any assumption about these subspaces; in particular, they need be neither convex nor connected as shown in Figure 2.2. Handling infeasible chromosomes is far from trivial. From the figure we can know that infeasible solution *b* is much near to optima *a* than infeasible solution *d* and feasible

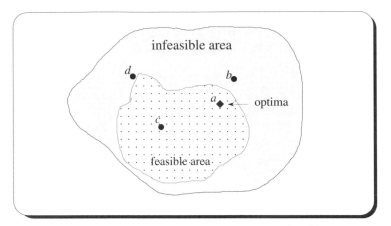

Figure 2.2. Solution space: feasible area and infeasible area.

solution c. We may hope to give less penalty to b than to d even though it is a little farther from the feasible area than d. We also can believe that b contains much more information about optima than c even though it is infeasible. However, we have no *a priori* knowledge about optima, so generally it is very hard for us to judge which one is better than others. The main issue of penalty strategy is how to design the penalty function $p(x)$ which can effectively guide genetic search toward the promising area of solution space. The relationship between infeasible chromosome and the feasible part of the search space plays a significant role in penalizing infeasible chromosomes: The penalty value corresponds to the "amount" of its infeasibility under some measurement. There is no general guideline on designing penalty function, and constructing an efficient penalty function is quite problem-dependent.

Evaluation Function with Penalty Term. Penalty techniques transform the constrained problem into an unconstrained problem by penalizing infeasible solutions. In general, there are two possible ways to construct the evaluation function with penalty term. One is to take the addition form expressed as follows:

$$eval(x) = f(x) + p(x) \tag{2.5}$$

where x represents a chromosome, $f(x)$ the objective function of problem, and $p(x)$ the penalty term. For maximization problems, we usually require that

$$\begin{aligned} p(x) &= 0 && \text{if } x \text{ is feasible} \\ p(x) &< 0 && \text{otherwise} \end{aligned} \tag{2.6}$$

Let $|p(x)|_{\max}$ and $|f(x)|_{\min}$ be the maximum of $|p(x)|$ and minimum of $|f(x)|$

among infeasible solutions in current population, respectively. We also require that

$$|p(x)|_{max} \leq |f(x)|_{min} \qquad (2.7)$$

to avoid negative fitness value. For minimization problems, we usually require that

$$\begin{array}{ll} p(x) = 0 & \text{if } x \text{ is feasible} \\ p(x) > 0 & \text{otherwise} \end{array} \qquad (2.8)$$

The second way is to take the multiplication form expressed as follows:

$$eval(x) = f(x)p(x) \qquad (2.9)$$

In this case, for maximization problems we require that

$$\begin{array}{ll} p(x) = 1 & \text{if } x \text{ is feasible} \\ 0 \leq p(x) < 1 & \text{otherwise} \end{array} \qquad (2.10)$$

and for minimization problems we require that

$$\begin{array}{ll} p(x) = 1 & \text{if } x \text{ is feasible} \\ p(x) > 1 & \text{otherwise} \end{array} \qquad (2.11)$$

Note that for the minimization problems, the fitter chromosome has the lower value of $eval(x)$. For some selection methods, it is required to transform the objective values into fitness values in such way that the fitter one has the larger fitness value.

Classification on Penalty Function. Several handling infeasibility techniques have been proposed in the area of genetic algorithms. In general, we can classify them into two classes:

- Constant penalty
- Variable penalty

The constant penalty approach is known to be less effective for the complex problem, and most recent studies pay particular attention to the variable penalty.
    In general, the variable penalty approach contains two components:

- Variable penalty ratio
- Penalty amount for the violation of constraints

The variable penalty ratio can be adjusted according to

- The degree of violation of constraints
- The iteration number of genetic algorithms

The first approach increases the penalty pressure as the violation becomes severe, which leads to the class of *static penalty*, and the second approach increases the penalty pressure along with the growing of evolutionary process, which leads to the class of *dynamic penalty* as discussed by Michalewicz [289].

Essentially, penalty is a function of the distance from feasible area. This can be given in three possible ways:

- The function of absolute distance of a single infeasible solution
- The function of relative distance of all infeasible solutions in current population
- The function of the adaptive penalty term

Most methods take the first approach. For highly constrained problem, the ratio of infeasible to feasible solutions is relatively high at each generation. In such cases, the second and third approaches are hopeful to make a good balance between the preservation of information and the pressure for infeasibility.

The penalty approaches can be further distinguished as

- Problem-dependent
- Problem-independent

Most penalty techniques belong to the class of problem-dependent approach.

The penalty approaches also can be distinguished as

- With parameter
- Without parameter

Most penalty techniques belong to the class of parameterized approach. It seems that the parameterized penalty functions tend to be problem-dependent.

Below, we introduce several penalty function methods used in genetic algorithms for solving nonlinear programming problems.

Homaifar, Qi, and Lai's Method. Homaifar et al. have considered the following nonlinear programming problem [222]:

$$\min \quad f(x)$$
$$\text{s.t.} \quad g_i(x) \geq 0, \qquad i = 1, 2, \ldots, m$$

and take the addition form as the evaluation function:

$$eval(x) = f(x) + p(x)$$

The penalty function is constructed with two components: (1) variable penalty factor and (2) penalty for the violation of constraints as follows:

$$p(x) = \begin{cases} 0 & \text{if } x \text{ is feasible} \\ \sum_{i=1}^{m} r_i g_i^2(x) & \text{otherwise} \end{cases} \tag{2.12}$$

where r_i is a variable penalty coefficient for the ith constraint. For each constraint, they create several levels of violation. Depending on the level of violation, r_i varies accordingly. However, determining the level of violation for each constraint and choosing suitable values of r_i is not an easy task and is problem-dependent.

Recent experiments of Michalewicz indicated that the quality of solution heavily depends on the values of these penalty coefficients [288]. If the penalty coefficients are moderate, the algorithm may converge to an infeasible solution; on the other hand, if the penalty coefficients are too large, the method is equivalent to rejecting strategy.

Joines and Houck's Method. Joines and Houck considered the following nonlinear programming problem [237]:

$$\begin{aligned} \min \quad & f(x) \\ \text{s.t.} \quad & g_i(x) \geq 0, \quad i = 1, 2, \ldots, m_1 \\ & h_i(x) = 0, \quad i = m_1 + 1, \ldots, m(= m_1 + m_2) \end{aligned}$$

and took the addition form as the evaluation function:

$$eval(x) = f(x) + p(t, x)$$

The penalty function is also constructed with (1) a variable penalty factor and (2) a penalty for the violation of constraints, as follows:

$$p(t, x) = \rho_t^{\alpha} \sum_{i=1}^{m} d_i^{\beta}(x) \tag{2.13}$$

where t is the iteration of genetic algorithm, and α and β are parameters used to adjust the scale of penalty value. The penalty term for single constraint $d_i(x)$

and the variable penalty factor ρ_t are given as follows:

$$d_i(x) = \begin{cases} 0 & \text{if } x \text{ is feasible} \\ |g_i(x)| & \text{otherwise for } 1 \leq i \leq m_1 \\ |h_i(x)| & \text{otherwise for } m_1 + 1 \leq i \leq m \end{cases} \tag{2.14}$$

$$\rho_t = C \times t \tag{2.15}$$

where C is a constant. The penalty on infeasible chromosomes is increased along with the evolutionary process due to the term of ρ_t. The variable penalty factor varies with the iteration of the genetic algorithm for Joines and Houck's method, while it varies according to the level of violation for Homaifar, Qi, and Lai's method.

The experiment result of Joines and Houck indicated that the quality of the solution was very sensitive to the values of the three parameters. How to determine the variable penalty factor is problem-dependent, and it is necessary to design the component with a proper dynamic property suitable for a given problem because this component is constantly increased along with the growth of generations, thereby giving infeasible chromosomes the *death penalty* at the later generation of genetic algorithms. In most of the experiments of Michalewicz, the method converged in early generations due to this reason [289].

Michalewicz and Attia's Method. Michalewicz and Attia considered the following nonlinear programming problem [289]:

$$\begin{aligned} \min \quad & f(x) \\ \text{s.t.} \quad & g_i(x) > 0, \quad i = 1, 2, \ldots, m_1 \\ & h_i(x) = 0, \quad i = m_1 + 1, \ldots, m(= m_1 + m_2) \end{aligned}$$

and took the addition form as the evaluation function:

$$eval(x) = f(x) + p(\tau, x)$$

The penalty function is also constructed with (1) a variable penalty factor and (2) a penalty for the violation of constraints, as follows:

$$p(\tau, x) = \frac{1}{2\tau} \sum_{i \in A} d_i^2(x) \tag{2.16}$$

where A is the set of active constraints, which consists of all nonlinear equations and violated nonlinear inequalities. A constraint $g_i(x)$ is violated at point x if and only if $g_i(x) > \delta$ ($i = 1, \ldots, m_1$), where δ is a parameter to decide whether a constraint is active. τ is the variable penalty component, called *temperature*. The penalty term for single constraint $d_i(x)$ is given as follows:

$$d_i(x) = \begin{cases} \max\{0, g_i(x)\} & \text{for } 1 \leq i \leq m_1 \\ |h_i(x)| & \text{for } m_1 + 1 \leq i \leq m \end{cases} \qquad (2.17)$$

They built a system called Genocop II with the technique [290]. Note that Genocop II is realized with the mechanism of SA-alike rather than GA. The variable component τ, beginning with a *starting* temperature τ_0 and ending at a *freezing* temperature τ_f, decreases in steps of the main loop according to a given cooling scheme. Genocop I is embedded in the main loop used to find an improved point. Within each execution of Genocop I, the temperature τ is fixed as a constant. The method is quite sensitive to the values of the parameters. The question of how to settle these parameters for a particular problem remains open.

Smith, Tate, and Coit's Method. The adapting penalty function was first proposed by Smith and Tate [382]; this function can alter the magnitude of the penalty dynamically by scaling according to the fitness of the best solution yet found. As better feasible and infeasible solutions are found, the penalty imposed on a given infeasible solution will change. Coit and Smith [82] further extended their previous studies and proposed the concept of *near-feasibility threshold* (NFT). Exterior penalty functions are characterized as being nondecreasing functions of the "distance" of a given solution from the feasible region. The NFT is the threshold distance from the feasible region at which the user would consider the search as "getting warm." The penalty function will encourage the genetic algorithm to explore within the feasible region and the NFT neighborhood of the feasible region and will discourage search beyond that threshold.

These authors considered the following nonlinear programming problem:

$$\begin{aligned} \max \quad & f(x) \\ \text{s.t.} \quad & g_i(x) \leq b_i, \qquad i = 1, 2, \ldots, m \end{aligned}$$

and took the addition form as the evaluation function:

$$eval(x) = f(x) + p(x)$$

The penalty function is constructed with (1) relative penalty coefficients for the violation of constraints and (2) an adaptive penalty term, as follows:

$$p(x) = - \sum_{i=1}^{m} \left(\frac{\Delta b_i(x)}{\Delta b_i^{\text{nef}}} \right)^{\alpha} (f_{\text{all}}^* - f_{\text{feas}}^*) \tag{2.18}$$

where α is a parameter which is used to adjust the severity of the penalty function. $\Delta b_i(x)$ is the value of violation for constraint i. Δb_i^{nef} is the NFT for constraint i. How to give a proper nef is problem-dependent. f_{feas}^* is the objective function value of the best feasible solution yet found, and f_{all}^* is the unpenalized objective function value of the best overall solution yet found.

Note that the adaptive term $(f_{\text{all}}^* - f_{\text{feas}}^*)$ may cause this penalty approach two kinds of danger: (1) zero-penalty and (2) over-penalty. For the case that $f_{\text{all}}^* = f_{\text{feas}}^*$, even though there exist infeasible solutions, this approach will give zero-penalty to all infeasible solutions; for the case that an infeasible one with very large value f_{all}^* occurred at the early stage of evolutionary process, this approach will give over-penalty to all infeasible solutions.

Yokota, Gen, Ida, and Taguchi's Method. Yokota, Gen, Ida, and Taguchi considered the same nonlinear programming problem as Smith, Tate, and Coit did, and they took the multiplication form as the evaluation function [428]:

$$eval(x) = f(x)p(x)$$

The penalty function is constructed as follows:

$$p(x) = 1 - \frac{1}{m} \sum_{i=1}^{m} \left(\frac{\Delta b_i(x)}{b_i} \right)^{\alpha} \tag{2.19}$$

$$\Delta b_i(x) = \max \{0, g_i(x) - b_i\} \tag{2.20}$$

where $\Delta b_i(x)$ is the value of violation for constraint i. This penalty function can be viewed as a special case of Smith, Tate, and Coit's method where the NFT for constraint i is set to be as $\Delta b_i^{\text{nef}} = b_i$. Thus we obtain a penalty that is relatively milder than that of Smith, Tate, and Coit's method. Note that the penalty function is designed with the nonparameterized approach and is problem-independent.

Gen and Cheng's Method. Gen and Cheng further refined their work in order to give a much severer penalty to infeasible solutions [151]. Let x be a chromosome in the current population $P(t)$. The penalty function is constructed as follows:

$$p(x) = 1 - \frac{1}{m} \sum_{i=1}^{m} \left(\frac{\Delta b_i(x)}{\Delta b_i^{\max}} \right)^{\alpha} \tag{2.21}$$

where

$$\Delta b_i(x) = \max \{0, g_i(x) - b_i\} \tag{2.22}$$

$$\Delta b_i^{\max} = \max \{\epsilon, \Delta b_i(x) | x \in P(t)\} \tag{2.23}$$

where $\Delta b_i(x)$ is the value of violation for constraint i for the chromosome x, $\Delta b_i^{\max}$ is the maximum of violation for constraint i among current population, and ϵ is a small positive number used to avoid penalty from zero-division. For highly constrained optimization problems, the infeasible solutions take a relatively large portion among population at each generation. This penalty approach adjusts the ratio of penalties adaptively at each generation in order to achieve a balance between the preservation of information and the selective pressure for infeasibility and avoidance of over-penalty.

2.2.3 Genetic Operators

For most applications of genetic algorithms to constrained optimization problems, the *real coding* technique is used to represent a solution to a given problem. In real coding implementation, each chromosome is encoded as a vector of real numbers, of the same lengths as the solution vector. Such coding is also known as *floating point representation* [287], *real number representation* [101], or *continuous representation* [302]. For example, for the optimization problem (2.1)–(2.4), the real vector $x = [x_1, x_2, \ldots, x_n]$ is used as the chromosome to represent a solution.

In recent years, several genetic operators have been proposed for such codings, which can be roughly classified into three classes:

- Conventional operators
- Arithmetical operators
- Direction-based operators

The conventional operators are made by extending the operators for binary representation into the real coding case. The arithmetic operators are constructed by borrowing the concept of linear combination of vectors from the area of convex sets theory. The direction-based operators are formed by introducing the approximate gradient (subgradient) direction or negative direction into genetic operators.

Conventional Operators

Simple Crossover. This kind of crossover operators is analogous to that of the binary implementation [101]. The basic one is *one-cut-point crossover*. Let two parents be $x = [x_1, x_2, \ldots, x_n]$ and $y = [y_1, y_2, \ldots, y_n]$. If they are crossed after the kth position, the resulting offspring are

$$x' = [x_1, x_2, \ldots, x_k, y_{k+1}, y_{k+2}, \ldots, y_n]$$
$$y' = [y_1, y_2, \ldots, y_k, x_{k+1}, x_{k+2}, \ldots, x_n]$$

The further generations of one-point crossover are *two-cut-point, multi-cut-point,* and *uniform* crossover. For details, see Spears and De Jong [385] and Syswerda [391].

Random Crossover. Essentially, this kind of crossover operator creates off-spring randomly within a hyper-rectangle defined by the parent points. The basic one is given by Radcliffe [343], called *flat crossover*, which produces an offspring by uniformly picking a value for each gene from the range formed by the values of two corresponding parents' genes. Eshelman and Schaffer presented a generalized crossover of Radcliffe's work [121], called *blend crossover*, to introduce more variance. It uniformly picks values that lie between two points containing the two parents.

Mutation. Mutation operators are quite different from the traditional one. A gene, being a real number, is mutated in a specific range. The basic one is *uniform mutation*. This one simply replaces a gene (real number) with a randomly selected real number within a specified range. Let a chromosome to be mutated be $x = [x_1, x_2, \ldots, x_n]$. We first select a random number $k \in [1, n]$ and then produce an offspring $x' = [x_1, \ldots, x'_k, \ldots, x_n]$, where x'_k is a random value (uniform probability distribution) from the range $[x_k^L, x_k^U]$. The values of x_k^L and x_k^U are typically the lower and upper bounds on the variable x_k, which can be determined by *domain constraint*. This range also can be calculated dynamically from the set of constraints (inequalities) [287].

The gene x'_k can be replaced by either x_k^L or x_k^U, each with equal probability. This kind variation is called *boundary mutation* [292]. Instead of the lower and upper bounds, this range also can be formed as $[x_{k-1}, x_{k+1}]$. This kind variation is called *plain mutation* [234].

Arithmetical Operators

Crossover. The basic concept of this kind of operator is borrowed from the convex set theory [26]. Generally, the weighted average of two vectors x_1 and x_2 is calculated as follows:

$$\lambda_1 x_1 + \lambda_2 x_2 \tag{2.24}$$

If the multipliers are restricted as

$$\lambda_1 + \lambda_2 = 1, \qquad \lambda_1 > 0, \qquad \lambda_2 > 0$$

the weighted form (2.24) is known as *convex combination*. If the nonnegativity condition on the multipliers is dropped, the combination is known as an *affine combination*. Finally, if the multipliers are simply required to be in real space E^1, the combination is known as a *linear combination*.

Similarly, arithmetic operators are defined as the combination of two vectors (chromosomes) as follows:

$$x_1' = \lambda_1 x_1 + \lambda_2 x_2$$
$$x_2' = \lambda_1 x_2 + \lambda_2 x_1$$

According to the restriction on multipliers, it yields three kinds of crossovers, which can be called *convex crossover*, *affine crossover*, and *linear crossover*.

The convex crossover may be the most commonly used one [287]. When restricting that $\lambda_1 = \lambda_2 = 0.5$, it yields a special case, which is usually called *average crossover* by Davis [101], or *intermediate crossover* by Schwefel [372].

The affine crossover was first examined by Wright [421] for a special case where the multipliers were restricted as

$$\lambda_1 = 1.5 \quad \text{and} \quad \lambda_2 = -0.5$$

Another case of affine crossover was considered by Mühlenbein and Schlierkamp-Voosen [302], called *extended intermediate crossover*, where one multiplier was given as a random real number in the interval $[-d, 1 + d]$.

The linear crossover (we use the term in strict sense as distinguished from convex and affine crossover) was first introduced by Cheng and Gen [67]. They restricted the multipliers as follows:

$$\lambda_1 + \lambda_2 \leq 2, \qquad \lambda_1 > 0, \qquad \lambda_2 > 0$$

Let us see a geometric explanation about arithmetic operators. For two parents x_1 and x_2, the collection of all convex combination of them is called *convex hull*. Similarly, we can define the *affine hull* as the collection of all affine combination and define the *linear hull* as the collection of all linear combination. Figure 2.3 shows them in the simple case in two-dimensional space. The offspring generated with convex crossover lie within the solid line, the offspring from affine crossover lie in both solid and dashed lines, and the offspring from linear crossover lie in whole space.

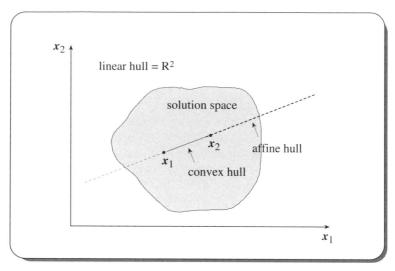

Figure 2.3. Illustration showing convex hull, affine hull, and linear hull.

One may be concerned that the affine and linear crossovers will yield infeasible offspring. This is not a critical problem right now because one can use very efficient penalty functions throughout the genetic algorithm search, which allows infeasible solutions to be engaged in evolutionary process. The key point is how to determine the proper values for multipliers. Generally, when applying affine and linear crossovers to a particular problem, the absolute values of multipliers should be restricted to be below an upper bound according to the domain constraint in order to force genetic search within a reasonable area.

Dynamic Mutation. This operator, also called *nonuniform mutation*, is given by Janilow and Michalewicz [234]. It is designed for fine-tuning capabilities aimed at achieving high precision. For a given parent x, if the element x_k of it is selected for mutation, the resulting offspring is $x' = [x_1, \ldots, x'_k, \ldots, x_n]$, where x'_k is randomly selected from following two possible choices:

$$x'_k = x_k + \Delta(t, x_k^U - x_k) \quad \text{or} \quad x'_k = x_k - \Delta(t, x_k - x_k^L)$$

The function $\Delta(t, y)$ returns a value in the range $[0, y]$ such that the value of $\Delta(t, y)$ approaches 0 as t increases (t is the generation number). This property causes the operator to search the space uniformly initially (when t is small) and very locally at later stages. The function $\Delta(t, y)$ is given as follows:

$$\Delta(t, y) = y \cdot r \cdot \left(1 - \frac{t}{T}\right)^b$$

where r is a random number from $[0, 1]$, T is the maximal generation number, and b is a parameter determining the degree of nonuniformity. It is possible for the operator to generate an offspring which is not feasible. In such a case, we can reduce the value of random number r.

Direction-Based Operators. With the genetic operators discussed above, there is no guarantee that the offspring are better than their parents. For direction-based operators, problem-specific knowledge is introduced into genetic operation in order to produce improved offspring.

Crossover. Direction-based crossover uses the values of objective function in determining the direction of genetic search [292]. The operator generates a single offspring x' from two parents x_1 and x_2 according to the following rule:

$$x' = r \cdot (x_2 - x_1) + x_2$$

where r is a random number between 0 and 1. It also assumes that the parent x_2 is not worse than x_1; that is, $f(x_2) \geq f(x_1)$ for maximization problems and $f(x_2) \leq f(x_1)$ for minimization problems.

Mutation. Direction-based mutation is given by Gen, Liu, and co-workers [163, 164]. It is well known that Taylor's expansion of a continuous differentiable function f is

$$f(x + \Delta x) \simeq f(x) + (\nabla f(x + \theta \Delta x))^T \Delta x, \qquad x \in R^n$$

where $0 \leq \theta \leq 1$, $\nabla f(x)$ denotes the gradient of the function f at the point x, and Δx is a small perturbation in R^n. For the maximizing problem, it may be better to choose the gradient direction as the mutation direction; for the minimizing problem, it may be better to choose the negative gradient direction as the mutation direction. Because we do not require that the problem have many good mathematical properties as the conventional methods do, we can simply calculate the ith component of the gradient approximately by

$$\frac{f(x_1, \ldots, x_i + \Delta x_i, \ldots, x_n) - f(x_1, \ldots, x_i, \ldots, x_n)}{\Delta x_i}$$

where Δx_i is a small real number. Let d be the approximate direction determined by this method; the offspring after mutation would be

$$x' = x + r \cdot d$$

where r is a random nonnegative real number, which is selected in such a way that the offspring is a feasible solution.

The direction can be given randomly as a free direction to avoid the chromosomes jamming into a *corner*. If the chromosome is near the boundary, the mutation direction given by some criteria might point toward the close boundary, and then jamming could occur. In such a case we can choose a random direction instead of the given direction.

2.2.4   Numerical Examples

Example 2.1. Consider the following problem:

$$\text{min}\quad f(x) = (x_1 - 2)^2 + (x_2 - 1)^2$$
$$\text{s.t.}\quad g_1(x) = x_1 - 2x_2 + 1 = 0$$
$$g_2(x) = \frac{x_1^2}{4} - x_2^2 + 1 \geq 0$$

This problem was originally given by Bracken and McCormick [47]. Homaifar, Qi, and Lai solved it with genetic algorithms [222]. The penalty function is given by $P(x) = r_1 g_1(x) + r_2 g_2(x)$, where r_1 and r_2 are penalty factors (for details refer to Homaifar et al. [222]). The GA parameters are chosen as follows: *chromosome length* $= 19, pop\_size = 400, p_c = 0.85$, and $p_m = 0.02$. The optimal solution is reached after 100 generations, and the precision is 0.0284. The solutions are given in Table 2.1. The results show that both constraints are satisfied by the genetic algorithm solutions, where GRG stands for generalized reduced gradient method [146].

Example 2.2. Consider the following problem:

Table 2.1. Solutions of Numerical Experimentation

Items	Reference Solution	Genetic Algorithm Solution	GRG Solution
$f(x)$	1.393	1.4339	1.3934
x_1	0.823	0.8080	0.8229
x_2	0.911	0.88544	0.9115
$g_1(x)$	1.0×10^{-3}	3.7×10^{-2}	1.00×10^{-4}
$g_2(x)$	7.46×10^{-4}	0.052	-5.18×10^{-5}

$$\min \quad f(x) = 5.3578547x_3^2 + 0.8356891x_1x_5$$
$$+ 37.293239x_1 - 40792.141$$
$$\text{s.t.} \quad 0 \leq 85.334407 + 0.0056858x_2x_5 + 0.00026x_1x_4$$
$$- 0.0022053x_3x_5 \leq 92$$
$$90 \leq 80.51249 + 0.0071317x_2x_5 + 0.0029955x_1x_2$$
$$+ 0.0021813x_3^2 \leq 110$$
$$20 \leq 9.300961 + 0.0047026x_3x_5 + 0.0012547x_1x_3$$
$$+ 0.0019085x_3x_4 \leq 25$$
$$78 \leq x_1 \leq 102$$
$$33 \leq x_2 \leq 45$$
$$27 \leq x_3 \leq 45$$
$$27 \leq x_4 \leq 45$$
$$27 \leq x_5 \leq 45$$

This problem is an interesting nonlinear optimization problem provided by Himmelblau [215]. For this problem, there are five independent variables, six nonlinear inequality constraints, and ten boundary conditions. Also, note that x_2 and x_4 are not included explicitly in the objective function. Homaifar, Qi, and Lai solved it with genetic algorithms [222]. The genetic algorithm parameters are selected as follows: chromosome length = 19, $pop\_size = 400$, $p_c = 0.8, p_m = 0.088$. The results are shown in Table 2.2; they indicate that the solution based on local reference is slightly better than the global reference. The boundary conditions are all satisfied by both methods.

Example 2.3. Consider the following problem:

Table 2.2. Solutions of Numerical Experimentation

Items	Reference Solution	Genetic Algorithm Solution Based on Global Reference	Genetic Algorithm Solution Based on Local Reference	GRG Solution
$f(x)$	−30665.5	−30175.804	−30182.269	−30373.950
x_1	78.00	80.61	81.49	78.62
x_2	33.00	34.21	34.09	33.44
x_3	29.995	31.34	31.24	31.07
x_4	45.00	42.05	42.20	44.18
x_5	36.776	34.85	34.37	35.22

$$\min \quad f(\boldsymbol{x}) = \sum_{j=1}^{10} x_j \left(c_j + \ln \frac{x_j}{x_1 + \cdots + x_{10}} \right)$$

$$\text{s.t.} \quad x_1 + 2x_2 + 2x_3 + x_6 + x_{10} = 2$$
$$x_4 + 2x_5 + x_6 + x_7 = 1$$
$$x_3 + x_7 + x_8 + 2x_9 + x_{10} = 1$$
$$x_i \geq 0.000001 \qquad (i = 1, \ldots, 10)$$

where

$$c_1 = -6.089, \quad c_2 = -17.164, \quad c_3 = -34.054, \quad c_4 = -5.914, \quad c_5 = -24.721$$
$$c_6 = -14.986, \quad c_7 = -24.100, \quad c_8 = -10.708, \quad c_9 = -26.662, \, c_{10} = -22.179$$

This problem was originally given by Hock and Schittkowski [219]. Michalewicz, Logan, and Swaminathan solved it with genetic algorithms [292]. The previously best-known solution was

$$\boldsymbol{x}* = [0.01773548, 0.08200180, 0.8825646, 0.0007233256, 0.4907851,$$
$$0.0004335469, 0.01727298, 0.007765639, 0.01984929, 0.05269826]$$

and $f(\boldsymbol{x}*) = -47.707579$. Michalewicz et al. found points with better value than the one above in all ten runs:

$$\boldsymbol{x}* = [0.04034785, 0.15386976, 0.77497089, 0.00167479, 0.48468539,$$
$$0.00068965, 0.02826479, 0.01849179, 0.03849563, 0.10128126]$$

for which the value of the objective function is equal to -47.760765. A single run of 500 iterations took 11 sec of CPU time.

Example 2.4. Consider the following problem:

$$\min \quad f(x, y) = 6.5x - 0.5x^2 - y_1 - 2y_2 - 3y_3 - 2y_4 - y_5$$
$$\text{s.t.} \quad x + 2y_1 + 8y_2 + y_3 + 3y_4 + 5y_5 \leq 16$$
$$-8x - 4y_1 - 2y_2 + 2y_3 + 4y_4 - y_5 \leq -1$$
$$2x + 0.5y_1 + 0.2y_2 - 3y_3 - y_4 - 4y_5 \leq 24$$
$$0.2x + 2y_1 + 0.1y_2 - 4y_3 + 2y_4 + 2y_5 \leq 12$$
$$-0.1x - 0.5y_1 + 2y_2 + 5y_3 - 5y_4 + 3y_5 \leq 3$$
$$y_3 \leq 1, \quad y_4 \leq 1, \quad \text{and} \quad y_5 \leq 2$$
$$x \geq 0, \quad y_i \geq 0 \qquad (i = 1, 2, \ldots, 5)$$

This problem was originally given by Floudas and Pardalos [127]. Michalewicz, Logan, and Swaminathan solved it with genetic algorithms [292]. The global solution is $[x, y^*] = [0, 6, 0, 1, 1, 0]$, and $f(x, y^*) = -11.005$. Michalewicz et al. approached the optimum quite closely in all ten runs; a typical optimum point found was

$$[0.000000, 5.976089, 0.005978, 0.999999, 1.000000, 0.000000]$$

for which the value of the objective function is equal to -10.988042. A single run of 1000 iterations took 29 sec of CPU time.

Example 2.5. Consider the following problem:

$$\min \quad f(x) = f_1(x) + f_2(x) + f_3(x)$$

$$\text{s.t.} \quad \frac{x_1}{\sqrt{3}} - x_2 \geq 0$$

$$-x_1 - \sqrt{3}x_2 + 6 \geq 0, \qquad 0 \leq x_1 \leq 6 \quad \text{and} \quad x_2 \geq 0$$

where

$$f_1(x) = x_2 + 10^{-5}(x_2 - x_1)^2 - 1.0, \qquad 0 \leq x_1 \leq 2$$

$$f_2(x) = \frac{1}{27\sqrt{3}}((x_1 - 3)^2 - 9)x_2^3, \qquad 2 \leq x_1 \leq 4$$

$$f_3(x) = \frac{1}{3}(x_1 - 2)^3 + x_2 - \frac{11}{3} \qquad 4 \leq x_1 \leq 6$$

This problem was originally given by Hock and Schittkowski [219]. Michalewicz, Logan, and Swaminathan solved it with genetic algorithms [292]. The problem was constructed from three separate problems. The function f has three global solutions:

$$x_1^* = [0, 0], \quad x_2^* = [3, \sqrt{3}], \quad \text{and} \quad x_3^* = [4, 0]$$

In all cases, $f(x_i^*) = -1, i = 1, 2, 3$. Michalewicz et al. performed three separate experiments. As a result, the global solution for the first experiment was $x_1^* = [0, 0]$, the global solution for the second experiment was $x_2^* = [3, \sqrt{3}]$, and the global solution for the third experiment was $x_3^* = [4, 0]$. They found global optima in all runs in all three cases; a single run of 500 iterations took 9 sec of CPU time.

2.3   STOCHASTIC OPTIMIZATION

One of the common problems in the practical application of mathematical programming is the difficulty for determining the proper values of model parameters. The true values of these parameters may not become known until after a solution has been chosen and implemented. This can sometimes be attributed solely to the inadequacy of investigation. However, the values of these parameters are often influenced by random events that are impossible to predict; that is, some or all of the model parameters may be *random variables*. What is needed is a way of formulating the problem so that the optimization will directly take the *uncertainty* into account. One such approach for mathematical programming under uncertainty is *stochastic programming* [214, 239].

2.3.1.   Mathematical Model

Let us now focus on stochastic programming. We know that a function $f(x, \xi)$, where x is a vector of real variables and ξ is a vector of random variables, is a random variable for any given x. Since it is meaningless to maximize a random variable, it is natural to use its expected value $E[f(x, \xi)]$ to replace it. Thus the stochastic programming problem can be generally written as follows:

$$\max \quad E[f(x, \xi)] \tag{2.25}$$

$$\text{s.t.} \quad E[g_i(x, \xi)] \leq 0, \qquad i = 1, 2, \ldots, m_1 \tag{2.26}$$

$$E[h_i(x, \xi)] = 0, \qquad i = m_1 + 1, \ldots, m(= m_1 + m_2) \tag{2.27}$$

where E denotes the expected operators, $x = [x_1, x_2, \ldots, x_n]$ is an n-dimensional real vector, $\xi = [\xi_1, \xi_2, \ldots, \xi_l]$ is an l-dimensional stochastic vector, and f; $g_i, i = 1, 2, \ldots, m_1$; $h_i, i = m_1 + 1, \ldots, m$, are real-valued functions defined on R^{n+l}.

If the distribution and density functions of stochastic vector ξ are $\Phi(\xi)$ and $\phi(\xi)$, respectively, then we have

$$E[f(x, \xi)] = \int_{R^m} f(x, \xi) \, d\Phi(\xi) = \int_{R^m} f(x, \xi)\phi(\xi) \, d\xi \tag{2.28}$$

$$E[g_2(x, \xi)] = \int_{R^m} g_i(x, \xi) \, d\Phi(\xi) = \int_{R^m} g_i(x, \xi)\phi(\xi) \, d\xi,$$

$$i = 1, 2, \ldots, m_1 \tag{2.29}$$

$$E[h_i(x, \xi)] = \int_{R^m} h_i(x, \xi) \, d\Phi(\xi) = \int_{R^m} h_i(x, \xi)\phi(\xi) \, d\xi,$$

$$i = m_1 + 1, \ldots, m \tag{2.30}$$

Most of the effort required to solve this problem is spent in the multivariate integration. The domain of the integration can become unmanageable, especially when the dimension is high and/or the constraints are complicated.

2.3.2   Monte Carlo Simulation

Monte Carlo simulation is now widely used to solve certain problems in statistics that are not analytically tractable. Monte Carlo simulation is a scheme employing random numbers—that is, independent and identically distributed random variables—to approximate the solutions of problems [262]. Although this technique is referred to as *simulation*, it does not necessarily have any interpretation as the mechanical imitation of a real process. One of its standard applications is to evaluate the integral

$$I = \int_a^b g(x)\,dx$$

where $g(x)$ is a real-valued function that is not analytically integrable. To approach this deterministic problem by Monte Carlo simulation, let Y be the random variable $(b - a)g(X)$, where X is a continuous random variable distributed uniformly on $[a, b]$, denoted by $U(a, b)$. Then the expected value of Y is

$$E[Y] - E[(b - a)g(X)]$$
$$= (b - a)E[g(X)]$$
$$= (b - a)\int_a^b g(x)f_X(x)\,dx = \int_u^b g(x)\,dx = I$$

where $f_X(x) = 1/(b - a)$ is the probability density function of a $U(a, b)$ random variable. Thus the problem of evaluation of the integral has been reduced to one of estimating the expected value of $E(Y)$. In particular, we can estimate the value by the sample mean

$$\overline{Y}(n) = \frac{1}{n}\sum_{j=1}^n Y_j = \frac{b - a}{n}\sum_{j=1}^n g(X_j)$$

where $X_1, X_2, \ldots, X_n$ are independent and identically distributed $U(a, b)$ random variables. Furthermore, it can be shown that $\overline{Y}(n)$ is an unbiased estimator of I; that is, $E[\overline{Y}(n)] = I$, and $\mathrm{Var}[\overline{Y}(n)] = \mathrm{Var}[Y]/n$. Assuming that $\mathrm{Var}[Y]$ is finite, it follows that $\overline{Y}(n)$ will be arbitrarily close to I for sufficiently large n (with probability 1).

In practice, Monte Carlo simulation would probably not be used to evaluate a single integral, since there are more efficient numerical analysis techniques for this purpose. Now we show how to evaluate a stochastic integral. For example, we need to calculate the following stochastic integral:

$$E[f(x, \xi)] = \int_D f(x, \xi)\phi(\xi)\, d\xi$$

For a fixed vector x, let Y be the random variable $|D| f(x, \Xi)\phi(\Xi)$, where Ξ is a random vector uniformly distributed over a bounded domain $D \in R^m$ and $|D|$ is the volume of the bounded domain. It is easy to verify that the expectation $E[Y]$ is the value of the integral. We can estimate $E(Y)$ by the sample mean as follows:

$$\overline{Y}(n) = \frac{|D|}{n} \sum_{j=1}^{n} f(x, \Xi_j)\phi(\Xi_j)$$

where $\Xi_1, \Xi_2, \ldots, \Xi_n$ are independent random vectors uniformly distributed over the bounded domain. $\overline{Y}(n)$ is an unbiased estimator of the value of the integral and $\mathrm{Var}[\overline{Y}(n)] = \mathrm{Var}[Y]/n$. In contrast to the classical method of multiple integral by iteration, the number of sampling points required to obtain a given degree of accuracy is independent of the number of dimensions, which makes the Monte Carlo simulation very attractive for large numbers of dimensions.

Procedure: Monte Carlo Simulation

begin
objective ← 0;
for j ← 1 to *number_simulation* **do**
  Ξ_j ← *random_vector* ( );
  objective ← *objective* $+ f(x, \Xi_j)\phi(\Xi_j)$;
end
objective ← *objective* $\times |D|/$*number_simulation*;
end

For a more detailed discussion of Monte Carlo simulation, see Hammersley and Handscombe [205], Halton [203], Rubinstein [361], and Morgan [298].

2.3.3   Evolution Program for Stochastic Optimization Problems

Gen, Liu, and Ida have proposed an evolution program for the following stochastic optimization problem [163]:

$$\max \quad E[f(x, \xi)]$$
$$\text{s.t.} \quad g_i(x) \leq 0, \qquad i = 1, 2, \ldots, m$$

They termed their algorithm the *evolution program*. The concept of evolution program was first introduced by Michalewicz [287]. It is based entirely on the idea of genetic algorithms; the difference is that it suggests the use of any data structure suitable for a problem together with any set of meaningful genetic operators.

Representation and Initialization. For numerical optimization problems, floating point representation has many advantages over binary representation [287]. In floating point representation, each chromosome is encoded as a vector of floating numbers, with the same length as the vector of decision variables. For a given problem with n decision variables, we can use a vector $[x_1, x_2, \ldots, x_n]$ as a chromosome to represent a solution to the optimization problem.

Initialization procedure begins with a predetermined interior point V_0 in the constraint set and a large positive number M_0. Then it produces *pop_size* chromosomes by the following method:

Step 1. Select a direction d in R^n randomly.

Step 2. Let $M = M_0$. Make a new chromosome as $v_0 + M \cdot d$ if it is feasible for the constraints; otherwise, set M by a random number in $(0, M)$ until $v_0 + M \cdot d$ is feasible.

Step 3. Repeat the above steps *pop_size* times and produce *pop_size* initial feasible solutions.

A feasible solution for the inequality constraints can be found in finite iterations by step 2 since V_0 is an interior point and the value of random number M decreases along with iterations.

Evaluation Function. Goldberg [171] has argued that with a fitness-proportional reproduction scheme, at the beginning of the run the extraordinary individuals would take over a significant proportion of the finite population in a single generation; this is undesirable, because it is a leading cause of premature convergence. Late in the run, there may still be significant diversity within the population; however, the population average fitness may be close to the population best fitness, and the survival of the fittest becomes a random walk among the mediocre. To overcome these problems, some scale procedures have been proposed, such as the linear-fitness scaling scheme and the power-law scaling scheme [287]. Gen, Liu, and Ida presented an exponential-fitness scaling scheme, which is a combination of ranking and scaling schemes.

First, the objective function for each chromosome is evaluated with Monte Carlo simulation and then chromosomes are ranked so as to the worst one in position 1 and the best one in position *pop_size* according to their objective function values. For the maximizing problem, chromosomes are sorted with the ascending order of the values; while for the minimizing problem, chromosomes are sorted with the descending order of the values.

Second, three preference parameters p_1, p_0, and $p_2 (0 < p_1 < p_0 < p_2 < 1)$ are defined; these are used to determine three critical chromosomes with ranking u_1, u_0, and u_2, respectively, such that $u_1 = \lfloor p_1 \cdot pop\_size \rfloor, u_0 = \lfloor p_0 \cdot pop\_size \rfloor$, and $u_2 = \lfloor p_2 \cdot pop\_size \rfloor$.

Finally, the chromosome with the ranking u_1 is assigned a fitness value of $e^{-1} \approx 0.37$, the one with ranking u_0 is assigned a fitness value of 1, and the one with ranking u_2 is assigned a fitness value of $2 - e^{-1} \approx 1.63$. For a chromosome $\boldsymbol{v}$ with ranking u, the relation between the ranking u and the exponential fitness $eval(\boldsymbol{v})$ is given as follows:

$$eval(\boldsymbol{v}) = \begin{cases} \exp\left[-\dfrac{u - u_0}{u_1 - u_0}\right], & u < u_0 \\[4mm] 2 - \exp\left[-\dfrac{u - u_0}{u_2 - u_0}\right], & u \geq u_0 \end{cases} \tag{2.31}$$

which is shown in Figure 2.4.

Then a roulette wheel can be made by the fitness values $eval(\boldsymbol{v}_k), k = 1, 2, \ldots, pop\_size$, and new population can be generated by spanning a roulette wheel which is made by these fitness values.

Crossover. Arithmetical crossover, which is defined as a convex combination of two vectors [287], is adopted. If the constraint set is convex, this operation ensures that both children are feasible if both parents are. For each pair of parents $\boldsymbol{v}_1$ and $\boldsymbol{v}_2$, the crossover operator will produce two offspring $\boldsymbol{v}'$ and $\boldsymbol{v}''$ as follows:

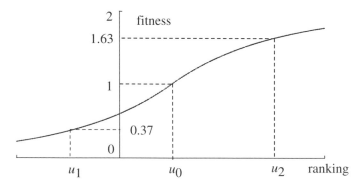

Figure 2.4. Exponential fitness.

$$v' = c_1 \cdot v_1 + c_2 \cdot v_2$$
$$v'' = c_2 \cdot v_1 + c_1 \cdot v_2$$

where $c_1, c_2 \geq 0$ and $c_1 + c_2 = 1$.

Mutation. Mutation operators use the same procedure as initialization to mutate chromosomes in a free direction. Let a parent be $v = [x_1, x_2, \ldots, x_n]$ and let a randomly generated direction of mutation be d. An offspring is made as follows:

$$v' = v + M \cdot d$$

If the offspring is not feasible, then set M by a random real number in $(0, M)$ until $v + M \cdot d$ is feasible.

Algorithm. The evolution program for stochastic optimization problems is summarized as follows:

Step 0. Parameter setting

> number of generations: max_gen
>
> population size: pop_size
>
> probability of crossover: p_c
>
> probability of mutation: p_m
>
> preference parameters: p_1, p_0, p_2
>
> a large positive number: M_0
>
> current generation: $gen \leftarrow 0$

Step 1. Initialization process

> give an interior point v_0;
> **for** $k \leftarrow$ **to** pop_size **do**
> $M \leftarrow M_0$;
>   produce a random direction d;
> $v_k \leftarrow v_0 + M \cdot d$;
>   **while**(v_k is not feasible) **do**
> $M \leftarrow random(M)$;
> $v_k \leftarrow v_0 + M \cdot d$;
>   **end**
> **end**

Step 2. Evaluation

> **for** $k \leftarrow 1$ **to** *pop_size* **do**
> compute the objectives f_k for $\boldsymbol{v}_k$ by Monte Carlo simulation;
> **end**
> rank chromosomes according to the objectives;
> **for** $k \leftarrow 1$ **to** *pop_size* **do**
> compute the exponential-fitness $eval(\boldsymbol{v}_k)$ based on the ranking;
> **end**

Step 3. Selection operation

> **for** $k \leftarrow 1$ **to** *pop_size* **do**
> compute selective probabilities $p_k = eval(\boldsymbol{v}_k)/\Sigma_{j=1}^{pop-size} eval(\boldsymbol{v}_j)$;
> **end**
> **for** $k \leftarrow 1$ **to** *pop_size* **do**
> compute the cumulative probabilities $q_k = \Sigma_{j=1}^{k} p_j$;
> **end**
> **for** $k \leftarrow 1$ **to** *pop_size* **do**
> **if** $q_{k-1} < random(\ ) \leq q_k$ **then**
> select $\boldsymbol{v}_k$;
> **end**
> **end**

Step 4. Crossover operation

> **for** $k \leftarrow 1$ **to** *pop_size*/2 **do**
> **if** $random(\ ) \leq p_c$ **then**
> $j \leftarrow random(pop\_size)$;
> $l \leftarrow random(pop\_size)$;
> $\alpha \leftarrow random(\ )$;
> $\boldsymbol{v}' \leftarrow \alpha\boldsymbol{v}_j + (1-\alpha)\boldsymbol{v}_l$;
> $\boldsymbol{v}'' \leftarrow \alpha\boldsymbol{v}_l + (1-\alpha)\boldsymbol{v}_j$;
> **end**
> **end**

Step 5. Mutation operation

> **for** $k \leftarrow$ **to** *pop_size* **do**
> **if** $(random(\ ) \leq p_m)$ **then**
> $M \leftarrow M_0$;
> produce a random direction $\boldsymbol{d}$;
> $\boldsymbol{v}'_k \leftarrow \boldsymbol{v}_k + M \cdot \boldsymbol{d}$;

while (v_k is not feasible) **do**
    $M \leftarrow random(M)$;
    $v'_k \leftarrow v_k + M \cdot d$;
end
end
end

Step 6. Termination test

$gen \leftarrow gen + 1$;
if ($gen < max\_gen$) **then**
    goto step 2;
else
    stop;
end

where *random*( ) means to return a random real number in (0, 1) and *random(num)* means to return a random real number in (0, *num*).

Example 2.6. Consider the following stochastic optimization problem with three decision variables and three stochastic variables:

$$\max \quad E[f(x, \xi)] = \int_{R^3} [(x_1 - \xi_1) \cdot \sin(\pi x_1) + (x_2 - \xi_2) \cdot \sin(4\pi x_2)$$
$$+ (x_3 - \xi_3) \cdot \sin(10\pi x_3)]\phi(\xi_1, \xi_2, \xi_3) \, d\xi_1 d\xi_2 d\xi_3$$
$$\text{s.t.} \quad 0 \le x_i \le 5, \quad i = 1, 2, 3$$

where $\phi(\xi_1, \xi_2, \xi_3)$ is the joint normal density function which has the following form:

$$\phi(\xi_1, \xi_2, \xi_3) = (2\pi)^{-3/2}(\det C)^{1/2}\exp\left\{-\frac{1}{2}\sum_{i=1}^{3}\sum_{j=1}^{3} c_{ij}(\xi_i - u_i)(\xi_j - u_j)\right\}$$

(2.32)

where $u_1 = 1, u_2 = 2, u_3 = 3$, and the positive definite matrix $C = [c_{ij}]$ is given by

$$c_{ij} = \begin{cases} 2, & i = j \\ 0, & i \ne j \end{cases}$$

(2.33)

and det C is the determinant of C. The best solution found by the genetic algorithm is $[x_1, x_2, x_3] = [4.5065, 4.6279, 0.1459]$ with the objective 8.9906.

2.4   NONLINEAR GOAL PROGRAMMING

In recent years, genetic algorithms have received a great deal of attention regarding their potential as optimization techniques for multicriteria optimization problems. Schaffer did the pioneering experiments in the area and proposed the *vector evaluated genetic algorithm* (VEGA) [369]. Tamaki, Mori, and Araki proposed the Pareto solution scheme for VEGA [396]. Horn, Hafpliotis, and Goldberg proposed the *niched Pareto genetic algorithm* [223]. Fonseca and Fleming described a rank-based fitness assignment method for multiple objective genetic algorithm [135]. Osyczka and Kundu proposed a distance method for the multiple objective genetic algorithm [324]. Tanino, Tanaka, and Hojo proposed an interactive approach for the multiple objective genetic algorithm [407]. Recently, Tamaki, Kita, and Kobayashi reviewed multiobject optimization by genetic algorithms [459].

Sakawa, Kato, and Shibano combined genetic algorithms and fuzzy programming technique to solve multiobjective 0-1 programming problems [364, 455, 456]. Gen and Liu investigated the application of genetic algorithms to solve the nonlinear goal programming problem [162].

Goal programming is one of the powerful techniques for solving multicriteria optimization problems. The concept of goal programming (GP) was developed by Charnes and Cooper [55]. Since then, many researchers have devoted their efforts to this area, such as Ijiri [230], Lee [268], Ignizio [329], Gen and Ida [153, 154], and Schniederjan [457]. The basic approach of goal programming is to establish a specific numeric goal for each objective, formulate an objective function for each objective, and then seek a solution that minimizes the (weighted) sum of deviations of these objective functions from their respective goals. Goal programming has been applied in a wide variety of real-world problems.

There are two kinds of goal programming problems. One is *nonpreemptive goal programming*, where all of the goals are of *roughly comparable importance*. The other is *preemptive goal programming*, where there is a hierarchy of priority levels for the goals, so that the goals of primary importance receive first-priority attention, those of secondary importance receive second-priority attention, and so forth. The term goal programming today usually refers to the preemptive approach. It has become an essential and widely recognized approach for solving multiple criteria decision-making problems.

Preemptive nonlinear goal programming, or simply nonlinear goal programming, is a very important type of engineering design problem involving nonlinear objectives and nonlinear constraints. The techniques of nonlinear goal programming were discussed by Ignizio [228], Hwang and Masud [224], and Weistroffer [412]. Four major approaches of nonlinear goal programming are reviewed and discussed by Saber and Ravindran [362]:

- Simplex-based approach
- Direct search approach
- Gradient search approach
- Interactive approach

Since the nonlinear goal programming problems can have different structures and degrees of nonlinearity, the performance of nonlinear goal programming techniques varies according to the nonlinear goal programming problem to be solved with respect to reliability, precision, convergence, and preparational and computational efforts. How to select a nonlinear goal programming technique is problem-dependent.

2.4.1 Formulation of Nonlinear Goal Programming

Nonlinear goal programming is one of the mathematical programming techniques developed to solve the problems with conflicting nonlinear objectives and nonlinear constraints wherein the user provides levels or targets of achievement for each objective and prioritizes the order in which the goals have to be achieved. It finds an optimal solution that satisfies as many of the goals as possible in the order specified. The generalized formulation of nonlinear goal programming is given as follows:

$$\min \quad z_0 = \sum_{k=1}^{q} \sum_{i=1}^{m_0} P_k(w_{ki}^+ d_i^+ + w_{ki}^- d_i^-) \tag{2.34}$$

$$\text{s.t.} \quad f_i(\boldsymbol{x}) + d_i^- - d_i^+ = b_i, \qquad i = 1, 2, \ldots, m_0 \tag{2.35}$$

$$g_i(\boldsymbol{x}) \leq 0, \qquad i = m_0 + 1, \ldots, \overline{m}_1(= m_0 + m_1) \tag{2.36}$$

$$h_i(\boldsymbol{x}) = 0, \qquad i - \overline{m}_1 + 1, \ldots, m(= \overline{m}_1 + m_2) \tag{2.37}$$

$$d_i^-, d_i^+ \geq 0, \qquad i = 1, 2, \ldots, m \tag{2.38}$$

where

P_k is the kth preemptive priority ($P_k \gg P_{k+1}$ for all k)

d_i^+ is the positive deviation variable representing the overachievements of goal i

d_i^- is the negative deviation variable representing the underachievements of goal i

w_{ki}^+ is the positive weight assigned to d_i^+ at priority P_k

w_{ki}^- is the negative weight assigned to d_i^- at priority P_k

$\boldsymbol{x}$ is n-dimensional decision vector

f_i is a function: $R^n \rightarrow R^1$ in goal constraints

g_i is a function: $R^n \rightarrow R^1$ in real inequality constraints

h_i is a function: $R^n \rightarrow R^1$ in real equality constraints

b_i is the target value or aspiration level of goal i

q is the number of priorities

m_0 is the number of goal constraints

m_1 is the number of real inequality constraints

m_2 is the number of real equality constraints

We sometimes rewrite the objective function (2.34) as follows:

$$\text{lexmin}\left\{ z_1 = \sum_{i=1}^{m_0} (w_{1i}^+ d_i^+ + w_{1i}^- d_i^-), \ldots, z_q = \sum_{i=1}^{m_0} (w_{qi}^+ d_i^+ + w_{qi}^- d_i^-) \right\} \quad (2.39)$$

where lexmin means lexicographically minimizing the objective.

2.4.2 Genetic Algorithms for Nonlinear Goal Programming

Gen and Liu [162] have proposed an alternative approach: a genetic algorithm for solving nonlinear goal programming problem because genetic algorithms belong to the class of problem-independent and probabilistic algorithms and can handle any kind of objective functions and any kind of constraints. Due to its evolutionary nature, genetic algorithms perform multiple directional and robust search in complex spaces by maintaining a population of potential solutions. Thus it provides us much ability to deal with much complex real-world nonlinear goal programming problem.

Representation and Initialization. For a given problem with n decision variables, we use a vector $[x_1, x_2, \ldots, x_n]$ as a chromosome to represent a solution to the problem. If equality constraints $h_i(x) = 0, i = 1, \ldots, r$ are linear, we can easily eliminate r variables by representing them with the remaining variables. In this case, a chromosome becomes $[x_1, x_2, \ldots, x_{n-r}]$.

The initialization procedure begins with a predetermined interior point V_0 in the constraint set and a large positive number M_0. Then we can produce *pop_size* chromosomes by the following method:

Step 1. Select a direction d in R^n randomly.

Step 2. Let $M = M_0$. Make a new chromosome represented as $v_0 + M \cdot d$ if it is feasible for the constraints; otherwise, set M by a random number in $(0, M)$ until $v_0 + M \cdot d$ is feasible.

Step 3. Repeat the above steps *pop_size* times and produce *pop_size* initial feasible solutions.

A feasible solution for the inequality constraints can be found in finite iterations by step 2 since $\boldsymbol{v}_0$ is an interior point and the value of random number M decreases along with iterations.

Evaluation Function. A kind of rank-based evaluation function is used to assess the merit of each chromosome. The evaluation procedure consists of mainly three steps:

Step 1. Calculate the objective values according to objective (2.39). There are q objective values associated with each chromosome; that is,

$$\left\{ \sum_{i=1}^{m_0} (w_{1i}^+ d_i^+ + w_{1i}^- d_i^-), \sum_{i=1}^{m_0} (w_{2i}^+ d_i^+ + w_{2i}^- d_i^-), \ldots, \sum_{i=1}^{m_0} (w_{qi}^+ d_i^+ + w_{qi}^- d_i^-) \right\}$$

Step 2. Sort chromosomes on the value of the first priority objective $\sum_{i=1}^{m_0} (w_{1i}^+ d_i^+ + w_{1i}^- d_i^-)$. If some chromosomes have the same value of the objective, then sort them on the second priority objective $\sum_{i=1}^{m_0} (w_{2i}^+ d_i^+ + w_{2i}^- d_i^-)$, and so forth. The tie is broken randomly. Thus chromosomes are rearranged in the ascending order according to the values of objectives.

Step 3. Assign each chromosome a rank-based fitness value. Let r_k be the rank of chromosome $\boldsymbol{v}_k$. For a parameter $a \in (0, 1)$ specified by user, the rank-based fitness function is defined as follows:

$$eval(\boldsymbol{v}_k) = a(1 - a)^{r_k - 1}$$

where $r_k = 1$ means the best chromosome and $r_k = pop\_size$ means the worst chromosome. We have

$$\sum_{k=1}^{pop\_size} eval(\boldsymbol{v}_k) \approx 1$$

The selection process is based on spinning a roulette wheel *pop_size* times; each time, a single chromosome is selected into a new population.

Crossover. Arithmetical crossover is adopted, in which it is defined as a convex combination of two vectors [287]. If the constraint set is convex, this operation ensures that both children are feasible if both parents are. For each pair of parents $\boldsymbol{v}_1$ and $\boldsymbol{v}_2$, the crossover operator will produce two children $\boldsymbol{v}'$ and $\boldsymbol{v}''$ as follows:

$$v' = c_1 \cdot v_1 + c_2 \cdot v_2$$
$$v'' = c_2 \cdot v_1 + c_1 \cdot v_2$$

where $c_1, c_2 \geq 0$ and $c_1 + c_2 = 1$.

Mutation. Mutation operators use the same procedure as initialization to mutate chromosomes in a free direction. Let a parent be $v = [x_1, x_2, \ldots, x_n]$ and let a randomly generated direction of mutation be d. An offspring is made as follows:

$$v' = v + M \cdot d$$

If the offspring is not feasible, then set M by a random number in $(0, M)$ until $v + M \cdot d$ is feasible.

2.4.3   Numerical Examples

A set of numerical examples will be presented with the following parameters: The population size, *pop_size*, is 30; the probability of crossover, p_c, is 0.2; the probability of mutation, p_m, is 0.6; and the parameter a in the rank-based evaluation function is 0.1.

Example 2.7. The first example is from Ignizio [228].

$$\begin{aligned}
\text{lexmin} \quad & \{d_3^+, 2d_1^- + d_2^+\} \\
\text{s.t.} \quad & x_1 x_2 + d_1^- - d_1^+ = 16 \\
& (x_1 - 3)^2 + x_2^2 + d_2^- - d_2^+ = 9 \\
& x_1 + x_2 + d_3^- - d_3^+ = 6
\end{aligned}$$

The best solution from a random run of genetic algorithm with 300 generations is

$$x = [3.01858, 2.98136]$$

which is close to the known solution [3, 3] because the former has objectives $d_3^+ = 0$ and $2d_1^- + d_2^+ = 14.0010$ and the latter has objectives $d_3^+ = 0$ and $2d_1^- + d_2^+ = 14.0000$.

Example 2.8. The second example, presented by Van de Panne and Popp [408] and also resolved by Lee and Olson [269], is a feed mixture problem which is to select four materials to mix in order to design a cattle feed mix subject to protein and fat constraints.

lexmin $\{d_1^- + d_2^- + d_3^- + d_4^-, d_5^-, d_6^+, d_7^-, d_8^+, d_9^-, d_{10}^+, d_{11}^-, d_{12}^+\}$

s.t. $x_1 + d_1^- - d_1^+ = 0.01$

$x_2 + d_2^- - d_2^+ = 0.01$

$x_3 + d_3^- - d_3^+ = 0.01$

$x_4 + d_4^- - d_4^+ = 0.01$

$12.0x_1 + 11.9x_2 + 41.8x_3 + 52.1x_4 - 0.841(0.2809x_1^2 + 0.1936x_2^2$
$\quad + 20.25x_3^2 + 0.6241x_4^2)^{1/2} + d_5^- - d_5^+ = 21$

$24.55x_1 + 26.75x_2 + 39.00x_3 + 40.50x_4 + d_6^- - d_6^+ = 30$

$12.0x_1 + 11.9x_2 + 41.8x_3 + 52.1x_4 - 1.282(0.2809x_1^2 + 0.1936x_2^2$
$\quad + 20.25x_3^2 + 0.6241x_4^2)^{1/2} + d_7^- - d_7^+ = 21$

$24.55x_1 + 26.75x_2 + 39.00x_3 + 40.50x_4 + d_8^- - d_8^+ = 29.90$

$12.0x_1 + 11.9x_2 + 41.8x_3 + 52.1x_4 - 1.645$
$\quad \cdot (0.2809x_1^2 + 0.1936x_2^2 + 20.25x_3^2 + 0.6241x_4^2)^{1/2} + d_9^- - d_9^+ = 21$

$24.55x_1 + 26.75x_2 + 39.00x_3 + 40.50x_4 + d_{10}^- - d_{10}^+ = 29.80$

$12.0x_1 + 11.9x_2 + 41.8x_3 + 52.1x_4 - 2.323(0.2809x_1^2 + 0.1936x_2^2$
$\quad + 20.25x_3^2 + 0.6241x_4^2)^{1/2} + d_{11}^- - d_{11}^+ = 21$

$24.55x_1 + 26.75x_2 + 39.00x_3 + 40.50x_4 + d_{12}^- - d_{12}^+ = 0$

$x_1 + x_2 + x_3 + x_4 = 1$

$2.3x_1 + 5.6x_2 + 11.1x_3 + 1.3x_4 \geq 5$

$x_1, x_2, x_3, x_4 \geq 0$

The equality constraint $x_1 + x_2 + x_3 + x_4 = 1$ is eliminated by replacing x_4 with $1 - (x_1 + x_2 + x_3)$. Thus a chromosome can be represented as $[x_1, x_2, x_3]$. The interior point is given as $[0.2, 0.3, 0.4]$, which is needed by initialization process. The best solution by genetic algorithms at a run with 600 generations is

$$x = [0.62742, 0.01001, 0.30914, 0.05343]$$

which is very close to $x = [0.62743, 0.01000, 0.30914, 0.054343]$ reported by Lee and Olson [269].

Example 2.9. This example is from El-Sayed, Ridgely, and Sandgren [118].

$$\min \quad z_0 = d_1^- + d_2^- + d_2^+ + d_3^- + d_3^+ + d_4^- + d_4^+$$

$$\text{s.t.} \quad 5.28x_1^2 + 3.74x_2^2 + 5.28x_3^2 + d_1^- - d_1^+ = 3.575$$

$$\frac{0.178}{x_1^2} + d_2^- - d_2^+ = 1.0$$

$$\frac{0.255}{x_2^2} + d_3^- - d_3^+ = 1.0$$

$$\frac{0.178}{x_3^2} + d_4^- - d_4^+ = 1.0$$

$$x_1, x_2, x_3 \geq 0$$

which is a three-bar structural optimization problem. The algorithm in El-Sayed et al. uses the linear goal programming techniques with successive linearization for the nonlinear equations to obtain the solution for the problem, and it yields a solution $x = [0.4239, 0.5047, 0.4239]$ with minimal objective value $z = 0.725$. The best solution of genetic algorithms at a run with 300 generations is

$$x = [0.4219, 0.6733, 0.4219], \qquad z_0 = 0.4376$$

which is much better than the result reported by El-Sayed et al. [118].

Example 2.10. Here we consider a complex nonlinear goal programming shown as follows:

$$\text{lexmin} \quad \{z_1 = d_1^-, z_2 = d_2^-, z_3 = d_3^+, z_4 = d_4^- + d_4^+\}$$

$$\text{s.t.} \quad x_1 \sin(\pi x_1) + x_2 \sin(2\pi x_2) + x_3 \sin(3\pi x_3) + x_4 \sin(4\pi x_4)$$
$$+ x_5 \sin(5\pi x_5) + d_1^- - d_1^+ = 18$$
$$x_1 \sin(\pi x_1) + x_2 \sin(2\pi x_2) + x_3 \sin(3\pi x_3) + x_4 \sin(4\pi x_4)$$
$$+ d_2^- - d_2^+ = 15$$
$$x_1 \sin(\pi x_1) + x_2 \sin(2\pi x_2) + x_3 \sin(3\pi x_3) + d_3^- - d_3^+ = 10$$
$$x_1 \sin(\pi x_1) + x_2 \sin(2\pi x_2) + d_4^- - d_4^+ = 0$$
$$x_1^2 + x_2^2 + x_3^2 + x_4^2 + x_5^2 \leq 100$$

This example is more complex than the former GP examples because the functions are multimodal; its one term $x \sin(5\pi x)$ is drawn in Figure 2.5. The functions made with the sum of $x \sin(i\pi x)$ are much more complex, so the traditional methods can do nothing about such an example. However, genetic algorithms are proven effective for this kind of GP problem. The best solution for a genetic

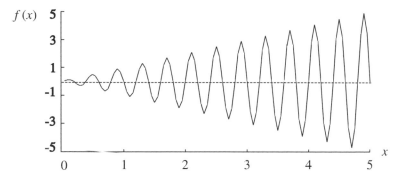

Figure 2.5. Graph of the function $f(x) = x \sin(5\pi x)$.

algorithm at a run with 6000 iterations is

$$x = [-1.860, 0.745, 6.823, 6.685, 1.995]$$

The objectives' variations with generations are shown in Figure 2.6.

2.5   INTERVAL PROGRAMMING

In the past decade, two different approaches have been proposed for interval optimization:

- Problem with interval variables and normal coefficients
- Problem with interval coefficients and normal variables

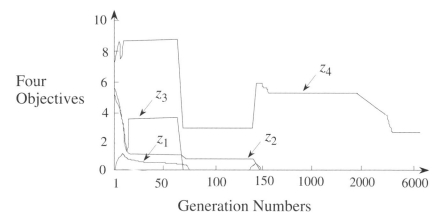

Figure 2.6. Four objectives in evolution process of the problem.

Interval analysis has been developed for the first-class problem [207], and several works have been reported which take a hybrid approach by combining interval analysis and genetic algorithms to deal with the global optimization problem [4, 306]. Interval programming techniques are developed for the second-class problem [232, 311]. When using mathematical programming methods to solve practical problems, it is usually not so easy for decision makers to determine the proper values of model parameters; on the contrary, such uncertainty can be roughly represented as an *interval of confidence*. This is the motivation for developing the interval programming technique.

In this section, we will show how to solve the interval programming problem with genetic algorithms. The basic idea is that we must first transform the interval programming model into an equivalent bicriteria programming model and then find out the Pareto solutions of the bicriteria programming problem using genetic algorithms.

2.5.1   Introduction

The following subsections will give a brief introduction to the basic definitions of interval arithmetic, the degree of inequality holding true, the order relation between intervals, and some related theorems.

Interval Arithmetic. An interval is defined as an ordered pair of real numbers as follows [6]:

$$A = [a^L, a^R]$$
$$= \{x | a^L \leq x \leq a^R; \ x \in R^1\} \tag{2.40}$$

where a^L and a^R are the left bound and right bound of interval A, respectively. The interval also can be defined as follows:

$$A = < a^C, a^W >$$
$$= \{x | a^C - a^W \leq x \leq a^C + a^W; \ x \in R^1\} \tag{2.41}$$

where a^C and a^W are the center and width of interval A, respectively. They are calculated as follows:

$$a^C = \tfrac{1}{2}(a^R + a^L) \tag{2.42}$$

$$a^W = \tfrac{1}{2}(a^R - a^L) \tag{2.43}$$

Figure 2.7 illustrates an interval A.

The generalization of ordinary arithmetic to closed intervals is known as interval arithmetic. For two intervals $A = [a^L, a^R]$ and $B = [b^L, b^R]$, the basic

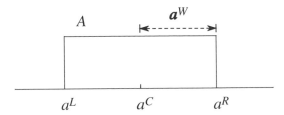

Figure 2.7. Illustration of an interval.

definitions of interval arithmetic are given as follows [6, 207]:

$$A + B = [a^L + b^L, a^R + b^R] \tag{2.44}$$

$$A - B = [a^L - b^R, a^R - b^L] \tag{2.45}$$

$$kA = \begin{cases} [ka^L, ka^R] & \text{if } k \geq 0 \\ [ka^R, ka^L] & \text{if } k < 0 \end{cases} \tag{2.46}$$

$$A \times B = [a^L \times b^L, a^R \times b^R] \qquad \text{if } a^L \geq 0, b^L \geq 0 \tag{2.47}$$

$$\frac{A}{B} = \left[\frac{a^L}{b^R}, \frac{a^R}{b^L} \right] \qquad \text{if } a^L \geq 0, b^L \geq 0 \tag{2.48}$$

$$\log(AB) = \log(A) + \log(B) \quad \text{if } a^L \geq 0, b^L \geq 0 \tag{2.49}$$

The proof for operation (2.49) can be found in Nakahara et al. [312].

Interval Inequality. Consider the following constraint with interval coefficients:

$$\sum_{j=1}^{n} A_j x_j \leq B \tag{2.50}$$

where

$$B = [b^L, b^R], A_j = [a_j^L, a_j^R], \qquad x_j \geq 0, j = 1, \ldots, n \tag{2.51}$$

How to determine the feasible region for the constraint is a fundamental problem for interval programming. Several definitions of feasible region have been proposed in past years. The definition given by Ishibuchi and Tanaka is based on the concept of degree of inequality holding true for two intervals [231].

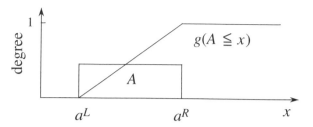

Figure 2.8. Illustration of the degree of inequality holding true.

Definition 2.1. For an interval A and a real number x, the degree for inequality $A \leq x$ holding true is given as follows:

$$g(A \leq x) = \max\left\{0, \min\left\{1, \frac{x - a^L}{a^R - a^L}\right\}\right\} \tag{2.52}$$

The definition is illustrated in Figure 2.8.

Definition 2.2. For two intervals A and B, the degree for inequality $A \leq B$ holding true is given as follows:

$$g(A \leq B) = g(A - B \leq 0) = \max\left\{0, \min\left\{1, \frac{b^R - a^L}{a^R - b^L + b^R - a^L}\right\}\right\} \tag{2.53}$$

According to Definition 2.2, the feasible region of interval constraint (2.50) can be determined by the following theorem:

Theorem 2.1 (Ishibuchi and Tanaka [231]). For a given degree of inequality holding true q, the interval inequality constraint (2.5) can be transformed equivalently into the following crisp inequality constraint:

$$\sum_{j=1}^{n} (q a_j^R x_j + (1 - q) a_j^L x_j) \leq (1 - q) b^R + q b^L \tag{2.54}$$

For more detailed discussion on the feasible region of interval constraint, see Nakahara, Sasaki, and Gen [311].

Order Relation between Intervals. Consider the following interval programming problem:

$$\max\left\{Z(x) = \sum_{j=1}^{n} C_j x_j \,\middle|\, x \in S \subset R_+^n\right\} \qquad (2.55)$$

where S is a feasible region of x and C_j is an interval coefficient which represents the uncertain unit profit from x_j. For a given x, the total profit $Z(x)$ is an interval. We need to make a decision based on such interval profits. The order relation of intervals is proposed to help us to do this, which represents the decision-maker's preference among interval profits. The following order relations are defined for the maximization problem:

Definition 2.3 For two intervals A and B, the order relation $\leq_{LR}$ is defined as follows:

$$A \leq_{LR} B \qquad \text{iff } a^L \leq b^L \text{ and } a^R \leq b^R \qquad (2.56)$$

The order relation represents the decision-maker's preference for the alternative with the higher minimum profit and higher maximum profit.

Definition 2.4 For two intervals A and B, the order relation $\leq_{CW}$ is defined as follows:

$$A \leq_{CW} B \qquad \text{iff } a^C \leq b^C \text{ and } a^W \leq b^W \qquad (2.57)$$

The order relation represents the decision-maker's preference for the alternative with the higher expected value and less uncertainty.

Definition 2.5 For two intervals A and B, the order relation $\leq_{LC}$ is defined as follows:

$$A \leq_{LC} B \qquad \text{iff } a^L \leq b^L \text{ and } a^C \leq b^C \qquad (2.58)$$

Note that the orders given above are partial order, which are transitive, reflexive, and antisymmetric.

Using Definition 2.5, the solution set of problem (2.55) is defined as the following nondominated solutions:

Definition 2.6 A vector $x \in S$ is a solution of problem (2.55) if and only if there is no $x' \in S$ which satisfies

$$Z(x) \leq_{LC} Z(x') \qquad (2.59)$$

According to Definition 2.6, the interval programming problem (2.55) can be transformed into an equivalent crisp bicriteria programming problem.

Theorem 2.2 (Ishibuchi and Tanaka [231]). The solution set of problem (2.55) defined by the Definition 2.5 can be obtained as the Pareto solutions of the following bicriteria programming problem:

$$\max\{z^L(x), z^C(x)|x \in S \subset R_+^n\} \tag{2.60}$$

Now consider the following nonlinear interval programming problem:

$$\max\left\{Z(x) = \prod_{j=1}^n C_j x_j \,\middle|\, x \in S \subset R_+^n\right\} \tag{2.61}$$

where the objective takes a product form. By the following theorem, this kind of nonlinear interval programming problem can be transformed into an equivalent linear interval programming problem.

Theorem 2.3 (Nakahara and Gen [312]). For the given order $\leq_{LC}$ and two positive intervals A and B, the following statement holds true:

$$A \leq_{LC} B \Leftrightarrow \log(A) \leq_{LC} \log(B) \tag{2.62}$$

Corollary 1. For the nonlinear interval programming problem (2.61), if $c_j^L \geq 0, \forall_j$, it is equivalent to the following linear interval programming problem:

$$\max\left\{Z(x) = \sum_{j=1}^n \log(C_j x_j) \,\middle|\, x \in S \subset R_+^n\right\} \tag{2.63}$$

Maximization Problem. There are two key steps when transforming interval programming to bicriteria linear programming:

- Using the definition of the degree of inequality holding true for two intervals, transform interval constraints into equivalent crisp constraints.
- Using the definition of the order relation between intervals, transform interval objectives into two equivalent crisp objectives.

Let us consider the following maximization interval programming problem:

$$\max \quad Z(\boldsymbol{x}) = \sum_{j=1}^{n} C_j x_j \tag{2.64}$$

$$\text{s.t.} \quad G_i(\boldsymbol{x}) = \sum_{j=1}^{n} A_{ij} x_j \leq B_i, \qquad i = 1, 2, \ldots, m \tag{2.65}$$

$$x_j^L \leq x_j \leq x_j^U: \quad \text{integer}, \qquad j = 1, 2, \ldots, n \tag{2.66}$$

where $C_j = [c_j^L, c_j^R], A_{ij} = [a_{ij}^L < a_{ij}^R]$, and $B = [b^L, b^R]$, respectively, and x_j^L and x_j^U are the lower and upper bounds for x_j, respectively. Problems (2.64)–(2.66) can be transformed into the following problem:

$$\max \quad z^L(\boldsymbol{x}) = \sum_{j=1}^{n} c_j^L x_j \tag{2.67}$$

$$\max \quad z^C(\boldsymbol{x}) = \sum_{j=1}^{n} \frac{1}{2} (c_j^L + c_j^R) x_j \tag{2.68}$$

$$\text{s.t.} \quad g_i(\boldsymbol{x}) = \sum_{j=1}^{n} a_{ij} x_j \leq b_i, \qquad i = 1, 2, \ldots, m \tag{2.69}$$

$$x_j^L \leq x_j \leq x_j^U: \quad \text{integer}, \qquad j = 1, 2, \ldots, n \tag{2.70}$$

where $a_{ij} = q a_{ij}^R + (1 - q) a_{ij}^L$ and $b_i = (1 - q) b_i^R + q b_i^L$.

In bicriteria context, two objectives usually conflict with other in nature, and the concept of optimal solution gives rise to the concept of nondominated solutions (or efficient solutions, or Pareto optimal solutions, or noninferior solutions), for which no improvement in any objective function is possible without sacrificing at least one of the other objective functions. Let F denote the set of feasible solutions; the formal definition of nondominated solution can be stated as follows:

Definition 2.7. A feasible solution $\bar{\boldsymbol{x}} \in F$ is said to be a nondominated solution if and only if

$$\forall \boldsymbol{x} \in F, \qquad z(\boldsymbol{x}) \leq z(\bar{\boldsymbol{x}}) \Rightarrow z(\boldsymbol{x}) = z(\bar{\boldsymbol{x}}) \tag{2.71}$$

where

$$z(\boldsymbol{x}) = (z_1(\boldsymbol{x}), z_2(\boldsymbol{x}), \ldots, z_q(\boldsymbol{x}))^t$$

There are two special points in criteria space: *positive ideal solution* and *neg-*

ative ideal solution. They stand for two extreme cases of the best objective and the worst objective. Even though infeasible, they provide some suggestive information for decision makers.

Definition 2.8. The positive ideal solution (PIS) is composed of all the best objective values attainable; this is denoted as $z^+ = \{z_1^+, z_2^+, \ldots, z_q^+\}$, where z_q^+ is the best value for the qth objective without considering other objectives.

Definition 2.9 The negative ideal solution (NIS) is composed of all the worst objective values attainable; this is denoted as $z^- = \{z_1^-, z_2^-, \ldots, z_q^-\}$, where z_q^- is the worst value for the qth objective without considering other objectives.

There are three primary approaches or philosophies that form the basis for nearly all the candidate multiobjective techniques that have been proposed [229]:

- Weighting or utility methods
- Ranking or prioritizing methods
- Efficient solution or generating methods

The weighting method refers to those approaches that attempt to express all objectives in terms of a single measure. It is attractive from a strictly computational point of view. However, the obvious drawback is that associated with actually developing truly credible weights.

The ranking or prioritizing methods try to circumvent the heady problems indicated above. They assign priorities to each objective according to their perceived importance. Most decision makers can do this.

The final method avoids the problems of finding weights and satisfying the ranking. It generates the entire set of nondominated solutions or an approximation of this set and then allows to the decision makers to select the nondominated solution which best represents their tradeoff among the objectives. Cohon refers to this solution as the best compromise solution [80].

2.5.2   Genetic Algorithm

Gen and Cheng have investigated how to solve interval programming problem with genetic algorithms [151]. The basic idea of the proposed approach is first to transform interval programming model into an equivalent bicriteria programming model and then to find out the Pareto solutions of the bicriteria programming problem using genetic algorithms. When working in multiobjective context with a genetic algorithm, a crucial issue is how to evaluate the credit of chromosomes and then how to represent the credit as fitness values. They proposed a weighted-sum approach which can force genetic search toward exploiting the set of Pareto solutions.

Chromosome Representation and Initial Population. A chromosome is defined as follows:

$$\boldsymbol{x}^k = [x_1^k \; x_2^k \; \cdots \; x_n^k]$$

where the subscript k is the index of chromosome. Initial population is randomly generated within the range $[x_j^L, x_j^U]$ for all x_j.

Crossover and Mutation. Crossover is implemented with uniform crossover operator [391], which has been shown to be superior to traditional crossover strategies for combinatorial problem. Uniform crossover first generates a random crossover mask and then exchanges relative genes between parents according to the mask. A crossover mask is simply a binary string with the same size of chromosome. The parity of each bit in the mask determines, for each corresponding bit in an offspring, which parent it will receive that bit form. This is illustrated in Figure 2.9.

Mutation is performed as random perturbation within the permissive range of integer variables.

Selection. Deterministic selection should be used; that is, delete all duplicate among parents and offspring, sort them in descending order, and then select the first *pop_size* chromosomes as the new population.

Evaluation of Chromosome. There are two main tasks involved in this phase: (1) how to handle infeasible chromosomes and (2) how to determine fitness values of chromosomes according to bicriteria. Let $\boldsymbol{x}^k$ be the kth chromosome in current generation, and the evaluation function is defined as follows:

$$eval(\boldsymbol{x}^k) = (w_1 z^L(\boldsymbol{x}^k) + w_2 z^C(\boldsymbol{x}^k)) p(\boldsymbol{x}^k) \tag{2.72}$$

where w_1 and w_2 are weights corresponding to the importance of the objectives, respectively. The evaluation function contains two terms: *weighted-sum objective* and *penalty*. The weighted-sum objective term tries to give selection pressure to force genetic search toward exploiting the set of Pareto solutions, and the penalty term tries to force genetic search to approach Pareto solutions from both feasible and infeasible regions.

Weighted-Sum Objective. Compared with conventional genetic algorithms, a distinctive feature of the implementation for bicriteria programming is that it is necessary to maintain a set of nondominated points along with the evolutionary process. Let E denote the set of nondominated solutions examined so far. Two special points in E interest us. One point contains the maximum of $z^L(\boldsymbol{x}^k)$ among others in E, and another contains the maximum of $z^C(\boldsymbol{x}^k)$. Denote these two points as (z_{min}^C, z_{max}^L) and $z_{max}^C, z_{min}^L)$, respectively, where

a random mask:

0	0	1	0	0	1	1	0

parents under mask:

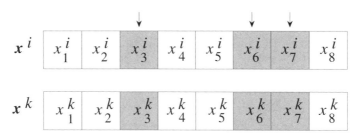

exchange relative genes:

Figure 2.9. Illustration of uniform crossover operation.

$$z^C_{\min} = \min \{z^C(x^k)|x^k \in E\}$$
$$z^C_{\max} = \max \{z^C(x^k)|x^k \in E\}$$
$$z^L_{\min} = \min \{z^L(x^k)|x^k \in E\}$$
$$z^L_{\max} = \max \{z^L(x^k)|x^k \in E\}$$

Then we can make a new objective function based on these two special points as follows:

$$w_1 z^L(x^k) + w_2 z^C(x^k)$$

where

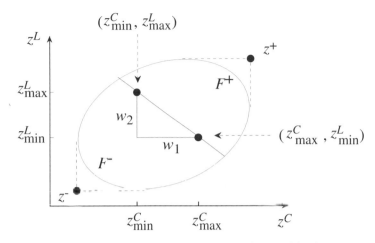

Figure 2.10. Illustrative explanation of new objective.

$$w_1 = z_{max}^C - z_{min}^C$$
$$w_2 = z_{max}^L - z_{min}^L$$

Figure 2.10 gives an illustrative explanation of the objective. The line formed with points (z_{min}^C, z_{max}^L) and (z_{max}^C, z_{min}^L) divides the criteria space into two half-spaces: One contains a positive ideal solution denoted by Z^+, and the other contains a negative ideal solution denoted by Z^-. The feasible solution space F is correspondingly divided into two parts: One is $F^- = F \cap Z^-$, and the other is $F^+ = F \cap Z^+$. It is easy to verify that a solution in F^+ has higher fitness value than those in F^-. Thus chromosomes in the half-space F^+ have a relatively larger chance to enter the next generation. At each generation, the Pareto set E is updated and the two special points may be renewed. This means that along with the evolutionary process, the line formed with this two points will move gradually from a negative ideal point to a positive ideal point. In other words, this fitness function gives such selection pressure to force a genetic search toward exploiting the nondominated points in the criteria space.

Penalty. The penalty term for the kth chromosome is defined as follows:

$$p(x^k) = 1 - \frac{1}{m} \sum_{i=1}^{m} \frac{\delta(g_{ki})(g_{ki} - b_i)}{g_i^{max} - b_i}$$

where

$$g_{ki} = \begin{cases} 0, & g_i(x^k) \le b_i \\ g_i(x^k); & \text{otherwise} \end{cases}$$

$$g_i^{\max} = \max \{g_{ki} | k = 1, 2, \ldots, pop\_size\}$$

$$\delta(g_{ki}) = \begin{cases} 0, & g_{ki} = 0 \\ 1, & \text{otherwise} \end{cases}$$

The penalty term can be viewed as a measure of the degree of infeasibility for a chromosome. Chromosomes generated from either the initial phase or the reproduction phase may violate system constraints. This special measure is used to evaluate how far an infeasible chromsome separates from the feasible region. Usually optimum occurs at the boundary between feasible and infeasible area. When we simply give a large and constant penalty to each infeasible area, it will be rejected from the evolutionary process; genetic search will approach optimum only from the feasible side. The proposed penalty approach can force the genetic search to approach the Pareto solutions from both feasible and infeasible regions as shown in Figure 2.11.

Overall Procedure. The overall procedure is summarized as follows:

Step 0. Set population size *pop_size*, mutation rate p_m, crossover rate p_c, and the maximum number of generations *max_gen*. Let $t \leftarrow 0$ and $E \leftarrow \emptyset$.

Step 1. Initialization. Randomly generate initial population x^k, $k = 1, 2, \ldots,$ *pop_size*.

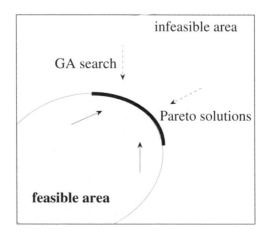

Figure 2.11. Genetic search direction.

Step 2. Crossover. Perform uniform crossover.

Step 3. Mutation. Perform random perturbation mutation.

Step 4. Update set E.

    1. Compute objective function values of bicriteria for each chromosome.

    2. Update set E by adding new nondominated points into E and deleting dominated points from E.

    3. Determine new special points (z^C_{min}, z^L_{max}) and (z^C_{max}, z^L_{min}).

Step 5. Evaluation. Compute fitness values for each chromosome using equation (2.72).

Step 6. Selection.

    1. Delete all duplicate chromosomes.

    2. Sort them in descending order.

    3. Select the first *pop_size* chromosomes as the new population.

Step 7. Terminal test. If $t = gen\_max$, then stop. Otherwise, let $t \leftarrow t + 1$ and go to Step 2.

2.5.3 Numerical Example

Let us consider the following example with the interval objective function.

$$\max \quad Z(x) = [15, 17]x_1 + [15, 20]x_2 + [10, 30]x_3 \tag{2.73}$$
$$\text{s.t.} \quad g_1(x) = x_1 + x_2 + x_3 \leq 30 \tag{2.74}$$
$$g_2(x) = x_1 + 2x_2 + x_3 \leq 40 \tag{2.75}$$
$$g_3(x) = x_1 + 4x_3 \leq 60 \tag{2.76}$$
$$x_j \geq 0: \quad \text{integer}, \quad j = 1, 2, 3 \tag{2.77}$$

This problem can be transformed into the following bicriteria programming problem:

$$\max \quad z^L(x) = 15x_1 + 15x_2 + 10x_3 \tag{2.78}$$
$$\max \quad z^C(x) = 16x_1 + 17.5x_2 + 20x_3 \tag{2.79}$$
$$\text{s.t.} \quad \text{constraints} \quad (2.74)\text{–}(2.77)$$

The 13 Pareto solutions found by genetic algorithms are listed in Table 2.3 and shown in Figure 2.12. The corresponding interval objectives are given in Figure 2.13. From the figure we can see that these intervals cannot be compared with each other under the order relation $\leq_{LC}$.

Table 2.3. Pareto Solutions

n	$z^L(\boldsymbol{x})$	$z^C(\boldsymbol{x})$	x_1	x_2	x_3	n	$z^L(\boldsymbol{x})$	$z^C(\boldsymbol{x})$	x_1	x_2	x_3
1	450	492	22	8	0	8	415	520	15	8	7
2	445	496	21	8	1	9	410	524	14	8	8
3	440	500	20	8	2	10	405	531	11	10	9
4	435	505	18	9	3	11	400	533	11	9	10
5	430	508	18	8	4	12	395	534	12	7	11
6	425	510	18	7	5	13	390	543	8	10	12
7	420	519	14	10	6						

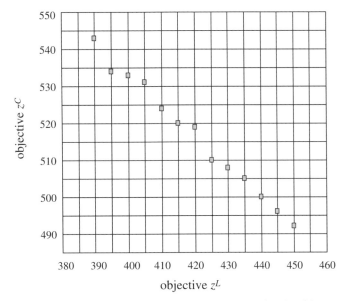

Figure 2.12. Pareto solutions obtained by genetic algorithms.

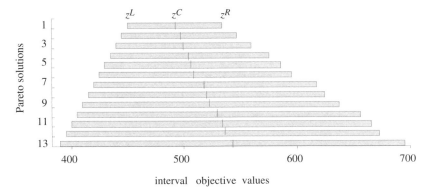

Figure 2.13. Interval objectives for Pareto solutions.

3

COMBINATORIAL OPTIMIZATION PROBLEMS

3.1 INTRODUCTION

Combinatorial optimization studies problems which are characterized by a finite number of feasible solutions. Such problems abound in everyday life, particularly in engineering design. An important and widespread area of applications concerns the efficient use of scarce resources to increase productivity. Typical engineering design problems include knapsack, quadratic 0-1 integer programming, machine scheduling, vehicle routing, network design, facility location and layout, traveling salesman problem, and so on. Recently, a group of young researchers from Europe gave their personal views on the current status of, and prospect for, combinatorial optimization problems [37].

Although, in principle, the optimal solution to such a finite problem can be found by a simple enumeration, in practice it is frequently impossible, especially for practical problems of realistic size where the number of feasible solutions can be extremely high. One of the most challenging problems in combinatorial optimization is to deal effectively with the *combinatorial explosion*. A major trend in solving such difficult problems is to utilize heuristic search. In fact, within every area of mathematical problem-solving, heuristics are of central importance. In combinatorial optimization, heuristics play an even more obviously central role.

In the past few years, genetic algorithms have received a rapidly growing interest in the combinatorial optimization community and have shown great power with very promising results from experimentation and practice of many industrial engineering areas. Since these meta-heuristics provide the user a great flexibility and do not require much problem-specific knowledge in order to get good solutions, it is expected that significant strides will be made in this area in the next few years.

In this chapter, we will explain how to solve knapsack problems, quadratic assignment problems, minimum spanning tree problems, traveling salesman problems, and film-copy deliverer problems with genetic algorithms. All techniques developed for them are applicable to other combinatorial problems. In particular, a traveling salesman problem is usually used as a paradigm for combinatorially difficult problems, and many new ideas have been tested on it. The other problems, such as job-shop scheduling problems, machine scheduling problems, vehicle routing problems, facility layout problems, and so on, will be discussed in subsequent chapters.

3.2   KNAPSACK PROBLEM

Knapsack problems have been intensively studied in the last decade, attracting both theorists and practicians. The theoretical interest arises mainly from their simple structure, which, on the one hand, allows exploitation of a number of combinatorial properties and, on the other hand, allows more complex optimization problems to be solved through a series of knapsack-type subproblems. From the practical point of view, these problems can model many industrial situations (e.g., capital budgeting, cargo loading, cutting stock) as well as most classical applications [282, 305].

Suppose that we want to fill up a knapsack by selecting some objects among various objects (generally called *items*). There are n different items available and each item j has a *weight* of w_j and a *profit* of p_j. The knapsack can hold a weight of at most W. The problem is to find an optimal subset of items so as to maximize the total profits subject to the knapsack's weight capacity. The profits, weights, and capacity are positive integers.

Let x_j be binary variables given as follows:

$$x_j = \begin{cases} 1 & \text{if item } j \text{ is selected} \\ 0 & \text{otherwise} \end{cases}$$

The knapsack problem can be mathematically formulated as follows:

$$\max \quad \sum_{j=1}^{n} p_j x_j \tag{3.1}$$

$$\text{s.t.} \quad \sum_{j=1}^{n} w_j x_j \leq W \tag{3.2}$$

$$x_j = 0 \text{ or } 1, \qquad j = 1, 2, \ldots, n \tag{3.3}$$

This is known as the *0-1 knapsack problem*, which is a pure integer programming with single constraint and forms a very important class of integer programming. There are many variations of the knapsack problem, such as the multiple-choice knapsack problem, the bounded knapsack problem, and the unbounded knapsack problem. It has been proven that knapsack problems are NP-hard. Several researchers have solved the knapsack problem using genetic algorithms [179, 216, 321]. The proposed genetic algorithms can be roughly classified into three classes:

- Binary representation approach
- Order representation approach
- Variable-length representation approach

We will explain them in turn.

3.2.1 Binary Representation Approach

Binary string is a natural representation of solutions to the 0-1 knapsack problem, where one means the inclusion and zero the exclusion of one of the n items from the knapsack. For example, a solution for the 10-item problem can be represented as the following bit string:

$$x = [x_1 \ x_2 \cdots x_{10}]$$
$$[0 \ 1 \ 0 \ 1 \ 0 \ 0 \ 0 \ 0 \ 1 \ 0]$$

It means that items 2, 4, and 9 are selected to be filled in the knapsack. This representation may yield an infeasible solution. In other words, setting too many bits might exceed the capacity of the knapsack. The following two methods have been proposed to handle the infeasibility:

- Penalty method
- Decoder method

The *penalty* method assigns a penalty to each infeasible solution. The *decoder* method generates a solution from a chromosome using the *greedy approximation* heuristic.

Penalty Method. Gordon and Whitley gave a simple penalty for each infeasible solution. The penalty value is equal to the amount in excess of the capacity of the knapsack [179]. Olsen has examined three kinds of penalty methods for the 0-1 knapsack problem [321]. The first one of his fitness functions takes the multiplication form as follows:

$$eval(x) = f(x)p(x) \tag{3.4}$$

The penalty function for maximization problem is constructed as follows:

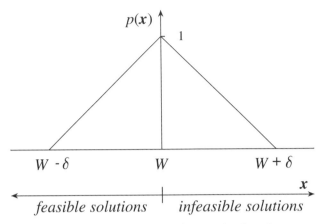

Figure 3.1. Illustration to the penalty function.

$$p(x) = 1 - \frac{\left| \sum_{j=1}^{n} w_j x_j - W \right|}{\delta} \tag{3.5}$$

where

$$\delta = \min \left\{ W, \left| \sum_{j=1}^{n} w_j - W \right| \right\} \tag{3.6}$$

The penalty term is depicted in Figure 3.1. From the figure we can see that the penalty function penalizes not only infeasible solutions for the excess, but also feasible solutions for the underutilization of the capacity of knapsack. This is a desirable property for such problems to attain the optima which will straddle or approach the border of feasible and infeasible areas. For the linear relaxation problem of the knapsack problem, the optima usually straddles the border.

Decoder Method. This method was given by Gordon and Whitley [179]. The decoding procedure contains the following two steps:

Procedure: Decoding

Step 1. Sort items with the value $x_j = 1$ in descending order by the *ratio of profit to weight.*

Step 2. Select items using the *first fit heuristic* until the knapsack cannot be filled any further.

Gordon and Whitley have tested these two approaches on two problems: a relatively difficult one with 20 items and a relative easy one with 80 items. The penalty approach worked better on the 20-item problem, and the decoder approach worked better on the 80-item problem.

3.2.2   Order Representation Approach

This representation was examined by Hinterding [216]. For an n-item problem, a chromosome contains n genes and each gene represents an item with an identical integer number. The order of an item in the permutation representation can be viewed as the priority for it to be filled into the knapsack. The *first fit heuristic* is used to generate a feasible solution from the ordering of items.

Let us look at the example given in Table 3.1. A chromosome of [1 6 4 7 3 2 5] would be decoded to the feasible solution of [1 6 4 5] with a profit of 73 under the weight restriction 100. Uniform crossover and swapping mutation were used for the method.

Hinterding used the following fitness function:

$$eval(x) = \frac{\sum_{j=1}^{n} p_j x_j}{\sum_{j=1}^{n} p_j} \tag{3.7}$$

It gives a fitness value in the range 0 to 1. Usually the value is always less than 1 except that the knapsack can contain all items.

The mapping from chromosome to solution is many to one. What appears to happen for this representation is that this mapping feature reduces effectively the diversity of the population. Hence a large population size must be used to increase the population diversity.

3.2.3   Variable-Length Representation Approach

This representation belongs to the *direct encoding* approach [216]. The genes in the chromosome represent a legal knapsack. Because the number of items in a legal knapsack is not fixed, the length of a chromosome is variable. Also the order of items in a chromosome has no meaning. The initialization procedure contains the following two steps:

Table 3.1. Seven-Item Knapsack Problem

Item Number:	1	2	3	4	5	6	7
Weight:	40	50	30	10	10	40	30
Profit:	40	60	10	10	3	20	60

Procedure: Initialization

Step 1. Generate a random sequence of items.

Step 2. Make a legal knapsack with the *first fit heuristic* based on the sequence.

Injection crossover, a simplification of Falkenauer and Delchambre's grouping crossover [123], was used to manipulate the chromosome of different lengths. It works in the following way:

Procedure: Crossover

Step 1. Choose an insertion point in the first parent, and choose a segment in the second parent.

Step 2. Inject the segment into the first parent at the insertion point.

Step 3. Delete duplicate genes outside the segment to get a proto-child.

Step 4. Cull genes from the proto-child with the *first fit heuristic* to yield a legal knapsack.

The procedure is illustrated in Figure 3.2. Mutation was based on Falkenauer and Delchambre's group mutation operator. It works in the following way:

1. *injecting a segment*

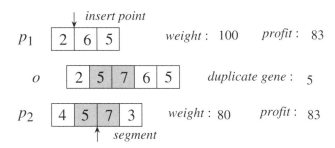

2. *deleting duplicate genes*

3. *culling genes with first fit heuristic*

Figure 3.2. Illustration to injection crossover.

Procedure: Mutation

Step 1. Delete a number of genes at random.

Step 2. Append the items in random order that were not present in the chromosome.

Step 3. Cull genes from the proto-child with the *first fit heuristic* to yield a legal knapsack.

For this representation, because chromosomes only contain items that are in the knapsack, mutation is essential to bring in new items and a large number of generations are needed to explore selections among items.

3.3   QUADRATIC ASSIGNMENT PROBLEM

The assignment problem is the special type of linear programming problem where the resources are being allocated to the activities on a *one-to-one basis*. Thus each resource or *assignee* (e.g., an employee, a machine, or a time slot) is to be assigned uniquely to a particular activity or *assignment* (e.g., a task, a site, or an event). There is a cost c_{ij} associated with assignee i ($i = 1, 2, \ldots, n$) and performing assignment j ($j = 1, 2, \ldots, n$), so that the objective is to determine how all the assignments should be made in order to minimize the total cost.

We use binary decision variables x_{ij} to designate the relationship between assignees and assignments as follows:

$$x_{ij} = \begin{cases} 1 & \text{if assignee } i \text{ performs assignment } j \\ 0 & \text{otherwise} \end{cases}$$

The assignment problem can be mathematically formulated as follows:

$$\min \quad \sum_{i=1}^{n} \sum_{j=1}^{n} c_{ij} x_{ij} \tag{3.8}$$

$$\text{s.t.} \quad \sum_{j=1}^{n} x_{ij} = 1, \qquad i = 1, 2, \ldots, n \tag{3.9}$$

$$\sum_{i-1}^{n} x_{ij} = 1, \qquad j = 1, 2, \ldots, n \tag{3.10}$$

$$x_{ij} = 0 \text{ or } 1, \qquad i, j = 1, 2, \ldots, n \tag{3.11}$$

We now introduce a new kind of cost into the assignment problem: the relative cost between assignees being specified to assignments. Let a_{ijkl} be the relative cost when assignee i performs assignment j and assignee k performs assignment l. The objective then can be rewritten as follows:

$$\min \quad \sum_{i=1}^{n} \sum_{j=1}^{n} c_{ij} x_{ij} + \sum_{i=1}^{n} \sum_{j=1}^{n} \sum_{k=1}^{n} \sum_{l=1}^{n} a_{ijkl} x_{ij} x_{kl} \qquad (3.12)$$

This extension yields the *quadratic assignment problem*. The quadratic assignment problem is a well-known classical combinatorial optimization problem. Since Koopmans and Beckman [250] first modeled the facility layout problem as a quadratic assignment problem, many articles have appeared in the OR literature proposing heuristics procedures and enumerative schemes for this problem. Among the more influential of these have been Hillier and Connors [213], Burkard and Bonninger [52], Bazaraa and Kirca [27], Picone and Wilhelm [337], Heragu and Kusiak [211], and Kaku and Thompson [238].

This problem is known to be NP-hard. Tate and Smith [348] have reported their results of an investigation of genetic algorithms for this problem and have shown that genetic algorithms performed consistently the same as, or better than, previously known heuristics without undue computational overhead.

3.3.1   Encoding

An order representation was used in Tate and Smith's work. For the test problems considered, the possible sites for assignment were a rectangular lattice of locations in the plane, with the intersite distance computed by either Euclidean norm or Manhattan norm. Sites were ordered in boustrophedon sequence by rows, establishing a one-to-one correspondence between sequences of $\{1, 2, \ldots, n\}$ and feasible assignments. The number of rows and columns was prespecified and fixed. Figure 3.3 shows an example of a 9-site problem. The chromosome for this example is a linear list of [7 3 1 2 9 5 6 8 4], where the integers represent objects and the positions in the list represent sites. Any permutation gives a feasible assignment of objects to sites.

7	3	1
2	9	5
6	8	4

Figure 3.3. An example of 9-site problem.

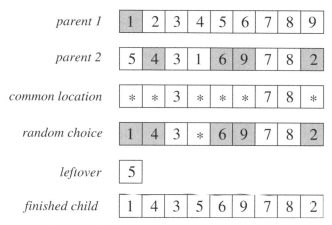

Figure 3.4. Illustration of crossover operator.

3.3.2 Genetic Operators

The crossover operator given by Tate and Smith works as follows [398]:

Procedure: Crossover

Step 1. Any object assigned to the same site in both parents occupies the site in the offspring.

Step 2. For the remaining sites, one or other objects assigned to that site in one parent is chosen at random, working left to right through the sequence of sites.

Step 3. Any unassigned objects are then matched with the remaining unassigned sites.

The procedure is illustrated in Figure 3.4.

Mutation takes the form of selecting two sites at random and reversing the order of all sites within the subsequence bounded by the two selected sites. This is usually called *inverse mutation.*

The basic flow of their implementation of genetic algorithms is given as follows:

Procedure: Mutation

Step 1. Create an initial population, usually feasible and random.

Step 2. Repeat

    1. Select parents.

    2. Breed offspring and add them to the population.

3. Mutate certain members of the current population.

4. Cull certain members of the current population.

Step 3. Until (some termination criterion is met).

One of the well-known quadratic assignment problems with the Euclidean distance is a 36-site and 34-object computer component placement problem first documented in Steinberg [387]. The 34 objects were set in a 4-by-9 grid of sites, and two dummy departments were introduced. The following three mixes of reproduction and mutation parameters were used in their experiments: 25% children and 75% probability of mutation, 50% children and 50% probability of mutation, and 75% children and 25% probability of mutation. The percent of children refers to the number of children created in each generation as a percent of population. The probability of mutation is the probability that any one solution will undergo mutation during a generation. Then the genetic algorithm was run under the three parameter mixes and ten random number seeds. Each run consisted of a fixed population size of 100 solutions, feasible and randomly generated. The search was allowed to continue until 2000 generations had passed. Table 3.2 shows the objective function values of the best and mean solutions as well as the variation over ten solutions for each parameter mix. Table 3.3 shows the least number of solutions searched to find a single run's best solutions (the column entitled "Quick"), the mean number and the variation of solutions searched over the ten seeds, and the coefficient of variation of solutions searched over the ten seeds. Table 3.4 shows the best found solution over the ten runs as a percent above the best-known solution. Tate and Smith also tested genetic algorithms with different replacement strategies (using pure mutation and no reproduction). In the case of 100% culling, each generation represents only the mutated offspring of the previous generation's chromosomes. Convergence is slow and does not seem to be converging toward the best-known solutions At the other extreme, replacing only the bottom 25% of the population in any given generation leads to convergence almost as fast and deep as that of the runs with both reproduction and mutation.

The results of experiments of Tate and Smith indicated that the performance of their genetic algorithm is very encouraging [398]. A simple mutation and crossover schemes are able to match the best-known heuristics from previous research, without the careful selection of mutation rate, population size, or any other specific parameters of the implementation.

Table 3.2. Objective Function Values of Solutions Found

Problem	Best	Mean	Variation
25% C/75% M	4296.0	4473.1	0.023
50% C/50% M	4271.5	4382.2	0.030
75% C/25% M	4385.6	4490.5	0.016

C: crossover, M: mutation

Table 3.3. Number of Solutions Explored before Finding the Best

Problem	Quick	Mean	Variation
25% C/75% M	32,125	123,650	0.410
50% C/50% M	38,059	145,674	0.333
75% C/25% M	26,093	122,331	0.446

Table 3.4. The Genetic Algorithm Solutions as a Percent Over Best-Known Solution

Problem	Best	Worst
25% C/75% M	1.752	10.602
50% C/50% M	1.173	10.614
75% C/25% M	3.836	8.942

3.4   MINIMUM SPANNING TREE PROBLEM

The minimum spanning tree (MST) problem has a venerable history in combinatorial optimization. The problem was first formulated in 1926 by Boruvka, who is said to have learned about it during the rural electrification of Southern Moraria where he gave a solution to find the most economical layout of a power-line network [181]. Since then the MST formulation has been applied to many combinatorial optimization problems such as transportation problems, telecommunication network design, distribution systems, and so on [247]. At the same time, some polynomial–time algorithms for solving the MST problem were developed by Kruskal [255], Prim [340], Dijkstra [109], and Sollin [383].

However, what has been the most attractive to many researchers is the extension problems of MST in the last decade, such as the degree-constrained minimum spanning tree (dc-MST) by Narula and Ho [314], and probabilistic minimum spanning tree (p-MST) by Bertsimas [32], the stochastic minimum spanning tree (s-MST) by Ishii et al. [379], the quadratic minimum spanning tree (q-MST) by Xu [422], the generalized minimum spanning tree (g-MST) by Myung, Lee, and Tcha [309], the minimum Steiner tree (MStT) by Gilbert [168], and so on.

The extensions of MST are generally an NP-hard problem in which polynomial–time solutions for them do not exist. Because of their complexity, recently some researchers have used the genetic algorithms to solve them, including the MStT by Hesser et al. [212], the p-MST by Palmer [328] and Abuali et al. [1], and the dc-MST by Gen and Zhou [166]. Some satisfactory results have been achieved compared with the conventional algorithms.

The research on the extensions of MST problem using genetic algorithms has brought about new experiences for combinatorial optimization problems [467, 468]. In this section we show how to solve the dc-MST problem using genetic algorithms.

3.4.1 Problem Description

Consider a connected and undirected graph $G = (V, E)$, where $V = \{v_1, v_2, \ldots, v_n\}$ is a finite set of *vertices* representing terminals or telecommunication stations etc. and $E = \{e_{ij} | e_{ij} = (v_i, v_j), v_i, v_j \in V\}$ is a finite set of *edges* representing connections between these terminals or stations. Each edge has an associated positive real number denoted with $W = \{w_{ij} | w_{ij} = w(v_i, v_j), w_{ij} > 0, v_i, v_j \in V\}$ representing distance, cost, and so on. The vertices and edges are sometimes referred to as *nodes* and *links*, respectively.

A spanning tree is a minimal set of edges from E that connects all the vertices in V and therefore at least one spanning tree can be found in graph G. The minimum spanning tree, denoted as T^*, is the spanning tree whose total weight of all edges is minimal. It can be formulated as follows:

$$T^* = \min_{T} \sum_{e_{ij} \in E} w_{ij} \qquad (3.13)$$

where T is a set of the spanning trees of graph G.

The *degree* of a vertex is the number of edges connected to it. The *leaf vertex* has only one edge connected to it. So the degree of leaf vertex in a tree is one and the others are more than one.

In some real-world problems the vertex degree is not always arbitrary. The problems may arise for instance when designing a road system in which at most four roads are allowed to meet at a crossing. In communication networks, a degree constraint limits the vulnerability in case of a drop of a crossing. That is, determine a spanning tree of minimum total edge weight such that, at each vertex v_i, the degree d_i is at most a given value b_i. Let T^*_{dc} be the minimum spanning tree meeting degree constraints among all trees T of graph G; then the problem can be formulated as follows:

$$T^*_{dc} = \min_{T} \sum_{\substack{e_{ij} \in E \\ d_i \leq b_i}} w_{ij} \qquad (3.14)$$

For convenience we only consider symmetric case on a complete graph, that is, $w_{ij} = w_{ji}$ and identical degree constraints $b_i = b$ for all $v_i \in V$.

3.4.2   Tree Encodings

For the MST problem, how to encode a tree is critical for the genetic algorithm approach because any chromosome should be a tree. There are several encodings developed for representing a tree. They can be classified as follows:

- Edge encoding
- Vertex encoding
- Edge and vertex encoding

Whatever it is, an effective encoding for a tree should possess the following traits [329]:

1. It should be capable of representing all possible trees and only trees, to the extent that no tree must not be represented.
2. It should be unbiased in the sense that all trees are equally represented; that is, all trees should be represented by the same number of encodings.
3. It should be easy to go back and forth between the encoded representation of the tree and the tree's representation in a more conventional form suitable for evaluating the fitness function and constraints.
4. It should be capable of encouraging short schemata so that the population can evolve toward more fit chromosomes.
5. It should possess locality in the sense that small changes in the representation make small changes in the tree.

We will explain the three kinds of encoding according to the above criteria.

Edge Encoding. If we associate an index k with each edge—that is, $E = \{e_k\}, k = 1, 2, \ldots, K$, where K is the number of edges in a graph—a bit string can represent a candidate solution by indicating which edges are used in a spanning tree as illustrated in Figure 3.5.

This edge encoding is really an intuitive representation for a tree. However, for n-vertex graph, a tree should be either (1) a connected subgraph with $n - 1$ edges or (2) a nonloop subgraph with $n - 1$ edges. But the edge encoding cannot preserve this property. If the bit string contains other than $n - 1$ edges, it is not a tree. Even if the bit string contains exactly $n - 1$ edges, it unlikely represents a tree because there may exist *loops*. Indeed, the probability of a random bit string encoding for a spanning tree is infinitesimally small as the vertices n increase. Thus it is quite likely that none of them would be trees in the initial population or after the crossover and mutation operations.

Piggott and Suraweera tried this edge encoding for the MST problem starting with the correct number of edges and preserving this edge count through the

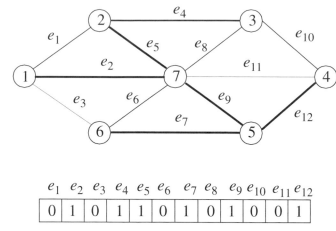

$$e_1\ e_2\ e_3\ e_4\ e_5\ e_6\ \ e_7\ e_8\ \ e_9\ e_{10}\ e_{11}\ e_{12}$$

0	1	0	1	1	0	1	0	1	0	0	1

Figure 3.5. A graph, with its edges encoding for a spanning tree.

genetic operations [338]. On the whole, however, this edge encoding is a poor representation for MST because of the extremely low probability of obtaining a tree.

Vertex Encoding. One of the classical theorems in graphical enumeration is Cayley's theorem that there are $n^{(n-2)}$ distinct labeled trees on a complete graph with n vertices. Prüfer provided a constructive proof of Cayley's theorem by establishing an one-to-one correspondence between such trees and the set of all strings of $n-2$ digits [342]. This means that we can use only $n-2$ digits permutation to uniquely represent a tree where each digit is an integer between 1 and n inclusive. This permutation is usually known as the *Prüfer number*.

Using the Prüfer number to encode a tree is one of vertex encoding. For any tree there are always at least two leaf vertices [381]. Based on this observation we can easily construct an encoding as follows:

Procedure: Encoding

Step 1. Let vertex i be the smallest labeled leaf vertex in a labeled tree T.

Step 2. Let j be the first digit in the encoding as the vertex j incident to vertex i is uniquely determined. Here we build the encoding by appending digits to the right, and thus the encoding is built and read from left to right.

Step 3. Remove vertex i and the edge from i to j; thus we have a tree with $n-1$ vertices.

Step 4. Repeat the above steps until one edge is left. We produce a Prüfer number or an encoding with $n-2$ digits between 1 and n inclusive.

It is also possible to generate a unique tree from a Prüfer number via the following procedure:

Procedure: Decoding

Step 1. Let P be the original Prüfer number and let $\overline{P}$ be the set of all vertices not included in P, which designates as eligible vertices for consideration in building a tree.

Step 2. Let i be the eligible vertex with the smallest label. Let j be the leftmost digit of P. Add the edge from i to j into the tree. Remove i from $\overline{P}$ and j from P. If j does not occur anywhere in P, put it into $\overline{P}$. Repeat the process until no digits are left in P.

Step 3. If no digits remain in P, there are exactly two vertices, r and s, still eligible for consideration. Add edge from r to s into the tree and form a tree with $n - 1$ edges.

An example is given to illustrate this kind of encoding. The Prüfer number (3 3 1 1) corresponds to a spanning tree on a six-vertex complete graph represented in Figure 3.6. The construction of the Prüfer number is described as follows: Locate the leaf vertex having the smallest label. In this case, it is vertex 2. Since vertex 3 (the only vertex) is adjacent to vertex 2 in the tree, assign 3 to the first digit in the Prüfer number, and then remove vertex 2 and the edge (2, 3). Repeat the process on the subtree until edge (1, 6) is left and the Prüfer number of this tree with four digits is finally produced.

Conversely, for the Prüfer number $P = (3\ 3\ 1\ 1)$, the vertices 2, 4, 5, and 6 are eligible and $\overline{P} = \{2, 4, 5, 6\}$. Vertex 2 is the eligible vertex with the smallest label. Vertex 3 is the leftmost digit in P. Add edge (2, 3) to the tree, remove vertex 2 from $\overline{P}$ for further consideration, and remove the leftmost digit 3 of P leaving $P = (3\ 1\ 1)$. Vertex 4 is now the eligible vector with the smallest label and the second vertex 3 is the leftmost digit in remaining P. Then add

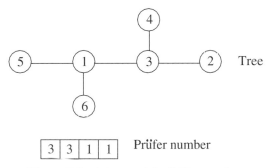

Figure 3.6. A tree and its Prüfer number.

edge (4, 3) to the tree, remove vertex 4 from $\overline{P}$ for further consideration, and remove the digit 3 from P leaving $P = (1\ 1)$. Because vertex 3 is now no longer in the remaining P, it becomes eligible and is put into $\overline{P} = \{3, 5, 6\}$. Indeed, vertex 3 is the eligible vertex with the smallest label, add edge (3, 1) to the tree, remove the leftmost digit 1 from P, and remove vertex 3 from $\overline{P}$ for further consideration. Now $P = (1)$ and only vertices 5 and 6 are eligible. Add edge (5, 1) to the tree, remove the last digit of P, and designate vertex 5 is not eligible and remove it from $\overline{P}$. Vertex 1 is now eligible as it is no longer in the remaining P and is put into $\overline{P} = \{1, 6\}$. P is now empty and only vertices 1 and 6 are eligible. Thus add edge (1, 6) to the tree and stop. The tree in Figure 3.6 is formed.

The Prüfer number is more suitable for encoding a spanning tree, especially in some extensions of the MST problem like dc-MST. But it has relatively little locality that changing even one digit of a Prüfer number can change the tree dramatically as Palmer and Kershenbaum pointed out [329]. Consider, for the Prüfer numbers 3341 and 3342, that just one digit is different in these two encodings, but only two edges are in common on the five-edge trees. It can be clearly seen in Figure 3.7.

There are also other vertex encodings introduced by Palmer [329] and Abuali et al. [1]. These encodings need complex algorithms to keep a tree.

Edge and Vertex Encoding. Edge and vertex encoding, also termed as the link and node biased encoding by Abuali et al. [1], was developed by Palmer [328]. In this encoding, the chromosome holds a bias value for each node and each link. Each link bias and each node bias are an integer in the range from 0 to 255. The spanning tree corresponding to the encoding is found by running Prim's minimum spanning tree algorithm [340] on the modified cost matrix $C' = [c'_{ij}]$.

$$c'_{ij} = c_{ij} + p_1 b_{ij} c_{max} + p_2 (b_i + b_j) c_{max} \tag{3.15}$$

c_{max} is the maximum link cost in the graph, b_{ij} is the link bias associated with edge from node i to node j, and b_i is the node bias associated with node

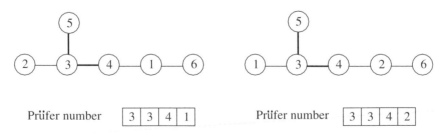

Prüfer number | 3 | 3 | 4 | 1   Prüfer number | 3 | 3 | 4 | 2

Figure 3.7. Illustration of Prüfer numbers 3341 and 3342.

i. p_1 and p_2 are control parameters. Palmer used $p_1 = 0$ and $p_2 = 1$ in his experiments.

In fact, this encoding does not directly encode a tree, but just a modified cost matrix. Based on the modified cost matrix, a tree is generated through Prim's algorithm [340]. This encoding has three evident disadvantages:

1. It requires a very long encoding (memory cost).
2. It needs a conventional minimum spanning tree algorithm to generate a tree from an encoding (computation cost).
3. It contains no useful information about a tree, such as degree, connection, and so on.

3.4.3 Genetic Algorithm Approach

Chromosome Representation. The dc-MST problem is the extension of MST with degree constraints, so the chromosome representation for a spanning tree should contain explicitly the degree on each vertex. Among the several tree encodings, only the Prüfer number encoding explicitly contains the information of vertex degree that any vertex with degree d will appear exactly $d - 1$ times in the encoding. Thus the Prüfer number encoding is adopted.

Crossover and Mutation. Prüfer number encoding can still represent a tree after any crossover or mutation operations. Simply, the one-cut-point crossover operator is used as illustrated in Figure 3.8. Mutation is performed as random perturbation within the permissive integer from 1 to n (n represents number of vertices in graph). An example is given in Figure 3.9.

Degree Modification. Because of the existence of degree constraint on each vertex, the chromosomes generated randomly in the initial population and the

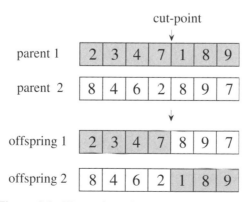

Figure 3.8. Illustration of crossover operation.

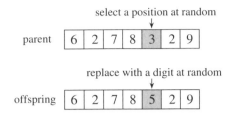

Figure 3.9. Illustration of mutation operation.

offspring produced by crossover and mutation may be illegal in the sense of violating the degree constraint. We need to modify the degree of each illegal vertex in a chromosome. By using the Prüfer number encoding, it is easy to modify the degree of each vertex.

Let $\overline{V}_{dc}$ be the set of vertices whose degree has not been checked and modified in a chromosome. If a vertex violates the degree constraint, which means that the number of this vertex in the chromosome is more than $d - 1$, then decrease the number of the vertex by checking the extra vertex and randomly replace it with another vertex from $\overline{V}_{dc}$. In a nine-vertex complete graph with degree value 3 on each vertex, the chromosome (2 6 6 2 6 8 1), in which the vertex 6 violates the degree constraint as the number of vertex 6 is three, can be modified as illustrated in Figure 3.10.

Evaluation and Selection. The evaluation procedure consists of two steps:

1. Convert a chromosome into a tree.
2. Calculate the total weight of the tree.

Let P be a chromosome, and let $\overline{P}$ be the set of eligible vertices. The fitness can be calculated according to the weight coefficient matrix $W = [w_{ij}]$. The evaluation procedure is illustrated as follows:

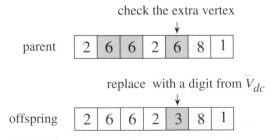

Figure 3.10. Modification of a chromosome with degree constraint.

Procedure: Evaluation

begin
  $T \leftarrow \{\phi\}$;
  $eval(T) \leftarrow 0$;
  define $\overline{P}$ according to P;
  repeat
    select the leftmost digit from P, say i;
    select the eligible vertex with the smallest label from $\overline{P}$, say j;
    $T \leftarrow T \cup \{e_{ij}\}$;
    $eval(T) \leftarrow eval(T) + w_{ij}$;
    remove i from P;
    remove j from $\overline{P}$;
    if i does not occur anywhere in remaining P **then**
     put i into $\overline{P}$;
    end
    $k \leftarrow k + 1$;
  until $k \leq n - 2$
  $T \leftarrow T \cup \{e_{rs}\}, r, s \in \overline{P}$;
  $eval(T) \leftarrow eval(T) + w_{rs}$;
end

As to selection procedure, a mixed strategy with $(\mu + \lambda)$ selection and *roulette wheel* selection is used. The $(\mu + \lambda)$ selection was suggested by Bäck and Hoffmeister [13], which selects μ best chromosomes from μ parents and λ offspring. If there are no μ different chromosomes available, the vacant pool of population is filled up with *roulette wheel* selection as follows:

Procedure: Selection

begin
  select μ' best different chromosomes;
  if $\mu' < \mu$ **then**
    select $\mu - \mu'$ chromosomes by *roulette wheel* selection;
  end
end

dc-MST Algorithm. Compared with conventional genetic algorithms, the major feature of this implementation is that the degree-modification step is embedded in the simple genetic algorithm to hold the degree constraints. The overall procedure for dc-MST problem is outlined as follows:

Procedure: Genetic Algorithm for dc-MST

  begin
    $t \leftarrow 0$:

```
    initialize P(t);
    modify degree P(t);
    evaluate P(t);
    while (not termination condition) do
     recombine P(t) to yield C(t);
     modify degree C(t);
     evaluate C(t);
     select P(t + 1) from P(t) and C(t);
     t ← t + 1;
    end
  end
```

Numerical Example. This example was given by Savelsbergh and Volgenant, who solved it using the branch and bound algorithm [368]; the optimal solution is 2256. It is a 9-vertex complete graph and is given in Table 3.5.

The parameters for the GA are as follows: population size pop_size = 50; crossover probability p_c = 0.5; mutation probability p_m = 0.01; maximum generation max_gen = 500; the constrained degree value for all vertices b = 3; and 30 run times.

By this GA approach, the optimal solution 2256 can be reached at most times. Figure 3.11 clearly illustrates that the GA approach requires neither heuristics nor knowledge of key properties of the problems to be solved; by emulating the biological process on chromosomes which represent the solutions, the optimal solutions can be reached with great probability.

Two kinds of selection strategies were compared. Table 3.6 shows that though both methods can obtain the optimal solution, the average value and the relative error by the GA with mixed strategy are much better than those of *roulette wheel* selection.

Figure 3.12 illustrates that the GA with mixed strategy has a high frequency to reach the optimal solution than does the GA with *roulette wheel* selec-

Table 3.5. Edge Weights of the Nine-Vertex dc-MST Problem

i	1	2	3	4	5	6	7	8	9
1	0	224	224	361	671	300	539	800	943
2	224	0	200	200	447	283	400	728	762
3	224	200	0	400	566	447	600	922	949
4	361	200	400	0	400	200	200	539	583
5	671	447	566	400	0	600	447	781	510
6	300	283	447	200	600	0	283	500	707
7	539	400	600	200	447	283	0	361	424
8	800	728	922	539	781	500	361	0	500
9	943	762	949	583	510	707	424	500	0

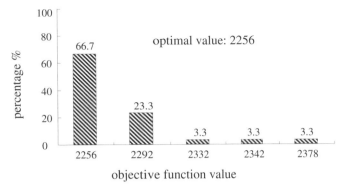

Figure 3.11. Solution distribution for dc-MST using genetic algorithms.

Table 3.6. Comparison of Results from Two Selection Strategies[a]

Methods	*min_val*	*ave_val*	*rel_err%*
Genetric algorithm with mixed strategy	2256	2270.7	0.6
Genetic algorithm with *roulette wheel*	2256	2365.2	4.8

[a]*min_val*, minimum value; *ave_val*, average value; *rel_err*, relative error.

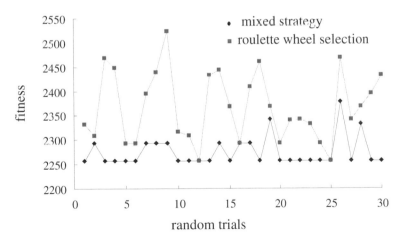

Figure 3.12. Solutions compared with two genetic algorithm selections.

tion. The former gets to the optimal solution 20 times in 30 run times, while the latter reaches it only 2 times in 30 run times. These results show that the mixed strategy is really more suitable for the combinatorial optimization than the roulette wheel selection.

3.5   TRAVELING SALESMAN PROBLEM

The *traveling salesman problem* (TSP) is one of the most widely studied combinatorial optimization problems. Its statement is deceptively simple: A salesman seeks the shortest tour through n cities. Since the mid-1950s, a vast amount of literature on the TSP has been published. The book edited by Lawer et al. provides an insightful and comprehensive survey of all major research results until that date [263]. In the past two decades, enormous progress has been made with respect to solving TSP to optimality. Landmarks in the research are a 48-city problem [90], a 120-city problem [193], a 318-city problem [87], a 532-city problem [326], a 666-city problem [194], and a 2392-city problem [327]. Reinelt has pointed out, in his recent monograph on TSP¡ that despite these achievements, the TSP is far from being solved. Many aspects of the problem still need to be considered, and questions are still left to be answered satisfactorily [352].

The study of this problem has attracted many researchers from different fields, for example, mathematics, operations research, physics, biology, or artificial intelligence. This is because TSP exhibits all aspects of combinatorial optimization and has served, and continues to serve, as the benchmark problem for new algorithmic ideas like simulated annealing, tabu search, neural networks, and evolutionary methods. The earlier studies using the genetic algorithm to solve TSP were those of Goldberg and Lingle [174] and Grefenstette et al. [185]. Since then, TSP has become a target for the genetic algorithm community [287]. The studies on genetic algorithms/TSP provide rich experiences and a sound basis for combinatorial optimization problems. Roughly speaking, major efforts have been made to achieve the following:

1. Give a proper representation to encode a tour.

2. Devise applicable genetic operators to keep building blocks and avoid illegality.

3. Prevent premature convergence.

For the first two topics, we will give a detailed explanation in subsequent sections. For the third topic, we refer to the work of Tsujimura, Gen, and Cheng, who introduced a measure of diversity of populations using the concept of information entropy to escape from falling into local optimum [404].

3.5.1   Representation

Traditionally, chromosomes are a simple binary string. This simple representation is not well suited for TSP and other combinatorial problems. During the past decade, several representation schemes have been proposed for TSP.

Among them, *permutation* representation and *random keys* representation have an appeal not only for the TSP but also for other combinatorial optimization problems.

Permutation Representation. This representation is perhaps the most natural representation of a TSP tour, where cities are listed in the order in which they are visited [101, 287]. The search space for this representation is the set of permutations of the cities. For example, a tour of a 9-city TSP

$$3 - 2 - 5 - 4 - 7 - 1 - 6 - 9 - 8$$

is simply represented as follows:

$$[3\ 2\ 5\ 4\ 7\ 1\ 6\ 9\ 8]$$

This representation is also called a *path representation* or *order representation*. This representation may lead to illegal tours if the traditional one-point crossover operator is used. Many crossover operators have been investigated for it.

Random Keys Representation. Random keys representation was first introduced by Bean [29]. This representation encodes a solution with *random numbers* from (0, 1). These values are used as sort *keys* to decode the solution. For example, a chromosome to a 9 city problem may be

$$[0.23\ 0.82\ 0.45\ 0.74\ 0.87\ 0.11\ 0.56\ 0.69\ 0.78]$$

where position i in the list represents city i. The random number in position i determines the visiting order of city i in a TSP tour. We sort the random keys in ascending order to get the following tour:

$$6 - 1 - 3 - 7 - 8 - 4 - 9 - 2 - 5$$

Random keys eliminate the infeasibility of the offspring by representing solutions in a soft manner. This representation is applicable to a wide variety of sequencing optimization problems including machine scheduling, resource allocation, vehicle routing, quadratic assignment problem, and so on.

3.5.2   Crossover Operators

During the past decade, several crossover operators have been proposed for permutation representation, such as partial-mapped crossover (PMX), order crossover (OX), cycle crossover (CX), position-based crossover, order-based crossover, heuristic crossover, and so on.

Roughly, these operators can be classified into two classes:

- Canonical approach
- Heuristic approach

The canonical approach can be viewed as an extension of two-point or multi-point crossover of binary strings to permutation representation. Generally, permutation representation will yield illegal offspring by two-point or multipoint crossover in the sense of that some cities may be missed while some cities may be duplicated in the offspring. Repairing procedure is embedded in this approach to resolve the illegitimacy of offspring.

The essence of the canonical approach is the blind random mechanism. There is no guarantee that an offspring produced by this kind of crossover is better than their parents. The application of heuristics in crossover intends to generate an improved offspring.

Partial-Mapped Crossover (PMX). PMX was proposed by Goldberg and Lingle [174]. PMX can be viewed as an extension of two-point crossover for binary string to permutation representation. It uses a special repairing procedure to resolve the illegitimacy caused by the simple two-point crossover. Thus the essentials of PMX are a simple two-point crossover plus a repairing procedure. PMX works as follows:

Procedure: PMX

Step 1. Select two positions along the string uniformly at random. The substrings defined by the two positions are called the *mapping sections*.

Step 2. Exchange two substrings between parents to produce proto-children.

Step 3. Determine the *mapping relationship* between two mapping sections.

Step 4. Legalize offspring with the mapping relationship.

This procedure is illustrated in Figure 3.13. Cities 1, 2, and 9 are duplicated in *proto-child* 1, while cities 3, 4, and 5 are missing from it. According to the mapping relationship determined in step 3, the excessive cities 1, 2, and 9 should be replaced by the missing cities 3, 5, and 4, respectively, while keeping the swapped substring unchanged.

Order Crossover (OX). OX was proposed by Davis [99]. It can be viewed as a kind of variation of PMX with a different repairing procedure. OX works as follows:

Procedure: OX

Step 1. Select a substring from one parent at random.

Step 2. Produce a proto-child by copying the substring into the corresponding positions of it.

1. select the substring at random

2. exchange substrings between parents

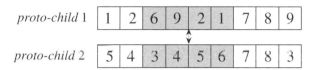

3. determine mapping relationship

4. legalize offspring with mapping relationship

Figure 3.13. Illustration of the PMX operator.

Step 3. Delete the cities which are already in the substring from the second parent. The resulted sequence of cities contains the cities that the proto-child needs.

Step 4. Place the cities into the unfixed positions of the proto-child from left to right according to the order of the sequence to produce an offspring.

The procedure is illustrated in Figure 3.14. It gives an example of making one offspring. With the same steps, we can produce the second offspring as [2 5 4 9 1 3 6 7 8] from the same parents.

Position-Based Crossover. Position-based crossover was proposed by Sys-

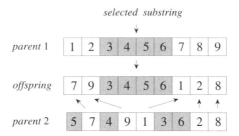

Figure 3.14. Illustration of the OX operator.

werda [99]. It is essentially a kind of uniform crossover for permutation representation together with a repairing procedure. It also can be viewed as a kind of variation of OX in which the cities are selected inconsecutively. Position-based crossover works as follows:

Procedure: Position-Based Crossover

Step 1. Select a set of positions from one parent at random.

Step 2. Produce a proto-child by copying the cities on these positions into the corresponding positions of the proto-child.

Step 3. Delete the cities which are already selected from the second parent. The resulting sequence of cities contains the cities the proto-child needs.

Step 4. Place the cities into the unfixed positions of the proto-child from left to right according to the order of the sequence to produce one offspring.

This procedure is illustrated in Figure 3.15. With the same steps, we can produce the second offspring as [2 4 3 5 1 9 6 7 8].

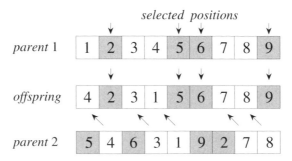

Figure 3.15. Illustration of the position-based crossover operator.

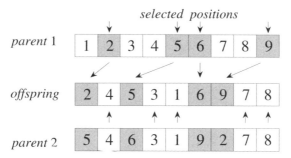

Figure 3.16. Illustration of the order-based crossover operator.

Order-Based Crossover. Order-based crossover was also proposed by Syswerda [99]. It is a slight variation of position-based crossover in which the order of cities in the selected position in one parent is imposed on the corresponding cities in the other parent, as illustrated in Figure 3.16. With the same steps, we can produce the second offspring as [4 2 3 1 5 6 7 9 8].

Cycle Crossover (CX). CX was proposed by Oliver, Smith, and Holland [320]. As with the position-based crossover, it takes some cities from one parent and selects the remaining cities from the other parent. The difference is that the cities from the first parent are not selected randomly and only those cities are selected which define a cycle according to the corresponding positions between parents. CX works as follows:

Procedure: CX

Step 1. Find the cycle which is defined by the corresponding positions of cities between parents.

Step 2. Copy the cities in the cycle to a child with the corresponding positions of one parent.

Step 3. Determine the remaining cities for the child by deleting those cities which are already in the cycle from the other parent.

Step 4. Fulfill the child with the remaining cities.

This procedure is illustrated in Figure 3.17. With the same steps, we can produce the second offspring as [5 4 3 9 2 6 7 8 1].

Subtour Exchange Crossover. This crossover operator was proposed by Yamamura, Ono, and Kobayashi [425, 461]. It selects subtours from parents first. Subtours contain the common cities. Offspring are created by exchanging subtours as shown in Figure 3.18.

Heuristic Crossover. Heuristic crossover was first presented by Grefenstette

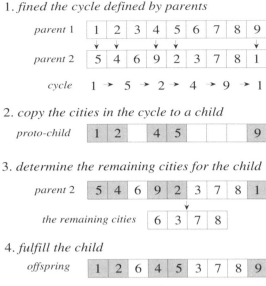

Figure 3.17. Illustration of the CX operator.

et al. [185]. In conventional heuristics for TSP, there are two basic construction approaches: *nearest neighbor* and *best insertion* heuristics. Grefenstette's crossover was implemented with the mechanism of nearest-neighbor heuristic. Cheng and Gen designed a crossover with the mechanism of best insertion for the vehicle routing and scheduling problem [65, 69]. The heuristic crossover works as follows:

Procedure: Heuristic Crossover

Step 1. For a pair of parents, pick a random city for the start.

Step 2. Choose the shortest edge (that is represented in the parents) leading

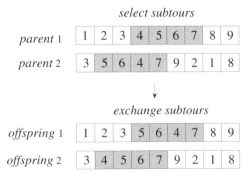

Figure 3.18. Illustration of the subtour exchange crossover operator.

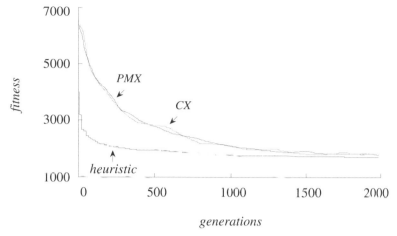

generations

Figure 3.19. Comparison of PMX, CX, and heuristic.

from the current city which does not lead to a cycle. If two edges lead to a cycle, choose a random city that continues the tour.

Step 3. If the tour is completed, stop; otherwise go to step 2.

Liepins et al. presented a slight modification on Grefenstette's version which let any tours always start from the same city [273]. Cheng and Gen also proposed a modified version of the heuristic crossover [61]. A *subtour rotation procedure* was introduced into the heuristic in order to pressure better subtours of parents in offspring. They compared the results of PMX, CX, and their modified heuristic in a 67-city problem from Nemhanser and Wolsey [316]. Figure 3.19 shows that the heuristic crossover is clearly superior to others.

3.5.3 Mutation Operators

It is relatively easy to produce some mutation operators for permutation representation. During the past decade, several mutation operators have been proposed for permutation representation, such as inversion, insertion, displacement, and reciprocal exchange mutation.

Inversion Mutation. Inversion mutation selects two positions within a chromosome at random and then inverts the substring between these two positions, as illustrated in Figure 3.20.

Insertion Mutation. Insertion mutation selects a city at random and inserts it in a random position as illustrated in Figure 3.21.

select a subtour at random

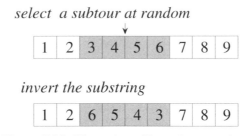

invert the substring

Figure 3.20. Illustration of inversion mutation.

select a city at random

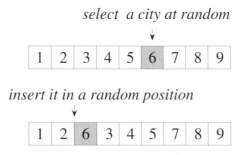

insert it in a random position

Figure 3.21. Illustration of insertion mutation.

select a subtour at random

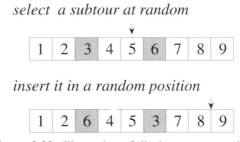

insert it in a random position

Figure 3.22. Illustration of displacement mutation.

Displacement Mutation. Displacement mutation selects a subtour at random and inserts it in a random position as illustrated in Figure 3.22. Insertion can be viewed as a special case of displacement in which the substring contains only one city.

Reciprocal Exchange Mutation. Reciprocal exchange mutation selects two positions at random and then swaps the cities on these positions as illustrated in Figure 3.23. This mutation is essentially the 2-opt heuristic for TSP.

Heuristic Mutation. Heuristic mutation was proposed by Cheng and Gen [62, 64]. It is designed with the neighborhood technique in order to produce an

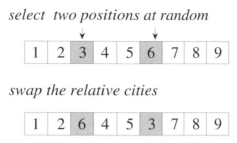

select two positions at random

| 1 | 2 | 3 | 4 | 5 | 6 | 7 | 8 | 9 |

swap the relative cities

| 1 | 2 | 6 | 4 | 5 | 3 | 7 | 8 | 9 |

Figure 3.23. Illustration of reciprocal exchange mutation.

improved offspring. A set of chromosomes transformable from a given chromosome by exchanging no more than λ genes are regarded as the neighborhood. The best one among the neighborhood is used as offspring produced by the mutation. The mutation operator works as follows:

Procedure: Heuristic Mutation

Step 1. Pick up λ genes at random.

Step 2. Generate neighbors according to all possible permutations of the selected genes.

Step 3. Evaluate all neighbors and select the best one as offspring.

This procedure is illustrated in Figure 3.24.

select three positions at random

| 1 | 2 | 3 | 4 | 5 | 6 | 7 | 8 | 9 |

the neighbors form with the cities

1	2	3	4	5	8	7	6	9
1	2	8	4	5	3	7	6	9
1	2	8	4	5	6	7	3	9
1	2	6	4	5	8	7	3	9
1	2	6	4	5	3	7	8	9

Figure 3.24. Illustration of the heuristic mutation operator.

3.6 FILM-COPY DELIVERER PROBLEM

The film-copy deliverer problem (FDP) is a new member of the combinatorial optimization community. It was first examined by Cheng and Gen [60]. The FDP can be formally stated as follows:

One film-copy is asked to be shown among n cinema. The number of times that each cinema is shown is predetermined and denoted by a non-negative integer d_i ($i = 1, 2, \ldots, n$). Each pair of cinema is connected by a link of length w_{ij}; and if cinema i and j are not directly linked, let $w_{ij} = +\infty$.

The problem is to find a tour for a deliverer, starting from and finally returning to a base cinema 1, so that the total length of the tour is minimized while being subjected to the constraint that the deliverer must bring the copy to the ith cinema exactly but not consecutively d_i times ($i = 1, 2, \ldots, n$). When all d_i ($i = 1, 2, \ldots, n$) are restricted to 1, FDP is converted into the TSP. From this point of view, FDP can be regarded as a new extension of TSP, but it may also be generalized to a wide variety of routing and scheduling problems. Zhang and Zheng investigated the transformation of FDP to TSP and solved the transformed problems with the heuristics known for TSP [430].

FDP can be easily represented with a graph. For a weighted and undirected graph we have $G = (V, E)$, where $V = \{v_1, v_2, \ldots, v_n\}$ is a finite set of vertices representing the cinema and $E = \{e_{ij} | e_{jj} = (v_i, v_j), v_i, v_j \in V\}$ is a finite set of edges representing the connection between cinema. Each edge has an associated non-negative real number denoted by $W = \{w_{ij} | w_{ij} = w(v_i, v_j), w_{ij} > 0, v_i, v_j \in V\}$ representing distances. Each vertex has an associated non-negative integer denoted by $D = \{d_i | d_i \text{ is non-negative integer}, i = 1, 2, \ldots, n\}$ representing the number of times that each cinema is shown. An example of graph representation of FDP is shown in Figure 3.25.

Two implementations of genetic algorithms for the FDP have been proposed. One was given by Cheng and Gen [66] and the other was given by Zhang and Zheng [431]. In general, the FDP is much harder to solve than the TSP.

3.6.1 Representation

A feasible deliverer tour is a walk on a graph. Recall that a *walk* of a graph G is a finite and alternating sequence of vertices and edges as shown below [56]:

$$v_{i_1}, e_{i_1 i_2}, v_{i_2}, e_{i_2 i_3}, \ldots, v_{i_{n-1}}, e_{i_{n-1} i_n}, v_{i_n}$$

There may exist repetitions of vertices and edges in a walk. A vertex v forms a $v - v$ *trivial walk*. A feasible deliverer tour is a walk on a given graph that contains exactly d_i ($i = 1, 2, \ldots, n$) repetitions of each vertex, but no trivial walk formed by any vertex.

Usually only the vertices of a walk are indicated since the edges present are then evident. Thus we have the following representatin of a walk:

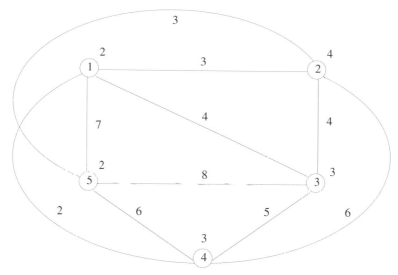

Figure 3.25. Graph representation of FDP.

$$\nu_{i_1}, \nu_{i_2}, \ldots, \nu_{i_n}$$

There may exist repetitions of vertices and edges in a walk. A vertex v forms a $v - v$ *trivial walk*. A feasible deliverer tour is a walk on a given graph that contains exactly d_i $(i = 1, 2, \ldots, n)$ repetitions of each vertex, but no trivial walk formed by any vertex.

Usually only the vertices of a walk are indicated since the edges present are then evident. Thus we have the following representation of a walk:

$$v_{i_1}, v_{i_2}, \ldots, v_{i_n}$$

This is a natural coding scheme for FDP. An example is given in Table 3.7. A feasible deliverer tour (or a chromsome) is represented as

$$[\,1\ 2\ 3\ 2\ 5\ 3\ 4\ 2\ 3\ 4\ 2\ 5\ 4\ 1\,]$$

Table 3.7. An Example

Cinema	1	2	3	4	5
Shows	2	4	3	3	2

which can be interpreted as follows: The film copy is first given to cinema 1 and then forwarded to cinema 2, and so on, according to the order determined in the sequence. This representation is a permutation representation. Generally for an n-cinema problem with d_i shows in the ith cinema, a legal chromosome contains exactly d_i repetitions of each cinema, resulting in the following total size:

$$l = \sum_{i=1}^{n} d_i \tag{3.16}$$

3.6.2 Genetic Operators

Crossover Operator. A *subtour preservation* crossover operator was proposed for this representation, because the subtour was considered to be a natural building block. The subtour is identified with the same cinema in the first and last positions of the tour. The proposed crossover performs with five main steps:

Procedure: Crossover

Step 1. Pick up a subtour from one parent randomly and pick up another subtour with the same first and last genes as the selected subtour.

Step 2. Generate offspring by exchanging the subtours. The offspring may have illegal length; that is, it may be longer or shorter than a legal chromosome because the two subtours usually have different length (contains different number of genes).

Step 3. Compare the subtours to determine the missed genes and exceeded genes for the offspring.

Step 4. Delete the exceeded genes from the offspring. The deletion may lead to some illegal positions; that is, some cinema are adjacent to themselves.

Step 5. Insert the missed genes into the offspring in the illegal positions to legalize them. If there is no such position, insert in a random legal position.

Let us see an example. First pick up a *subtour₁*, [2 3 5 3 4 2], from parent p_1 randomly. The *subtour₂* from parent p_2 is [2 4 3 1 2], which, as with the first one, begins from the second cinema 2 and ends at the third cinema 2 in the sequence of parent p_2. Replace *subtour₁* with *subtour₂* in parent p_1 to generate an offspring. Obviously it is illegal because cinema 5 and 3 are the missed ones and cinema 1 is the exceeded one. It is legalized by deleting the exceeded one and inserting the missed ones. Note that the deletion leads to an illegal position; that is, cinema 4 is adjacent to itself. Thus the missed gene 5

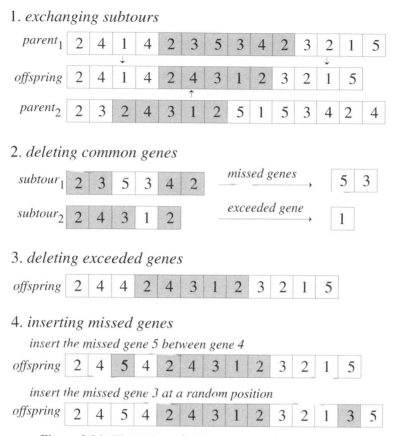

Figure 3.26. Illustration of subtour preservation crossover.

is inserted in the first position. Gene 3 is inserted in a random legal position because there is no illegal position at all. This is shown in Figure 3.26.

Mutation Crossover. The mutation operator was designed with the mechanism of neighborhood search in order to produce an improved offspring. Figure 3.27 shows an example where a subtour with length of $\lambda = 4$ is randomly selected. The legal permutations of the cinema (no cinema is adjacent to itself) in the subtour, together with remaining cinema of the chromosome, form the neighbor chromosomes.

The proposed procedure was tested in a randomly generated problem consisting of 30 cinema and 124 shows with different parameter settings to investigate how these parameters impacted on the performance of the genetic algorithm. The results showed that mutation played a critical role and that the value of mutation probability had significant influence on the performance of the genetic algorithm.

parent chromosome

2	4	1	4	2	3	5	3	4	2	3	2	1	5

neighbor chromosomes

2	4	1	4	3	2	5	3	4	2	3	2	1	5
2	4	1	4	3	2	3	5	4	2	3	2	1	5
2	4	1	4	3	5	2	3	4	2	3	2	1	5
2	4	1	4	3	5	3	2	4	2	3	2	1	5
2	4	1	4	5	3	2	3	4	2	3	2	1	5

Figure 3.27. Illustration of mutation.

4

RELIABILITY OPTIMIZATION PROBLEMS

4.1 INTRODUCTION

In the broadest sense, *reliability* is a *measure of performance* of systems. As systems have grown more complex, the consequences of their unreliable behavior have become severe in terms of cost, effort, lives, and so on, and the interest in assessing system reliability and the need for improving the reliability of products and systems have become very important.

The reliability of a system can be defined as the probability that the system has operated successfully over a specified interval of time under stated conditions. In the past few decades, the field of reliability has grown sufficiently large to include separate specialized subtopics, such as reliability analysis, failure modeling, reliability optimization, reliability growth and its modeling, reliability testing, reliability data analysis, accelerated testing, and life cycle cost [346].

One of the major goals is to find the best way to increase the system's reliability. *Reliability optimization* provides a means to help the reliability engineer achieve such a goal. Most methods of reliability optimization assume that systems have redundancy components in series and/or parallel systems and that alternative designs are available. Optimization concentrates on optimal allocation of redundancy components and optimal selection of alternative designs to meet system requirements. In the past two decades, numerous reliability optimization techniques have been proposed. Generally, these techniques can be classified as linear programming, dynamic programming, integer programming, geometric programming, heuristic method, Lagrangean multiplier method, and so on [345]. A good review of optimization techniques for system reliability can be found in the book by Tillman, Hwang, and Kuo [402].

4.1.1 Combinatorial Aspects of System Reliability

Complex systems are usually decomposed into functional entities composed of units, subsystems, or components for the purpose of reliability analysis. Network modeling techniques and combinatorial aspects of reliability analysis are employed to connect the components in series, parallel, or meshed structure or in any combination thereof. Probability concepts are then employed to compute the reliability of the system in terms of the reliabilities of its components.

Series Structure. From a reliability point of view, a set of n components is said to be *in series* if the success of the system depends on the success of all the components. The components themselves need not be physically or topologically in series; what is relevant is only the fact that all of them must succeed for the system to succeed. Such a system is also known as a *nonredundant* system. The block diagram of such a system is shown in Figure 4.1. Referring to this figure, let x_i denote the event that the ith unit is successful and let $\bar{x}_i$ denote the event that the ith unit is not successful. Then the system reliability R is defined as the probability of system success as follows:

$$R = P(x_1 x_2 \cdots x_n) = P(x_1)P(x_2|x_1)P(x_3|x_1 x_2) \cdots P(x_n|x_1 x_2 \cdots x_{n-1})$$

If the unit failures are independent, then we have

$$R = P(x_1)P(x_2) \cdots P(x_n) = \prod_{i=1}^{n} P(x_i) \equiv \prod_{i=1}^{n} R_i$$

and system unreliability R can be expressed as follows:

$$
\begin{aligned}
Q &= P(\bar{x}_1 + \bar{x}_2 + \cdots + \bar{x}_n) \\
  &= 1 - P(x_1, x_2 \cdots x_n) \\
  &= 1 - R \\
  &= 1 - \prod_{i=1}^{n} R_i \\
  &= 1 - \prod_{i=1}^{n} (1 - Q_i)
\end{aligned}
$$

Figure 4.1. Series structure.

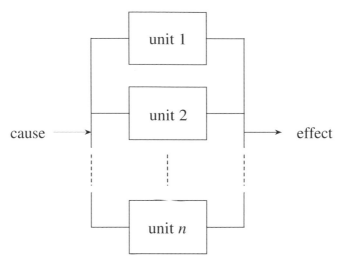

Figure 4.2. Parallel structure.

Parallel Structure. From a reliability point of view, a set of components is said to be *in parallel* if the system can succeed when at least one component succeeds. Such a system is also known as a *fully* or *completely redundant* system. The block diagram of such a system is shown in Figure 4.2. The system reliability is defined as the probability of system success as follows:

$$R = P(x_1 + x_2 + \cdots + x_n)$$
$$= 1 - P(\bar{x}_1 \bar{x}_2 \cdots \bar{x}_n)$$
$$= 1 - P(\bar{x}_1)P(\bar{x}_2|\bar{x}_1)P(\bar{x}_3|\bar{x}_1\bar{x}_2)\cdots P(\bar{x}_n|\bar{x}_1\bar{x}_2\cdots\bar{x}_{n-1})$$

If the unit failures are independent, then we have

$$R = 1 - \prod_{i=1}^{n} P(\bar{x}_i)$$

$$= 1 - \prod_{i=1}^{n} Q_i$$

and system unreliability can be expressed as follows:

$$Q = \prod_{i=1}^{n} Q_i$$

$$= \prod_{i=1}^{n} (1 - R_i)$$

k-Out-of-n Structure. Such a system consists of n identical independent components of which at least $k < n$ of the components should succeed in order for the system to succeed. If p is the probability of success of a component, the probability of exactly k successes and $(n - k)$ failures in n components should be

$$\binom{n}{k} p^k (1 - p)^{n-k}$$

Such a system is said to have *redundancy*. For $k = 1$, they become a truly parallel (fully redundant) system; and for $k = n$, they become a series (nonredundant) system.

4.1.2   Reliability Optimization Models with Several Failure Modes

Using redundant components is well accepted as a technique to improve the reliability of a system. Usually this problem can be formulated as a nonlinear integer program [24, 403].

Notation and Nomenclature

k-out-of-n:F	the system is failed if and only if at least k out of its n elements are failed
k-out-of-n:G	the system is good if and only if at least k out of its n elements are good
class O failure modes	those of which subsystem i is 1-out-of m_i: F
class A failure modes	those of which subsystem i is 1-out-of-m_i: G
m_i	number of elements in subsystem i and $\boldsymbol{m} = [m_1, m_2, \ldots, m_N]$
s_i	total number of failure modes in subsystem i
h_i	number of class O failure modes in subsystem i $(u = 1, 2, \ldots, h_i)$
$s_i - h_i$	number of class A failure modes in subsystem i $(u = h_i + 1 \ h_i + 2, \ldots, s_i)$
q_{iu}	probability of failure mode u for each element in subsystem i
u_i	upper bound of m_i elements in subsystem i

b_t — available amount of system resource t ($t = 1, 2, \ldots, T$)

$g_{ti}(m_i)$ — subsystem i requires this amount of resource t when it contains m_i elements

$Q_u^0(m_i), Q_u^A(m_i)$ — failure probability in subsystem i (m_i elements) for failure mode u of the class of O or A failure modes

$Q^0(m_i), Q^A(m_i)$ — failure probability in subsystem i (m_i elements) subject to the class of O or A failure modes

$Q_i(m_i)$ — unreliability in subsystem i (m_i elements)

$R(\mathbf{m})$ — reliability of the system when the element allocation is $\mathbf{m}$

$G_t(\mathbf{m})$ — the system requires this amount of resource t when the element allocation is $\mathbf{m}$

Mathematical Model. The following assumptions are made for the problem:

1. All elements in subsystem i are s-independent with respect to failure mode u, for each i, u combination taken by itself.
2. The system is 1-out-of-N: F with respect to its subsystems.
3. The failure modes are a partitioning (mutually exclusive and exhaustive) of the failure event. Thus failure probabilities for failure modes combine to obtain total failure probability.
4. Within a subsystem, the elements are 1-out-of-m_i: G for some failure modes while at the same time are 1 out-of-m_i: F for the other failure modes.

Failures occur when the entire subsystem is subjected to the failure condition. All elements in subsystem i can fail by only one of the s_i modes at any given time, and these failures are divided into the class O or class A failure modes.

The failure probability $Q_u^O(m_i)$ in subsystem i for failure mode u of class O failure modes is

$$Q_u^O(m_i) = 1 - (1 - q_{iu})^{m_i}, \qquad u = 1, 2, \ldots, h_i \tag{4.1}$$

The failure probability $Q^O(m_i)$ in subsystem i subject to the class O failure modes is

$$Q^O(m_i) = \sum_{u=1}^{h_i} Q_u^O(m_i) \tag{4.2}$$

The failure probability $Q_u^A(m_i)$ in subsystem i for failure modes u of class A is

$$Q_u^A(m_i) = (q_{iu})^{m_i}, \qquad u = h_i + 1, h_i + 2, \ldots, s_i \tag{4.3}$$

The failure probability $Q^A(m_i)$ in subsystem i subject to the class A failure modes is

$$Q^A(m_i) = \sum_{u=h_i+1}^{s_i} Q_u^A(m_i) \tag{4.4}$$

The unreliability $Q_i(m_i)$ of subsystem i is obtained by adding the failure probabilities for class O failure modes to those for class A failure modes; that is,

$$Q_i(m_i) = Q^O(m_i) + Q^A(m_i) \tag{4.5}$$

Then the reliability optimization of redundant system with several failure modes can be formulated as follows:

$$\max \quad R(\boldsymbol{m}) = \prod_{i=1}^{N} (1 - Q_i(m_i)) \tag{4.6}$$

$$\text{s.t.} \quad G_t(\boldsymbol{m}) = \sum_{i=1}^{N} g_{ti}(m_i) \le b_t, \qquad t = 1, 2, \ldots, T \tag{4.7}$$

$$1 \le m_i \le u_i : \quad \text{integer}, \qquad i = 1, 2, \ldots, N \tag{4.8}$$

The optimization problem is to determine the element allocation which maximizes the nonlinear system reliability (4.6) subject to (a) the T nonlinear constraints (4.7) and (b) the nonnegative conditions with upper bound (4.8). This is a nonlinear integer programming.

4.1.3 Reliability Optimization Models with Alternative Design

Another way to improve the reliability of a system is using an alternative design having greater reliability but perhaps higher cost. Fyffe et al. formulated the reliability optimization problem with redundant units and alternative design as a nonlinear integer programming problem [145].

Notation and Nomenclature

α_i	the design alternatives available for the ith subsystem
β_i	upper bound of alternative design for the ith subsystem
$R(\boldsymbol{m}, \boldsymbol{\alpha})$	reliability of the system with redundant units $\boldsymbol{m}$ and alternative design $\boldsymbol{\alpha}$

$C(\boldsymbol{m}, \boldsymbol{\alpha})$ cost function when system is subject to redundant units $\boldsymbol{m}$ and alternative design $\boldsymbol{\alpha}$

$W(\boldsymbol{m}, \boldsymbol{\alpha})$ weight function when system is subject to redundant units $\boldsymbol{m}$ and alternative design $\boldsymbol{\alpha}$

C overall constraint on the system cost

W overall constraint on the system weight

Mathematical Model. Assume that the system consists of N subsystems. Associated with each subsystem there exist several choices of design alternatives that the system designer can employ in order to meet the reliability requirement. It can be formulated as a nonlinear integer programming problem as follows:

$$\max \quad R(\boldsymbol{m}, \boldsymbol{\alpha}) = \prod_{i=1}^{N} R_i(m_i, \alpha_i) \qquad (4.9)$$

$$\text{s.t.} \quad C(\boldsymbol{m}, \boldsymbol{\alpha}) = \sum_{i=1}^{N} c_i(\alpha_i) m_i \leq C \qquad (4.10)$$

$$W(\boldsymbol{m}, \boldsymbol{\alpha}) = \sum_{i=1}^{N} w_i(\alpha_i) m_i \leq W \qquad (4.11)$$

$$1 \leq m_i \leq u_i \qquad (4.12)$$

$$1 \leq \alpha_i \leq \beta_i \qquad (4.13)$$

$$m_i \text{ and } \alpha_i \text{ are integers}, \qquad i = 1, 2, \ldots, N$$

The problem is to determine which design alternative to select and then decide how many redundant units to use in order to achieve the greatest reliability while keeping the total system cost and weight within the allowable amounts. This problem also can be thought of as an N-stage decision problem, where at each stage i the decisions determine m_i and α_i.

4.2 SIMPLE GENETIC ALGORITHM FOR RELIABILITY OPTIMIZATION

4.2.1 Problem Formulation

Gen, Ida, and Yokota first proposed a simple genetic algorithm to solve reliability optimization [159, 226, 427]. They considered the reliability optimization problem of a redundancy system with several failure modes given by Tillman [403], which is used as a benchmark problem by many researchers [149, 158]. The problem is to maximize the system reliability subject to three nonlinear constraints with parallel redundant units in subsystems that are subject to *A failures*, which occur when the entire subsystem is subjected to the failure condition. This can be mathematically stated as follows:

$$\max \quad R(\boldsymbol{m}) = \prod_{i=1}^{3} [1 - [1 - (1 - q_{i1})^{m_i + 1}] - \sum_{u=2}^{4} (q_{iu})^{m_i + 1}]$$

$$\text{s.t.} \quad G_1(\boldsymbol{m}) = (m_1 + 3)^2 + (m_2)^2 + (m_3)^2 \le 51$$

$$G_2(\boldsymbol{m}) = 20 \sum_{i=1}^{3} (m_i + \exp(-m_i)) \ge 120$$

$$G_3(\boldsymbol{m}) = 20 \sum_{i=1}^{3} (m_i \exp(-m_i/4)) \ge 65$$

$$1 \le m_1 \le 4, \quad 1 \le m_2, \quad m_3 \le 7$$

$$m_i \ge 0: \quad \text{integer}, \quad i = 1, 2, 3$$

where $\boldsymbol{m} = [m_1 \ m_2 \ m_3]$. The subsystems are subject to four failure modes ($s_i = 4$) with one O failure ($h_i = 1$) and three A failures, for $i = 1, 2, 3$. For each subsystem the failure probability is shown in Table 4.1.

Table 4.1. Failure Modes and Probabilities in Each Subsystem

Subsystem i	Failure Modes $s_i = 4, h_i = 1$	Failure Probabilities q_{iu}
1	O	0.01
	A	0.05
	A	0.10
	A	0.18
2	O	0.08
	A	0.02
	A	0.15
	A	0.12
3	O	0.04
	A	0.05
	A	0.20
	A	0.10

Reprinted from *IEEE Transactions on Reliability*, vol. R. 18, F. A. Tillman, "Optimization by integer programming of constrained reliability problems with several modes of failure," p. 51, 1969, with kind permission.

4.2.2 Genetic Algorithm and Numerical Example

Chromosome Representation. In conventional genetic algorithms, chromosome is represented as a binary string. For this problem, the integer value of each variable m_i is represented as a binary string. The length of the string depends on the upper bound u_i of redundant units. For instance, when upper bound u_i equals 4, we need three binary bits to represent m_i. In this example, the upper bounds of redundant units in each subsystem are $u_1 = 4$, $u_2 = 7$, $u_3 = 7$, so each decision variable m_i needs tree binary bits. This means that totally 9 bits are required. If $m_1 = 2$, $m_2 = 3$, and $m_3 = 3$, we have the following chromosome:

$$\boldsymbol{v} = [x_{33}\ x_{32}\ x_{31}\ x_{23}\ x_{22}\ x_{21}\ x_{13}\ x_{12}\ x_{11}]$$
$$= [0\ 1\ 1\ 0\ 1\ 1\ 0\ 1\ 0]$$

where x_{ij} is the symbol of the jth binary bit for variable m_i.

Initial Population. Initial population of chromosomes is generated randomly. Each chromosome is a binary string containing 9 bits. Such a random approach may yield illegal chromosomes in two aspects: violating the system constraint and/or violating the upper bound. Thus the *feasible checking* step is performed to guarantee that all chromosomes are legal.

Procedure: Initialization

begin
for $i \leftarrow 1$ **to** *pop_size* **do**
  produce a random chromosome $\boldsymbol{v}_i$;
  if ($\boldsymbol{v}_i$ is not feasible) **then**
    $i \leftarrow i - 1$;
  end
end
end

Let *pop_size* = 5; the randomly generated chromosomes are given as follows:

$$\boldsymbol{v}_1 = [1\ 1\ 0\ 0\ 1\ 0\ 0\ 0\ 1]$$
$$\boldsymbol{v}_2 = [1\ 0\ 0\ 0\ 0\ 1\ 0\ 1\ 1]$$
$$\boldsymbol{v}_3 = [1\ 0\ 1\ 0\ 0\ 1\ 0\ 1\ 0]$$
$$\boldsymbol{v}_4 = [0\ 1\ 1\ 0\ 1\ 0\ 0\ 1\ 1]$$
$$\boldsymbol{v}_5 = [1\ 0\ 0\ 0\ 1\ 0\ 0\ 0\ 1]$$

The corresponding integer values are

$$
\begin{array}{cccc}
 & m_3 & m_2 & m_1 \\
\boldsymbol{v}_1 = [& 6 & 2 & 1] \\
\boldsymbol{v}_2 = [& 4 & 1 & 3] \\
\boldsymbol{v}_3 = [& 5 & 1 & 2] \\
\boldsymbol{v}_4 = [& 3 & 2 & 3] \\
\boldsymbol{v}_5 = [& 4 & 2 & 1]
\end{array}
$$

Evaluation of Chromosome. When evaluating chromosomes, each legal one is assigned with the value of objective function $R(\boldsymbol{m})$ and each illegal chromosome is assigned a large penalty as follows:

$$
eval(\boldsymbol{v}_k) = \begin{cases} R(\boldsymbol{m}), & G_t(\boldsymbol{m}) \le b_t \quad \forall t \quad \text{and} \quad 1 \le m_i \le u_i \quad \forall i \\ -M, & \text{otherwise} \end{cases}
$$

where $\boldsymbol{v}_k$ represents the kth chromosome. M is a positive large integer.
The fitness values (system reliability) of the above chromosomes are

$$
\begin{aligned}
eval(\boldsymbol{v}_1) &= 0.543625 \\
eval(\boldsymbol{v}_2) &= 0.632703 \\
eval(\boldsymbol{v}_3) &= 0.610062 \\
eval(\boldsymbol{v}_4) &= 0.629119 \\
eval(\boldsymbol{v}_5) &= 0.589642
\end{aligned}
$$

Crossover. One-cut-point crossover is used here. Let the probability of crossover be $p_c = 0.4$ and let the sequence of random numbers be

$$0.550279 \quad 0.379650 \quad 0.243294 \quad 0.494583 \quad 0.771811$$

If random number r is less than p_c, we select the relative chromosome for crossover. This means that the chromsomes $\boldsymbol{v}_2$ and $\boldsymbol{v}_3$ are selected for crossover. The position of the cut-point is randomly generated from the range [1, 9]. Assume the position is 3; thus we have

$$
\boldsymbol{v}_2 = [1\ 0\ 0\ |0\ 0\ 1\ 0\ 1\ 1]
$$
$$
\updownarrow
$$
$$
\boldsymbol{v}_3 = [1\ 0\ 1|0\ 0\ 1\ 0\ 1\ 0]
$$

The offspring generated by exchanging the right parts of their parents would be

$$o_1 = [1\ 0\ 0\ 0\ 0\ 1\ 0\ 1\ 0]$$
$$o_2 = [1\ 0\ 1\ 0\ 0\ 1\ 0\ 1\ 1]$$

The corresponding integer values of offspring are

$$\begin{array}{ccc} m_3 & m_2 & m_1 \\ o_1 = [4 & 1 & 2] \\ o_2 = [5 & 1 & 3] \end{array}$$

The fitness values of the offspring are

$$eval(o_1) = 0.607591$$
$$eval(o_2) = 0.635277$$

Mutation. Mutation is performed on a bit-by-bit basis. Let the probability of mutation p_m be 0.1; this means that, on average, 10% of bits would undergo mutation. There are $9 \times 5 = 45$ bits in the whole population, and 4 or 5 bits would undergo mutation per generation. A random number r is generated from the range [0, 1]; if $r < 0.1$, we mutate the relative bit.

A total of 45 random numbers are generated, and four of them were smaller than 0.1; the bit number, relative chromosome, and the bit position within each chromosome are listed below:

Bit Number	Chromosome	Position	Random Number
14	2	5	0.018754
15	2	6	0.094512
18	2	9	0.065742
31	4	4	0.001257

This means that the chromosomes v_2 and v_4 were selected for mutation. The resulting offspring would be

$$v_2 = [1\ 0\ 0\ 0\ 0\ 1\ 0\ 1\ 1]$$
$$\updownarrow\updownarrow \quad \updownarrow$$
$$o_3 = [1\ 0\ 0\ 0\ 1\ 0\ 0\ 1\ 0]$$

$$v_4 = [0\ 1\ 1\ 0\ 1\ 0\ 0\ 1\ 1]$$
$$\updownarrow$$
$$o_4 = [0\ 1\ 1\ 1\ 1\ 0\ 0\ 1\ 1]$$

The corresponding integer values of the offspring are

$$m_3 \quad m_2 \quad m_1$$
$$o_3 = [4 \quad 2 \quad 2]$$
$$o_4 = [3 \quad 5 \quad 3]$$

The fitness values of the offspring are

$$eval(o_3) = -M$$
$$eval(o_4) = -M$$

which means that both of them are illegal.

Selection. A deterministic selection is adopted as the selection strategy; that is, sort all of parents and offspring in descending order and select the first *pop_size* chromosomes as the new population.

$$v'_1 = [1\ 0\ 1\ 0\ 0\ 1\ 0\ 1\ 1] \quad (o_2), \qquad eval(v'_1) = 0.635277$$
$$v'_2 = [1\ 0\ 0\ 0\ 0\ 1\ 0\ 1\ 1] \quad (v_2), \qquad eval(v'_2) = 0.632703$$
$$v'_3 = [0\ 1\ 1\ 0\ 1\ 0\ 0\ 1\ 1] \quad (v_4), \qquad eval(v'_4) = 0.629119$$
$$v'_4 = [1\ 0\ 1\ 0\ 0\ 1\ 0\ 1\ 0] \quad (v_3), \qquad eval(v'_3) = 0.610062$$
$$v'_5 = [1\ 0\ 0\ 0\ 0\ 1\ 0\ 1\ 0] \quad (o_1), \qquad eval(v'_1) = 0.607591$$

The best solution from a random run of genetic algorithm with 50 generations is

$$v* = [0\ 1\ 1\ 0\ 0\ 1\ 0\ 1\ 0], \qquad eval(v*) = 0.660685$$

The corresponding integer values of the offspring are

$$[m_3 \quad m_2 \quad m_1] = [3 \quad 1 \quad 2]$$

This result is identical with that given by Tillman [403] and Gen [149]. The statistical data over 30 runs are shown in Table 4.2.

4.3 RELIABILITY OPTIMIZATION WITH REDUNDANT UNIT AND ALTERNATIVE DESIGN

4.3.1. Problem Formulation

Gen, Yokota, Ida, and Taguchi further extended their work to the reliability optimization problem by considering both redundant units and alternative design [159, 428]. Because it is more complex than the problem discussed in the pre-

Table 4.2. Statistical Data over 30 Trials

Total runs:	30
Frequency for obtaining optima:	0.933
Earliest generation for obtaining optima:	1
Latest generation for obtaining optima:	44
Average generation for obtaining optima:	13.4
Standard deviation[a]:	0.00093

[a]Standard deviation: $(x^* - \sum x_i/N)/x^*$.

vious section, the classic approach of genetic algorithms cannot be applied without modification.

The example used here was first introduced by Fyffe et al. [145] as follows:

$$\max \quad R(\boldsymbol{m}, \boldsymbol{\alpha}) = \prod_{i=1}^{14} (1 - (1 - R_i(\alpha_i))^{m_i}) \tag{4.14}$$

$$\text{s.t.} \quad G_1(\boldsymbol{m}, \boldsymbol{\alpha}) = \sum_{i=1}^{14} c_i(\alpha_i) m_i \leq 130 \tag{4.15}$$

$$G_2(\boldsymbol{m}, \boldsymbol{\alpha}) - \sum_{i=1}^{14} w_i(\alpha_i) m_i \leq 170 \tag{4.16}$$

$$1 \leq m_i \leq u_i, \qquad \forall i \tag{4.17}$$

$$1 \leq \alpha_i \leq \beta_i, \qquad \forall i \tag{4.18}$$

$$m_i, \alpha_i \geq 0: \quad \text{integer} \quad \forall i$$

where α_i represents the design alternatives available for the ith subsystem, m_i represents the identical units used in redundancy for the ith subsystem, u_i is the upper bound of redundant units for the ith subsystem, and β_i is the upper bound of alternative design for the ith subsystem. The problem is to determine which design alternative to select and then decide how many redundant units to use in order to achieve the greatest reliability while keeping the total system cost and weight within the allowable amounts. Data are given in Table 4.3.

4.3.2 Genetic Algorithm and Numerical Example

Chromosome Representation. A gene is defined as an ordered couple of design alternatives α_{ki} and redundant units m_{ki} shown as follows:

$$v_{ki} = (\alpha_{ki}, m_{ki})$$

Table 4.3. Redundant Units and Alternative Design

Subsystem	Design Alternative											
	1			2			3			4		
i	R	c_i	w_i	R	c_i	w_i	R	c_i	w_i	R	c_i	w_i
1	0.90	1	3	0.93	1	4	0.91	2	2	0.95	2	5
2	0.95	2	8	0.94	1	10	0.93	1	9	*	*	*
3	0.85	2	7	0.90	3	5	0.87	1	6	0.92	4	4
4	0.83	3	5	0.87	4	6	0.85	5	4	*	*	*
5	0.94	2	4	0.93	2	3	0.95	3	5	*	*	*
6	0.99	3	5	0.98	3	4	0.97	2	5	0.96	2	4
7	0.91	4	7	0.92	4	8	0.94	5	9	*	*	*
8	0.81	3	4	0.90	5	7	0.91	6	6	*	*	*
9	0.97	2	8	0.99	3	9	0.96	4	7	0.91	3	8
10	0.83	4	6	0.85	4	5	0.90	5	6	*	*	*
11	0.94	3	5	0.95	4	6	0.96	5	6	*	*	*
12	0.79	2	4	0.82	3	5	0.85	4	6	0.90	5	7
13	0.98	2	5	0.99	3	5	0.97	2	6	*	*	*
14	0.90	4	6	0.92	4	7	0.95	5	6	0.99	6	9

$\beta_i = 4$ for $i = 1, 3, 6, 9, 12, 14$, $\beta_i = 3$ for $i = 2, 4, 5, 7, 8, 10, 11, 13$, and $u_i = 5 \; \forall i$.

Reprinted from *IEEE Transactions on Reliability*, vol. R. 17, D. Fyffe, W. Hines, and N. Lee, "System reliability allocation and a computational algorithm," p. 68, 1968 with kind permission.

where the subscript k is the index of chromosome the gene belongs to and the subscript i indicate the subsystem i. A chromosome is an ordered list of such genes as follows:

$$\boldsymbol{v}_k = [v_{k1} \; v_{k2} \; \cdots \; v_{k14}]$$

The representation can be rewritten as follows:

$$\boldsymbol{v}_k = [(\alpha_{k1}, m_{k1})(\alpha_{k2}, m_{k2}) \cdots (\alpha_{k14}, m_{k14})]$$

Initial Population. The initial population of chromosomes is generated randomly as follows:

Procedure: Initialization

begin
for $k \leftarrow 1$ **to** *pop_size* **do**
for $i \leftarrow 1$ **to** 14 **do**
$\alpha_{ki} \leftarrow random(1, \beta_i)$;
$m_{ki} \leftarrow random(1, u_i)$;
end

$$\boldsymbol{v}_k \leftarrow [(\alpha_{k1}, m_{k1}) \cdots (\alpha_{k14}, m_{k14})];$$
end
end

where *random*(1, *num*) means to return a random integer within the range [1, *num*]. Let *pop_size* = 8; the randomly generated chromosomes are given as follows:

$$\boldsymbol{v}_1 = [(2,1)(1,1)(4,1)(1,4)(3,4)(4,2)(2,1)(1,3)(3,2)(2,2)(2,1)(1,2)(3,4)(3,4)]$$
$$\boldsymbol{v}_2 = [(2,2)(1,3)(2,3)(1,1)(2,5)(2,1)(1,1)(3,1)(4,1)(3,2)(1,2)(1,2)(1,2)(1,1)]$$
$$\boldsymbol{v}_3 = [(1,3)(2,3)(3,1)(3,2)(2,3)(1,1)(2,1)(3,1)(4,1)(3,4)(1,2)(1,2)(3,4)(2,1)]$$
$$\boldsymbol{v}_4 = [(2,4)(3,1)(4,3)(3,1)(1,1)(2,3)(3,2)(2,1)(3,3)(2,2)(3,5)(1,1)(3,1)(1,2)]$$
$$\boldsymbol{v}_5 = [(2,2)(2,1)(2,4)(1,3)(1,2)(2,5)(1,2)(1,3)(2,1)(2,2)(2,2)(3,1)(2,1)(2,3)]$$
$$\boldsymbol{v}_6 = [(1,1)(3,4)(2,3)(3,1)(1,4)(1,1)(3,1)(1,4)(1,1)(3,1)(1,1)(2,5)(3,1)(2,1)]$$
$$\boldsymbol{v}_7 = [(4,1)(3,1)(2,4)(3,2)(3,4)(4,3)(2,1)(3,4)(4,2)(2,4)(1,1)(1,2)(1,1)(2,1)]$$
$$\boldsymbol{v}_8 = [(3,2)(3,1)(3,3)(2,2)(2,3)(4,2)(3,2)(3,3)(3,1)(2,1)(1,3)(3,2)(1,2)]$$

Evaluation of Chromosome. The random approach to generate initial population may yield illegal chromosomes, which violate system constraints. In fact a common and critical problem when applying genetic algorithms to large-scale integer optimization problem is that the chromosomes generated from either genetic operations or random initial procedures usually violate the constraints of the given problem. A ready way to handle such infeasibility is to give a large penalty to each illegal chromosome as discussed in the previous section. This approach will essentially narrow search space by eliminating all illegal ones from evolutionary process. In such a case, it is very difficult to find better candidates for the global optimum with any selection mechanisms, and genetic search tends to be inefficient.

To overcome such a problem, a special measure function is introduced to determine how far illegal chromosomes separate from the feasible region. Usually the optimum occurs at the boundary between the feasible area and the infeasible area. When we simply give a large and constant penalty to each illegal chromosome, they will be rejected from the evolutionary process, and the genetic search will approach the optimum only from the feasible area. If the information of the degree of infeasibility of solutions is embedded in fitness evaluation, the genetic search will approach the optimum from both the feasible area and the infeasible area; that is, it can be performed within more wider search space as shown in Figure 4.3. The measure of the degree of infeasibility for an illegal chromosome is given as follows:

$$d_{kt} = \begin{cases} 0, & G_t(\boldsymbol{m}, \boldsymbol{\alpha}) \le b_t \\ (G_t(\boldsymbol{m}, \boldsymbol{\alpha}) - b_t)/b_t, & \text{otherwise} \end{cases}$$

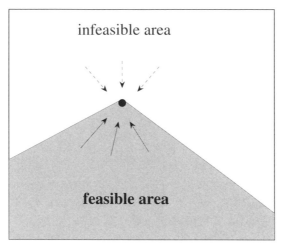

Figure 4.3. Genetic search direction.

where the subscript k indicates the kth chromosome and t indicates the tth constraint. Then the fitness function is given as follows:

$$eval(\boldsymbol{v}_k) = R(\boldsymbol{m}, \boldsymbol{\alpha}) \left(1 - \frac{1}{T} \sum_{t=1}^{T} d_{kt} \right)$$

where T is the total number of constraints.

The fitness values of initial chromosomes are

$$eval(\boldsymbol{v}_1) = 0.656578, \qquad eval(\boldsymbol{v}_5) = 0.741668$$
$$eval(\boldsymbol{v}_2) = 0.516839, \qquad eval(\boldsymbol{v}_6) = 0.520029$$
$$eval(\boldsymbol{v}_3) = 0.561119, \qquad eval(\boldsymbol{v}_7) = 0.637986$$
$$eval(\boldsymbol{v}_4) = 0.493847, \qquad eval(\boldsymbol{v}_8) = 0.753369$$

Crossover. The uniform crossover operator given by Syswerda is used here; it has been shown to be superior to traditional crossover strategies for the combinatorial problem [391]. Uniform crossover first generates a random crossover mask and then exchanges relative genes between parents according to the mask. A crossover mask is simply a binary string with the same size of chromosome. The parity of each bit in the mask determines, for each corresponding bit in an offspring, from which parent it will receive that bit. This is illustrated in Figure 4.4. Let the probability of crossover be $p_c = 0.4$ and let the sequence of random numbers be *a random mask*:

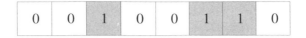

parents under mask:

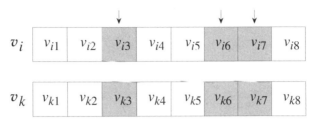

exchange relative genes:

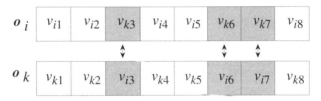

Figure 4.4. Illustration of the uniform crossover operation.

$$0.486129 \qquad 0.536149 \qquad 0.785178 \qquad 0.754631$$
$$0.632408 \qquad 0.084658 \qquad 0.105411 \qquad 0.771811$$

If a random number r is less than p_c, select a relative chromosome for crossover. This means that the chromosomes v_6 and v_7 were selected for crossover. The positions randomly generated are 12, 8, and 2; then we have

$$v_6 = [(1, 1)(3, 4)(2, 3)(3, 1)(1, 4)(1, 1)(3, 1)(1, 4)(1, 1)(3, 1)(1, 1)(2, 5)(3, 1)(2, 1)]$$
$$\updownarrow \qquad\qquad\qquad\qquad\qquad \updownarrow \qquad\qquad\qquad \updownarrow$$
$$v_7 = [(4, 1)(3, 1)(2, 4)(3, 2)(3, 4)(4, 3)(2, 1)(3, 4)(4, 2)(2, 4)(1, 1)(1, 2)(1, 1)(2, 1)]$$

The resulting offspring by exchanging relative genes of their parents would be

$$o_1 = [(1, 1)(3, 1)(2, 3)(3, 1)(1, 4)(1, 1)(3, 1)(3, 4)(1, 1)(3, 1)(1, 1)(1, 2)(3, 1)(2, 1)]$$
$$o_2 = [(4, 1)(3, 4)(2, 4)(3, 2)(3, 4)(4, 3)(2, 1)(1, 4)(4, 2)(2, 4)(1, 1)(2, 5)(1, 1)(2.1)]$$

The fitness values of the offspring are

$$eval(o_1) = 0.462970$$
$$eval(o_2) = 0.653214$$

Mutation. Mutation is performed as random perturbation. For a selected gene $v_{ki} = (\alpha_{ki}, m_{ki})$, α_{ki} will be replaced by a random integer within $[1, \beta_i]$ and m_{ki} will be replaced by a random integer within $[1, u_i]$.

Let the probability of mutation, p_m, be 0.05; this means that, on average, 14 $\times$ 8 $\times$ 0.05 = 5.6 bits would undergo mutation. A total of 112 random numbers were generated, and four of them were smaller than 0.05; the bit number, relative chromosomes, and the bit position within each chromosome are listed below:

Bit Number	Chromosome	Position	Random Number
22	2	8	0.031892
31	3	3	0.020447
43	4	1	0.041556
70	5	14	0.001257

This means that the chromosomes v_2, v_3, v_4, and v_5 were selected for mutation. The resulting offspring would be

$$o_3 = [(2,2)(1,3)(2,3)(1,1)(2,5)(2,1)(1,1)(1,4)(4,1)(3,2)(1,2)(1,2)(1,2)(1,1)]$$
$$o_4 = [(1,3)(2,3)(1,5)(3,2)(2,3)(1,1)(2,1)(3,1)(4,1)(3,4)(1,2)(1,2)(3,4)(2,1)]$$
$$o_5 = [(3,4)(3,1)(4,3)(3,1)(1,1)(2,3)(3,2)(2,1)(3,3)(2,2)(3,5)(1,1)(3,1)(1,2)]$$
$$o_6 = [(2,2)(2,1)(2,4)(1,3)(1,2)(2,5)(1,2)(1,3)(2,1)(2,2)(2,2)(3,1)(2,1)(4,1)]$$

The fitness values of offspring are

$$eval(o_3) = 0.567215$$
$$eval(o_4) = 0.613955$$
$$eval(o_5) = 0.493827$$
$$eval(o_6) = 0.734647$$

Selection. The deterministic selection is adopted as the selection strategy; that is, we sort all parents and offspring in descending order and select the first *pop_size* chromosomes as the new population.

$$v'_1 = [(3,2)(3,1)(3,3)(2,2)(2,3)(4,2)(3,2)(3,3)(3,3)(3,1)(2,1)(1,3)(3,2)(1,2)] \quad (v_8)$$
$$v'_2 = [(2,2)(2,1)(2,4)(1,3)(1,2)(2,5)(1,2)(1,3)(2,1)(2,2)(2,2)(3,1)(2,1)(2,3)] \quad (v_5)$$
$$v'_3 = [(2,2)(2,1)(2,4)(1,3)(1,2)(2,5)(1,2)(1,3)(2,1)(2,2)(2,2)(3,1)(2,1)(4,1)] \quad (o_6)$$
$$v'_4 = [(2,1)(1,1)(4,1)(1,4)(3,4)(4,2)(2,1)(1,3)(3,2)(2,2)(2,1)(1,2)(3,4)(3,4)] \quad (v_1)$$
$$v'_5 = [(4,1)(3,4)(2,4)(3,2)(3,4)(4,3)(2,1)(1,4)(4,2)(2,4)(1,1)(2,5)(1,1)(2,1)] \quad (o_2)$$
$$v'_6 = [(4,1)(3,1)(2,4)(3,2)(3,4)(4,3)(2,1)(3,4)(4,2)(2,4)(1,1)(1,2)(1,1)(2,1)] \quad (v_7)$$
$$v'_7 = [(1,3)(2,3)(1,5)(3,2)(2,3)(1,1)(2,1)(3,1)(4,1)(3,4)(1,2)(1,2)(3,4)(2,1)] \quad (o_4)$$
$$v'_8 = [(2,2)(1,3)(2,3)(1,1)(2,5)(2,1)(1,1)(1,4)(4,1)(3,2)(1,2)(1,2)(1,2)(1,1)] \quad (o_3)$$

The fitness values of new generation are

$$eval(v'_1) = 0.753369, \qquad eval(v'_2) = 0.741668$$
$$eval(v'_3) = 0.734647, \qquad eval(v'_4) = 0.656578$$
$$eval(v'_5) = 0.653214, \qquad eval(v'_6) = 0.637986$$
$$eval(v'_7) = 0.613955, \qquad eval(v'_8) = 0.567215$$

The best chromosome at the 623rd generation from a random run is

$$v^* = [(3,3)(1,2)(4,3)(3,3)(2,3)(2,2)(1,2)(1,4)(3,2)(2,3)(1,2)(1,4)(2,2)(3,2)]$$

The reliability of the system is 0.970015. The result is identical to the one given by Nakagawa et al. [310] and Gen et al. [156]. The statistical data over 30 runs are shown in Table 4.4.

4.4 RELIABILITY OPTIMIZATION WITH REDUNDANT MIXING COMPONENTS

4.4.1 Problem Formulation

Coit and Smith have examined the reliability optimization problem with mixing components redundant in subsystems. The overall system is partitioned into N

Table 4.4. Statistical Data over 30 Trials

Total Runs:	30
Frequency for getting optima:	0.170
Earliest generation for getting optima:	161
Latest generation for getting optima:	975
Average generation for getting optima:	423.8
Standard deviation[a]:	0.002375

[a]Standard deviation: $(x^* - \sum x_i/N)/x^*$.

subsystems in parallel [82, 83]. For each required function, there are several different component types available with different levels of cost, reliability, weight, and other characteristics. A subsystem is considered to be composed of mixing components and to have k-out-of-n: G redundancy. This kind of redundancy allocation problem is to select the optimal combination of components and levels of redundancy to collectively meet weight and cost constraints while maximizing system reliability.

This problem can be formulated as follows:

$$\max \quad \prod_{i=1}^{N} R_i(\boldsymbol{x}_i | k_i) \tag{4.19}$$

$$\text{s.t.} \quad \sum_{i=1}^{N} C_i(\boldsymbol{x}_i) \leq C \tag{4.20}$$

$$\sum_{i=1}^{N} W_i(\boldsymbol{x}_i) \leq W \tag{4.21}$$

$$k_i \leq \sum_{j=1}^{m_i} x_{ij} \leq u_i, \qquad i = 1, 2, \ldots, N \tag{4.22}$$

$$x_{ij}: \quad \text{integer} \quad \forall i, j$$

where

N is the number of subsystems

C is the cost constraint

W is the weight constraint

x_{ij} is the number of the jth component used in subsystem i

$\boldsymbol{x}_i$ is the vector $(x_{i1}, x_{i2}, \ldots, x_{im_i})$

m_i is the number of available component choices for subsystem i

k_i is the lower bound of components in parallel for subsystem i

u_i is the upper bound of components in parallel for subsystem i

$R_i(\boldsymbol{x}_i | k_i)$ is the reliability of subsystem i for a given k_i

$C_i(\boldsymbol{x}_i)$ is the total cost of subsystem i

$W_i(\boldsymbol{x}_i)$ is the total weight of subsystem i

The total cost and weight of subsystem i can be explicitly expressed as follows:

$$C_i(\boldsymbol{x}_i) = \sum_{j=1}^{m_i} c_{ij} x_{ij} \tag{4.23}$$

$$W_i(\boldsymbol{x}_i) = \sum_{j=1}^{m_i} w_{ij} x_{ij} \tag{4.24}$$

where

c_{ij} is the cost of component j available for subsystem i

w_{ij} is the weight of the component j available for subsystem i

The system reliability can be expressed as a function of the decision variables x_{ij} by the following equation:

$$\prod_{i=1}^{N} R_i(\boldsymbol{x}_i|k_i) = \prod_{i=1}^{N} \left(1 - \sum_{l=0}^{k_i-1} \sum_{t \in T_i} \prod_{j=1}^{m_i} \binom{x_{ij}}{t_j} r_{ij}^{t_j}(1 - r_{ij})^{x_{ij}-t_j} \right) \tag{4.25}$$

where

T_i is the set $\{(t_1, t_2, \ldots, t_{m_i})| \sum_{j=1}^{m_i} t_j = l\}$

t is the vector $(t_1, t_2, \ldots, t_{m_i})$

r_{ij} is the reliability of component j available for subsystem i

4.4.2   Genetic Algorithm and Numerical Example

Solution Encoding. Each possible solution to the redundancy allocation problem is a collection of n_i parts in parallel ($k_i \le n_i \le u_i$) for N different subsystems. The n_i parts can be chosen in any combination from among m_i available components. The m_i components are indexed in descending order according to their reliability (i.e., 1 representing the most reliable one, etc.). The solution encoding is an integer vector representation with $\sum u_i$ positions. Each of the N subsystems is represented by u_i positions, with each component listed according to their reliability index. An index of $m_i + 1$ is assigned to a position where an additional component is not used. The subsystem representations are then placed adjacent to each other to complete the vector representation. As an example, consider a system with $N = 3$, $m_1 = 5$, $m_2 = 4$, $m_3 = 5$, and u_i predetermined to be 5 for all i. The following example

$$\boldsymbol{v}_k = [1\ 1\ 6\ 6\ 6 \mid 2\ 2\ 3\ 5\ 5 \mid 4\ 6\ 6\ 6\ 6]$$

represents a prospective solution; two of the most reliable components are used in parallel for the first subsystem, two of the second most reliable and one of the third most reliable components are used in parallel for the second subsystem, and one of the fourth most reliable components is used for the third subsystem.

Initial Population. The initial population is determined by randomly selecting, for each solution vector, N integers between k_i and u_i to represent the number of parts in parallel (n_i) for each particular subsystem. Then, n_i parts are randomly and uniformly selected from among the m_i available components, assuming an unlimited available supply of each type. The chosen components are then sequenced in accordance with their reliability. A population size of 40 was suggested according to their experiences.

Crossover. The crossover operator used performs uniform crossover where common alleles of the parents are retained in the child and noncommon alleles are selected with equal probability from either parent.

Mutation. The mutation operator performs random perturbations. Each value with the solution vector (which is randomly selected to be mutated) is changed with probability equal to the mutation rate. A mutated component was changed to an index of $m_i + 1$ with 50% probability and to a randomly chosen component, from among m_i choices, with 50% probability.

Selection. A survival-of-the-fittest strategy is employed. After crossover breeding, the *pop_size* best solutions from among the previous generation and the new child vectors are retained to form the next generation. Mutation was then performed after culling an inferior solution from the population. The best solution within the population is never chosen for mutation to ensure that the optimal solution is never altered via mutation. This is a form of elitist selection [186].

Adaptive Penalty. The penalty function used here is based on the notion of a *near-feasibility threshold* (NFT) corresponding to a set of constraints. Exterior penalty functions are characterized as being nondecreasing functions of the "distance" of a given solution from the feasible region. The penalty function will encourage the genetic algorithm to explore within the feasible region and the NFT neighborhood of the feasible region and will discourage search beyond that threshold. The penalized objective function for the redundancy allocation problem is given as follows:

$$R_{ip} = R_i - \left(\left(\frac{\Delta w_i}{\text{NFT}_w} \right)^{\alpha} + \left(\frac{\Delta c_i}{\text{NFT}_c} \right)^{\alpha} \right) (R_{\text{all}} - R_{\text{feas}}) \qquad (4.26)$$

where

R_{ip} is the penalized objective function value of solution i

R_i is the unpenalized objective function value of solution i

R_{all} is the unpenalized value of the best solution yet found

R_{feas} is the value of the best feasible solution yet found

NFT_w is the near-feasible threshold for weight constraint

NFT_c is the near-feasible threshold for cost constraint

Δw_i is the value of over-weight for solution i

Δc_i is the value of over-cost for solution i

α is the user-specified severity parameter

Given a specific problem and constraints set, it may not be indicative to what extent the infeasible region should be searched to provide convergence to optimal or near-optimal feasible solutions. Although effective values of NFT could be found through experimentation for any particular redundancy allocation problem, this was considered to be undesirable for a penalty function intended to be robust regarding the specific problem instance. Different rules were proposed for approximating NFT including a dynamic way. The dynamic NFT is defined as follows:

$$NFT = \frac{NFT_0}{1 + \lambda g} \tag{4.27}$$

where NFT_0 is an upper bound or starting point for NFT, g is the generation number, and λ is an arbitrary constant which ensures that the entire region between NFT_0 and zero is searched.

Note that the adaptive term $(R_{all} - R_{feas})$ may cause this penalty approach two kinds of danger: (1) zero-penalty and (2) over-penalty. For the case that $R_{all} = R_{feas}$ even though there exist infeasible solutions, this approach will give zero-penalty to all infeasible solutions; for the case that an infeasible solution with very large value R_{all} occurred at the early stage of the evolutionary process, this approach will give over-penalty to all infeasible solutions.

Test Problems and Results. The test problems used are the original problems posed by Fyffe, Hines, and Lee [145] and the 33 problem variations from Nakagawa and Miyazaki [310]. The problem objective is to maximize reliability for a system with 14 subsystems, three or four component choices for each, and $k = 1$ for all subsystems. For each component alternative, there is a specified reliability, cost, and weight. For the original problem, only identical components could be placed in parallel. In Smith and Coit's studies, they tried to allocate redundant mixing components in subsystems to reach maximal reliability of the overall system.

The 33 problem variations were analyzed by genetic algorithms with different NFT alternatives including:

Table 4.5. Comparison of Feasibility and Performance

Percentage	Zero NFT	5% NFT	3% NFT	1% NFT	Dynamic NFT
Feasible	100.00%	1.21%	80.00%	100.00%	100.00%
Best > N&M	0.00%	63.64%	45.45%	9.09%	81.82%
Total > N&M	0.00%	27.27%	15.45%	0.91%	44.85%

Table 4.6. Comparison of Final Solution

Performanace	Zero NFT	5% NFT	3% NFT	1% NFT	Dynamic NFT
Best	0.97096	0.97337	0.97302	0.97180	0.97366
Average	0.96894	0.97239	0.97167	0.96956	0.97288
Variation (%)	0.14933	0.08218	0.11305	0.15847	0.06573

- Static NFT (5%, 3%, and 1% of each constraint)

- Dynamic NFT

- Zero NFT (only feasibles allowed)

Ten runs were made for each of the 33 problem variations and NFT alternatives. The results are summarized in Tables 4.5 and 4.6. The results in Table 4.5 present, for each of the five NFT alternatives, the percentage of trials where genetic algorithms converged to a final feasible solution. Additionally, the table presents the percentage of problems where the best feasible solution encountered during the GA search exceeded the results from Nakagawa and Miyazaki. Table 4.6 presents the best feasible solution, average feasible solution, and the standard deviation averaged over all 33 problems.

4.5   RELIABILITY OPTIMIZATION WITH FUZZY GOAL AND FUZZY CONSTRAINTS

4.5.1   Problem Formulation

Sasaki, Yokota, and Gen have examined reliability optimization problem under the consideration of fuzzy goal and fuzzy constraints [367]. The corresponding crisp problem is the optimal design of redundancy system with several failure modes as described in Section 4.1.2. The reliability optimization of redundant system with fuzzy goal and fuzzy constraints can be formulated as follows:

$$\max \quad R(\boldsymbol{m}) = \prod_{i=1}^{N} (1 - Q_i(m_i)) \gtrsim g_0 \tag{4.28}$$

$$\text{s.t.} \quad G_t(\boldsymbol{m}) = \sum_{t=1}^{N} g_{ti}(m_i) \lesssim b_t, \qquad t = 1, 2, \ldots, T \tag{4.29}$$

$$1 \le m_i \le u_i, \qquad i = 1, 2, \ldots, N \tag{4.30}$$

The symbols $\lesssim$ and $\gtrsim$ represent fuzzy inequality, g_0 represents the desired goal of the reliability $R(\boldsymbol{m})$ given by decision maker, and the other variables have been defined in the previous section.

Let z_0^- be the worst tolerable value of system reliability. The membership function $\mu_0(\boldsymbol{m})$ for objective (4.28) is defined as follows:

$$\mu_0(\boldsymbol{m}) = \begin{cases} 1, & R(\boldsymbol{m}) > g_0 \\[2mm] \dfrac{R(\boldsymbol{m}) - z_0^-}{g_0 - z_0^-}, & z_0^- \le R(\boldsymbol{m}) \le g_0 \\[2mm] 0, & R(\boldsymbol{m}) < z_0^- \end{cases} \tag{4.31}$$

The membership function $\mu_0(\boldsymbol{m})$ is depicted in Figure 4.5.

Let δ_t be the tolerable overuse of the tth resource. The membership function μ_t for constraint (4.29) is defined as follows:

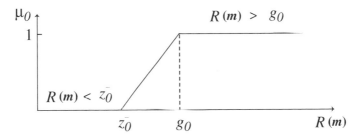

Figure 4.5. Illustration of the membership function $\mu_0(\boldsymbol{m}`$

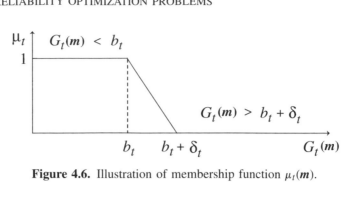

Figure 4.6. Illustration of membership function $\mu_t(m)$.

$$\mu_t(m) = \begin{cases} 1, & G_t(m) < b_t \\ 1 - \dfrac{G_t(m) - b_t}{\delta_t}, & b_t \leq G_t(m) \leq b_t + \delta_t \\ 0, & G_t(m) > b_t + \delta_t \end{cases} \qquad (4.32)$$

The membership function $\mu_t(m)$ is depicted in Figure 4.6. Now we can transform fuzzy reliability optimization problem (4.28)–(4.30) into equivalent crisp problem as follows:

$$\text{max} \quad f(m) = \sum_{t=0}^{T} w_t \mu_t(m) \qquad (4.33)$$

$$\text{s.t.} \quad R(m) - (g_0 - z_0^-)\mu_0(m) \geq z_0^- \qquad (4.34)$$

$$(\mu_t(m) - 1)\delta_t + G_t(m) \leq b_t, \qquad t = 1, 2, \ldots, T \qquad (4.35)$$

$$1 \leq m_i \leq u_i, \qquad i = 1, 2, \ldots, N \qquad (4.36)$$

where $w_t, t = 0, 1, \ldots, T$, are the weights of membership functions μ_t given by the decision maker with $\sum_{t=0}^{T} w_t = 1$. The fuzzy reliability optimization problem is to determine the best number of redundancy units for each subsystem to maximize the grades of system designers' satisfaction for both the achievement of objective of reliability and the best utilization of resources simultaneously.

Let us consider the following example, modified from the one given by Tillman [403], to maximize the nonlinear reliability objective subject to fuzzy nonlinear constraints:

$$\max \quad R(\boldsymbol{m}) = \prod_{i=1}^{6} \left[1 - (1 - (1 - q_{i1})^{m_i+1}) - \sum_{u=2}^{4} (q_{iu})^{m_i+1} \right] \geq 0.93$$

$$\text{s.t.} \quad G_1(\boldsymbol{m}) = (m_1 + 3)^2 + (m_2)^2 + (m_3)^2 + (m_4 + 3)^2 + (m_5)^2 + (m_6)^2 \leq 51$$

$$G_2(\boldsymbol{m}) = 20 \sum_{i=1}^{6} \{m_i + \exp(-m_i)\} \geq 260$$

$$G_3(\boldsymbol{m}) = 20 \sum_{i=1}^{6} \{m_i \exp(-m_i/4)\} \geq 140$$

$$1 \leq m_i \leq 3: \quad \text{positive integer}, \qquad i = 1, 2, \ldots, 6$$

where $\boldsymbol{m} = [m_1 \cdots m_6]$. The subsystems are subject to four failure modes ($s_i = 4$) with one O failure ($h_i = 1$) and three A failures, for subsystem $i = 1, \ldots, 6$. For each subsystem the failure probabilities are shown in Table 4.7.

Let $g_0 = 0.93$, $z_0^- = 0.86$, $\delta_1 = 14.0$, $\delta_2 = 8.0$, $\delta_3 = 8.0$, $w_0 = 0.85$, $w_1 = 0.05$, $w_2 = 0.05$, and $w_3 = 0.05$, respectively. The above problem can be transformed into the following equivalent crisp problem:

$$\max \quad f(\boldsymbol{m}) = 0.85\mu_0(\boldsymbol{m}) + 0.05\mu_1(\boldsymbol{m}) + 0.05\mu_2(\boldsymbol{m}) + 0.05\mu_3(\boldsymbol{m})$$

$$\text{s.t.} \quad R(\boldsymbol{m}) - (g_0 - z_0^-)\mu_0(\boldsymbol{m}) \geq z_0^-$$

$$(\mu_1(\boldsymbol{m}) - 1)\delta_1 + (m_1 + 3)^2 + (m_2)^2 + (m_3)^2 + (m_4 + 3)^2 + (m_5)^2$$
$$+ (m_6)^2 \leq 51$$

$$(1 - \mu_2(\boldsymbol{m}))\delta_2 + 20 \sum_{i-1}^{6} \{m_i + \exp(-m_i)\} \geq 260$$

$$(1 - \mu_3(\boldsymbol{m}))\delta_3 + 20 \sum_{i=1}^{6} \{m_i \exp(-m_i/4)\} \geq 140$$

$$m_i \geq 0: \quad \text{integer}, \qquad i = 1, 2, \ldots, 6$$

4.5.2 Genetic Algorithm and Numerical Example

Chromosome and Initial Population. A chromosome is defined as an ordered list of the number of redundant units m_{ki} shown as follows:

$$\boldsymbol{v}_k = [m_{k1} \; m_{k2} \cdots m_{k6}]$$

where $\boldsymbol{v}_k$ is the kth chromosome.

Table 4.7. Failure Modes and Probabilities in Each Subsystem

Subsystem i	Failure Modes $s_i = 4, h_i = 1$	Failure Probabilities q_{iu}
1	O	0.002
	A	0.05
	A	0.10
	A	0.18
2	O	0.004
	A	0.02
	A	0.15
	A	0.12
3	O	0.005
	A	0.05
	A	0.20
	A	0.10
4	O	0.003
	A	0.01
	A	0.18
	A	0.10
5	O	0.002
	A	0.02
	A	0.10
	A	0.07
6	O	0.002
	A	0.02
	A	0.10
	A	0.10

The initial population of chromosomes is generated randomly within the range [1, 3]. Let *pop_size* = 5; the initial chromosomes are given as follows:

$$v_1 = [1\ 2\ 3\ 3\ 1\ 3]$$
$$v_2 = [3\ 1\ 2\ 2\ 1\ 3]$$
$$v_3 = [1\ 2\ 3\ 3\ 3\ 2]$$
$$v_4 = [2\ 2\ 3\ 3\ 1\ 1]$$
$$v_5 = [3\ 1\ 2\ 2\ 3\ 2]$$

Evaluation of Chromosome. When evaluating chromosomes, each legal one is assigned the value of objective function $f(m)$ and each illegal chromosome is penalized as 0.

$$eval(v_k) = \begin{cases} f(m); & R(m) \geq z_0^-, G_t(m) \leq b_t + \delta_t, \text{ and } 1 \leq m_i \leq u_i \quad \forall i, t \\ 0 & \text{otherwise} \end{cases}$$

$$(4.37)$$

The fitness values of the above chromosomes are

$$eval(\boldsymbol{v}_1) = 0.271185$$
$$eval(\boldsymbol{v}_2) = 0.408735$$
$$eval(\boldsymbol{v}_3) = 0.416843$$
$$eval(\boldsymbol{v}_4) = 0.532714$$
$$eval(\boldsymbol{v}_5) = 0.555988$$

Crossover. One-cut-point crossover is used here. Let the probability of crossover, p_c, be 0.34 and let the sequence of random numbers be

0.089663   0.215613   0.345064   0.526048   0.641652

This means that the chromosomes $\boldsymbol{v}_1$ and $\boldsymbol{v}_2$ were selected for crossover. The position of the cut-point is randomly generated from the range [1, 6]. If we assume that the position is 3, then we have

$$\boldsymbol{v}_1 = [1\ 2\ 3\ |\ 3\ 1\ 3]$$
$$\updownarrow$$
$$\boldsymbol{v}_2 = [3\ 1\ 2\ |\ 2\ 1\ 3]$$

The offspring generated by exchanging the right parts of their parents would be

$$\boldsymbol{o}_1 = [1\ 2\ 3\ 2\ 1\ 3]$$
$$\boldsymbol{o}_2 = [3\ 1\ 2\ 3\ 1\ 3]$$

The fitness values of the offspring are

$$eval(\boldsymbol{o}_1) = 0.291535$$
$$eval(\boldsymbol{o}_2) = 0.388195$$

Mutation. Mutation is performed as random perturbation. For a selected gene m_{ki}, it will be replaced by a random integer within [1, 3]. Let the probability of mutation, p_m, be 0.2; this means that, on average, $6 \times 5 \times 0.2 = 6$ bits would undergo mutation. A total of 30 random numbers were generated, and six of them were smaller than 0.2. The bit number, the relative chromosome, and the bit position within each chromosome are listed below:

Bit Number	Chromosome	Position	Random Number
16	3	4	0.113498
18	3	6	0.038301
29	5	5	0.196936
39	7	3	0.034211
41	7	5	0.154180
42	7	6	0.134129

The offspring resulting from mutation would be

$$o_3 = [1\ 2\ 3\ 2\ 3\ 1]$$
$$o_4 = [3\ 1\ 2\ 2\ 2\ 2]$$
$$o_5 = [3\ 1\ 3\ 2\ 3\ 1]$$

The fitness values of offspring are

$$eval(o_3) = 0.227088$$
$$eval(o_4) = 0.565843$$
$$eval(o_5) = 0.375953$$

Selection. The deterministic selection is adopted as the selection strategy; that is, we delete all duplicate among parents and offspring and then sort them in descending order and select the first *pop_size* chromosomes as the new population.

$$v'_1 = [3\ 1\ 2\ 2\ 2\ 2]\quad (o_4)\qquad eval(v'_1) = 0.565843$$
$$v'_2 = [3\ 1\ 2\ 2\ 3\ 2]\quad (v_5)\qquad eval(v'_2) = 0.555988$$
$$v'_3 = [2\ 2\ 3\ 3\ 1\ 1]\quad (v_4)\qquad eval(v'_4) = 0.532714$$
$$v'_4 = [1\ 2\ 3\ 3\ 3\ 2]\quad (v_3)\qquad eval(v'_3) = 0.416843$$
$$v'_5 = [3\ 1\ 2\ 2\ 1\ 3]\quad (v_2)\qquad eval(v'_1) = 0.408735$$

The best solution from a random run of genetic algorithm with 300 generations is

$$v* = [3\ 3\ 3\ 2\ 2\ 2],\qquad R(m*) = 0.928987$$

The values of corresponding membership function are

$$\mu_0 = 0.986,\qquad \mu_1 = 1.0,\qquad \mu_2 = 1.0,\qquad \mu_3 = 1.0$$

4.6 RELIABILITY OPTIMIZATION WITH INTERVAL COEFFICIENTS

In the past three decades, three major approaches have been proposed for decision-making problems under uncertainties:

- Stochastic programming
- Fuzzy programming
- Interval programming

In the stochastic programming approach, the coefficients of the mathematical model are viewed as random variables and their probability distributions are assumed to be known. In the fuzzy programming approach, the coefficients are viewed as fuzzy sets and their membership functions are assumed to be known. However, in many applications of real-world problems, it is not so easy for decision makers to specify either probability distributions or membership functions; on the contrary, such uncertainty can be easily represented as an *interval of confidence*. This is the motivation to developing interval arithmetic and interval programming [232, 311].

Gen and Cheng have examined the optimal design problem of system reliability with uncertain coefficients. These coefficients can be roughly given as the intervals of confidence. Then this problem is formulated as an interval programming model. A solution procedure developed in Chapter 2 for interval programming problems was used to solve it [150]. The basic idea of the solution procedure is first to transform the nonlinear interval programming model into an equivalent nonlinear bicriteria programming model and then to use genetic algorithms to determine the set of Pareto solutions. The problem is characterized as

- Combinatorial nature
- Nonlinear in both objectives and constraints
- Multiple objectives

This makes the problem very difficult to solve in general. The adaptively moving line method in criteria space is used to construct a fitness function so as to force the genetic search to exploit the nondominated points in the criteria space. The adaptive penalty technique is used to guide the genetic search to approach the Pareto solutions from both the feasible and infeasible regions.

4.6.1 Problem Formulation

Traditional formulations on reliability optimization problems have assumed that the coefficients of models are fixed quantities, and reliability optimization problems are treated as deterministic optimization problems. Because the optimal design of system reliability is resolved in the same stage of overall system design, model coefficients are highly uncertain and imprecise during the design phase and it is usually very difficult to determine the precise values for them. However, these coefficients can be roughly given as the *intervals of confidence*. It means that we can utilize the interval programming technique to deal with such uncertainty in the optimal design problem of system reliability.

Let us consider the following problem in Section 4.1.3. Assume that the system consists of N subsystems. Associated with each subsystem there exist several choices of design alternatives. The problem is to determine which design alternative should be selected, and how many redundant units should be used to achieve the greatest reliability while keeping the total system cost and weight within the allowable limits. It was usually formulated as integer nonlinear programming problem. For the uncertain circumstances, the parameters are replaced by interval numbers and the problem can be formulated as the following interval programming model [150]:

$$\max \quad R(\boldsymbol{m}, \boldsymbol{\alpha}) = \prod_{i=1}^{N} (1 - (1 - R_i(\alpha_i))^{m_i}) \tag{4.38}$$

$$\text{s.t.} \quad G_1(\boldsymbol{m}, \boldsymbol{\alpha}) = \sum_{i=1}^{N} C_i(\alpha_i) m_i \leq C \tag{4.39}$$

$$G_2(\boldsymbol{m}, \boldsymbol{\alpha}) = \sum_{i=1}^{N} W_i(\alpha_i) m_i \leq W \tag{4.40}$$

$$1 \leq \alpha_i \leq \beta_i: \quad \text{integer,} \qquad i = 1, \ldots, N \tag{4.41}$$

$$1 \leq m_i \leq u_i: \quad \text{integer,} \qquad i = 1, \ldots, N \tag{4.42}$$

where

N is the total number of subsystems

m_i is the number of redundant units for subsystem i

α_i is the alternative design for subsystem i

$\boldsymbol{m}$ is a vector $[m_1 \ m_2 \cdots m_N]$

$\boldsymbol{\alpha}$ is a vector $[\alpha_1 \ \alpha_2 \cdots \alpha_N]$

β_i is the upper bound of α_i

u_i is the upper bound of m_i

$R_i(\alpha_i)$ is the interval reliability $R_i(\alpha_i) = [r_i^L(\alpha_i), r_i^R(\alpha_i)]$

$C_i(\alpha_i)$ is the interval coefficient $C_i(\alpha_i) = [c_i^L(\alpha_i), c_i^R(\alpha_i)]$

$W_i(\alpha_i)$ is the interval coefficient $W_i(\alpha_i) = [w_i^L(\alpha_i), w_i^R(\alpha_i)]$

C is the interval overall constraint on the system cost $C = [c^L, c^R]$

W is the interval overall constraint on the system weight $W = [w^L, w^R]$

The problem then can be transformed into an equivalent crisp bicriteria programming problem. There are two key steps when transforming interval programming to bicriteria programming: (1) Using the definition of the degree of inequality-holding-true for two intervals, transform interval constraints into equivalent crisp constraints, and (2) using the definition of the order relation between intervals, transform interval objectives into two equivalent crisp objectives.

First, taking natural logarithms to equation (4.38), we have

$$Z(\boldsymbol{m}, \boldsymbol{\alpha}) = \ln R(\boldsymbol{m}, \boldsymbol{\alpha}) \tag{4.43}$$

$$= \sum_{i=1}^{N} D_i(m_i, \alpha_i) \tag{4.44}$$

where $D_i(m_i, \alpha_i)$ is an objective function with the interval coefficient. It can be further written as follows:

$$D_i(m_i, \alpha_i) = \ln(1 - (1 - R_i(\alpha_i))^{m_i}) \tag{4.45}$$

$$= [d_i^L(m_i, \alpha_i), d_i^R(m_i, \alpha_i)] \tag{4.46}$$

where

$$d_i^L(m_i, \alpha_i) = \ln(1 - (1 - r_i^L(\alpha_i)^{m_i}) \tag{4.47}$$

$$d_i^R(m_i, \alpha_i) = \ln(1 - (1 - r_i^R(\alpha_i)^{m_i}) \tag{4.48}$$

According to Theorem 2.3 of Nakahara et al. [312], interval objective (4.44) is equivalent to the nonlinear interval objective (4.38) in the sense that they keep the same order relations among interval reliability. According to Theorem 2.2 of Ishibuchi and Tanaka [232], for the given values of h_1 and h_2, the degree of inequality holding true for constraints (4.39) and (4.40), the problem (4.38)–(4.42) can be transformed into the following problem:

$$\max \quad z^L(\boldsymbol{m}, \boldsymbol{\alpha}) = \sum_{i=1}^{N} d_i^L(m_i, \alpha_i) \tag{4.49}$$

$$\max \quad z^C(\boldsymbol{m}, \boldsymbol{\alpha}) = \sum_{i=1}^{N} \frac{d_i^L(m_i, \alpha_i) + d_i^R(m_i, \alpha_i)}{2} \tag{4.50}$$

$$\text{s.t.} \quad g_1(\boldsymbol{m}, \boldsymbol{\alpha}) = \sum_{i=1}^{N} c_i(\alpha_i) m_i \le c \tag{4.51}$$

$$g_2(\boldsymbol{m}, \boldsymbol{\alpha}) = \sum_{i=1}^{N} w_i(\alpha_i) m_i \le w \tag{4.52}$$

$$1 \le m_i \le u_i: \quad \text{integer}, \quad i = 1, 2, \ldots, N \tag{4.53}$$

$$1 \le \alpha_i \le \beta_i: \quad \text{integer}, \quad i = 1, 2, \ldots, N \tag{4.54}$$

where the coefficients in constraints are determined as follows:

$$c_i(\alpha_i) = h_1 c_i^R(\alpha_i) + (1 - h_1) c_i^L(\alpha_i) \tag{4.55}$$
$$w_i(\alpha_i) = h_2 w_i^R(\alpha_i) + (1 - h_2) w_i^L(\alpha_i) \tag{4.56}$$
$$c = (1 - h_1) c^R + h_1 c^L \tag{4.57}$$
$$w = (1 - h_2) w^R + h_2 w^L \tag{4.58}$$

4.6.2   Genetic Algorithm

The implementation of genetic algorithms for problem (4.49)–(4.54) is essentially the same as the one developed in Chapter 2 for interval programming.

Chromosome Representation and Initial Population. A chromosome is defined as follows:

$$\boldsymbol{v}_k = [(\alpha_{k1}, m_{k1})(\alpha_{k2}, m_{k2}) \cdots (\alpha_{kN}, m_{kN})]$$

where the subscript k is the chromosome index, α_{ki} is the design alternative for subsystem i, and m_{ki} represents the redundant units.

The initial population is generated randomly within the range $[1, u_i]$ for all m_{ki} and the range $[1, \beta_i]$ for all α_{ki}.

Crossover and Mutation. Crossover is implemented with the uniform crossover operator [391], and mutation is performed as random perturbation within the permissive range of integer variables.

Selection. Deterministic selection is used; that is, we delete all duplicate among parents and offspring and then sort them in descending order and select the first *pop_size* chromosomes as the new population.

Evaluation. There are two main tasks involved in this phase: (1) how to handle infeasible chromosomes and (2) how to determine fitness values of chromosomes according to bicriteria. If we let v_k be the kth chromosome in the current generation, the evaluation function is defined as follows:

$$eval(v_k) = (w_1 z^L(m_k, \alpha_k) + w_2 z^C(m_k, \alpha_k))p(m_k, \alpha_k) \tag{4.59}$$

The evaluation function contains two terms: *weighted-sum objective* and *penalty*. The weighted-sum objective term tries to exert selection pressure to force the genetic search to exploit the set of Pareto solutions, and the penalty term tries to force the genetic search to approach Pareto solutions from both the feasible and infeasible regions.

Weighted-Sum Objective. Let E denote the set of nondominated solutions examined so far. Two special points in E interest us. One point contains the maximum of $z^L(m_k, \alpha_k)$ among others in E and another contains the maximum of $z^C(m_k, \alpha_k)$. Denote these two points as $(z^C_{\min}, z^L_{\max})$ and $(z^C_{\max}, z^L_{\min})$, respectively, where

$$z^C_{\min} = \min\{z^C(m_k, \alpha_k)|m_k, \alpha_k \in E\}$$
$$z^C_{\max} = \max\{z^C(m_k, \alpha_k)|m_k, \alpha_k \in E\}$$
$$z^L_{\min} = \min\{z^L(m_k, \alpha_k)|m_k, \alpha_k \in E\}$$
$$z^L_{\max} = \max\{z^L(m_k, \alpha_k)|m_k, \alpha_k \subset E\}$$

The weighted-sum objective function then can be constructed as follows:

$$w_1 z^L(m_k, \alpha_k) + w_2 z^C(m_k, \alpha_k)$$

where

$$w_1 = z^C_{\max} - z^C_{\min}$$
$$w_2 = z^L_{\max} - z^L_{\min}$$

The line formed with points $(z^C_{\min}, z^L_{\max})$ and $(z^C_{\max}, z^L_{\min})$ divides the criteria space into two half-spaces: One contains positive ideal solution denoted with Z^+, and another contains negative ideal solution denoted with Z^-. The feasible solution space F is correspondingly divided into two parts: One is $F^- = F \cap Z^-$

and another is $F^+ = F \cap Z^+$. It is easy to verify that a solution in F^+ has higher fitness value than those in F^-. Thus chromosomes in the half-space F^+ have a relatively larger chance to enter the next generation. At each generation, the Pareto set E is updated and the two special points may be renewed. This means that along with the evolutionary process, the line formed with these two points will move gradually from the negative ideal point to the positive ideal point (adaptively moving line). In other words, this fitness function gives selection pressure to force the genetic search to exploit the nondominated points in the criteria space.

Penalty. The penalty function is constructed as follows:

$$p(\boldsymbol{m}_k, \boldsymbol{\alpha}_k) = 1 - \frac{1}{2} \sum_{t=1}^{2} \left(\frac{\Delta b_t(\boldsymbol{m}_k, \boldsymbol{\alpha}_k)}{\Delta b_t^{\max}} \right) \tag{4.60}$$

where

$$\Delta b_t(\boldsymbol{m}_k, \boldsymbol{\alpha}_k) = \max\{0, g_t(\boldsymbol{m}_k, \boldsymbol{\alpha}_k) - b_t\} \tag{4.61}$$

$$\Delta b_t^{\max} = \max\{\epsilon, \Delta b_t(\boldsymbol{m}_k, \boldsymbol{\alpha}_k) \mid k = 1, \ldots, pop\_size\} \tag{4.62}$$

where $\Delta b_t(\boldsymbol{m}_k, \boldsymbol{\alpha}_k)$ is the value of violation for constraint t for the kth chromosome, $\Delta b_t^{\max}$ is the maximum of violation for constraint t among the current population, and ϵ is a small positive number used to help the penalty avoid zero division. The penalty term can force the genetic search to approach the Pareto solutions from both feasible and infeasible regions.

Overall Procedure

Procedure: Genetic Algorithms

begin
  $t \leftarrow 0$;
  initialize $P(t)$;
  form Pareto solution set $E(t)$;
  evaluate $P(t)$;
  while (not termination condition) **do**
    recombine $P(t)$;
    update Pareto solution set $E(t)$;
    evaluate $P(t)$;
    select $P(t + 1)$ from $P(t)$;
    $t \leftarrow t + 1$;
  end
end

where the step of update set $E(t)$ works as follows:

Procedure: Update Set E

begin
    compute objective function values of bicriteria for each chromosome;
    add new nondominated points into E;
    delete dominated points from E;
    determine new special points $(z_{\min}^C, z_{\max}^L)$ and $(z_{\max}^C, z_{\min}^L)$;
end

4.6.3   Numerical Example

The numerical example is an extension of the problem given by Fyffe et al. [145]. The system consists of 14 subsystems. Each subsystem has three or four alternative designs, and the possible maximum of redundant units for each subsystem is six. The interval coefficients of the problem are shown in Table 4.8.

The parameters for genetic algorithm were set as follows: population size, pop_size = 40; crossover probability, p_c = 0.4; mutation probability, p_m = 0.4; maximum generation, max_gen = 2000. The values of h_1 and h_2 were set as h_1 = 0.5 and h_2 = 0.5.

The Pareto solutions found by the proposed algorithm are listed in Table 4.9. The corresponding solutions (the optimal designs) are given in Table 4.10. These solutions are depicted in solution space as shown in Figure 4.7, and the corresponding interval numbers (interval objectives) are given in Figure 4.8. We can see from the figure that these interval numbers cannot be compared with each other under the order relation $\leq_{LC}$.

Table 4.8. Interval Coefficients for the Test Problem

	Design Alternative					
	1			2		
Subsystem i	r_i^L, r_i^R	c_i^L, c_i^R	w_i^L, w_i^R	r_i^L, r_i^R	c_i^L, c_i^R	w_i^L, w_i^R
1	[0.86, 0,91]	[1, 4]	[3, 6]	[0.88, 0.94]	[1, 4]	[3, 8]
2	[0.92, 0.97]	[2, 6]	[7, 11]	[0.89, 0,94]	[1, 3]	[10, 13]
3	[0.80, 0.87]	[1, 3]	[4, 9]	[0.85, 0.91]	[3, 6]	[4, 9]
4	[0.78, 0.85]	[1, 5]	[4, 6]	[0.82, 0.88]	[4, 7]	[6, 9]
5	[0.89, 0.94]	[2, 5]	[4, 8]	[0.88, 0.93]	[2, 5]	[3, 7]
6	[0.93, 0.99]	[3, 9]	[5, 10]	[0.93, 0.99]	[5, 9]	[3, 8]
7	[0.86, 0.92]	[4, 7]	[7, 10]	[0.87, 0.93]	[4, 8]	[8, 11]
8	[0.76, 0.82]	[3, 6]	[4, 9]	[0.85, 0.91]	[5, 9]	[7, 10]
9	[0.92, 0.98]	[2, 7]	[8, 12]	[0.93, 0.99]	[3, 8]	[9, 14]
10	[0.78, 0.84]	[4, 9]	[6, 10]	[0.80, 0.86]	[4, 7]	[5, 8]
11	[0.89, 0.95]	[3, 7]	[5, 9]	[0.90, 0.95]	[4, 9]	[6, 10]
12	[0.74, 0.80]	[1, 4]	[4, 9]	[0.77, 0.83]	[3, 6]	[5, 8]
13	[0.93, 0.99]	[2, 6]	[5, 10]	[0.94, 0.98]	[3, 8]	[5, 10]
14	[0.85, 0.91]	[4, 7]	[6, 10]	[0.87, 0.92]	[4, 7]	[7, 10]

	Design Alternative					
	3			4		
Subsystem i	r_i^L, r_i^R	c_i^L, c_i^R	w_i^L, w_i^R	r_i^L, r_i^R	c_i^L, c_i^R	w_i^L, w_i^R
1	[0.87, 0.93]	[2, 6]	[2, 6]	[0.90, 0.96]	[4, 8]	[5, 9]
2	[0.88, 0.94]	[2, 5]	[9, 13]	—	—	—
3	[0.82, 0.88]	[1, 4]	[6, 10]	[0.87, 0.93]	[4, 7]	[4, 8]
4	[0.80, 0.86]	[5, 9]	[4, 9]	—	—	—
5	[0.90, 0.96]	[3, 8]	[5, 9]	—	—	—
6	[0.92, 0.98]	[2, 6]	[5, 10]	[0.91, 0.97]	[2, 7]	[4, 8]
7	[0.89, 0.95]	[5, 9]	[9, 13]	—	—	—
8	[0.86, 0.92]	[6, 9]	[6, 10]	—	—	—
9	[0.91, 0.96]	[4, 7]	[7, 11]	[0.86, 0.92]	[3, 7]	[8, 12]
10	[0.85, 0.90]	[4, 8]	[4, 7]	—	—	—
11	[0.91, 0.96]	[5, 10]	[6, 10]	—	—	—
12	[0.80, 0.86]	[4, 7]	[6, 9]	[0.85, 0.91]	[5, 8]	[7, 12]
13	[0.92, 0.97]	[3, 7]	[6, 10]	—	—	—
14	[0.90, 0.95]	[5, 10]	[5, 10]	[0.94, 0.99]	[6, 11]	[9, 14]

$C = [95, 130]$, $W = [163, 177]$, $u_i = 5$, $i = 1, \ldots, 14$, $\beta_i = 4$ for $i = 1, 3, 6, 9, 12, 14$, and $\beta_i = 3$ for $i = 2, 4, 5, 7, 8, 10, 11, 13$.

Table 4.9. Pareto Solution

n	z^L	z^R
1	0.97801	0.98758
2	0.97826	0.98660
3	0.97832	0.98647

Table 4.10. Optimal Design of System Reliability

Subsystem i	Pareto solutions					
	1		2		3	
	Design	Units	Design	Units	Design	Units
1	1	5	3	5	1	5
2	1	5	1	5	1	5
3	1	4	2	4	1	4
4	1	5	1	4	1	5
5	2	3	1	3	2	3
6	3	3	2	3	3	3
7	2	5	2	3	2	5
8	1	4	1	5	2	3
9	2	4	1	2	2	4
10	1	4	1	5	1	4
11	2	2	1	5	2	2
12	3	4	2	4	3	4
13	1	4	1	2	1	4
14	3	4	3	5	3	4

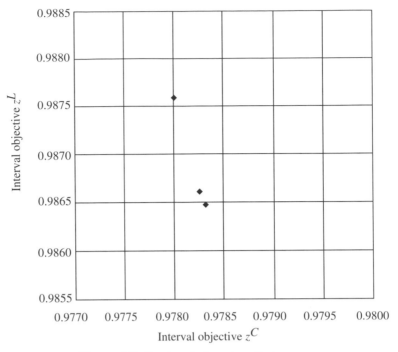

Figure 4.7. Pareto solutions in criteria space.

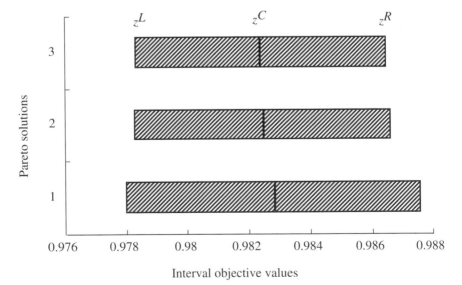

Figure 4.8. Interval objective values.

5

FLOW-SHOP SEQUENCING PROBLEMS

5.1 INTRODUCTION

Since Johnson published the first paper on flow-shop sequencing problem in 1954 [236], this problem has held the attention of many researchers. The flow-shop sequencing problem is generally described as follows: There are m machines and n jobs, each job consists of m operations, and each operation requires a different machine. n jobs have to be processed in the same sequence on m machines. The processing time of job i on machine j is given by t_{ij} ($i = 1, \ldots, n; j = 1, \ldots, m$). The objective is to find the sequence of jobs minimizing the maximum flow time, which is called *makespan*. The main assumptions for this problem are usually made as follows.

- Every job has to be processed on all machines in the order $1, 2, \ldots, m$.
- Every machine processes only one job at a time.
- Every job is processed on one machine at a time.
- The operations are not preemptable.
- The set-up times for the operations are sequence-independent and are included in the processing times.
- The operating sequences of the jobs are the same on every machine, and the common sequence has to be determined.

The flow-shop sequencing problems may be classified into the following three categories:

- Deterministic flow-shop problem
- Stochastic flow-shop problem
- Fuzzy flow-shop problem

Figure 5.1. Gantt chart for a flow-shop sequencing problem.

Deterministic problems assume that fixed processing times of jobs are known. In the stochastic cases, processing times vary according to chosen probability distributions [115]. In fuzzy decision context, a fuzzy due date is assigned to each job to represent the grade of satisfaction of decision makers for the completion time of the job [233, 405]. Furthermore, if any given job has a processing time equal to zero on one or more machines, the flow-shop is called a *general flow-shop problem*, or else it is a *pure flow-shop problem*. The majority of the research effort during the past 30 years has been devoted to pure deterministic flow-shop sequencing problem. Such a problem is usually labeled as $n/m/F/c_{\max}$, which means n-job/m-machine/flow-shop/maximum flow time.

If we let $c(j_i, k)$ denote the completion time of job j_i on machine k and let $\{j_1, j_2, \ldots, j_n\}$ denote a job permutation, then we can calculate the completion times for an n-job m-machine flow-shop problem as follows:

$$c(j_1, 1) = t_{j_1 1} \tag{5.1}$$

$$c(j_1, k) = c(j_1, k - 1) + t_{j_1 k}, \qquad k = 2, \ldots, m \tag{5.2}$$

$$c(j_i, 1) = c(j_{i-1}, 1) + t_{j_i 1}, \qquad i = 2, \ldots, n \tag{5.3}$$

$$c(j_i, k) = \max\{c(j_{i-1}, k), c(j_i, k - 1)\} + t_{j_i k}, \quad i = 2, \ldots, n, k = 2, \ldots, m \tag{5.4}$$

The makespan is calculated as follows:

$$c_{\max} = c(j_n, m) \tag{5.5}$$

Usually a *Gantt chart* is used to represent schedules for flow-shop problems. An example schedule for the 4-job 2-machine problem is shown in Figure 5.1.

The flow-shop problem has been intensively studied in the literature. A recent survey paper about the flow-shop problem is given by Dudek, Panwalkar, and Smith [115], and recent books have been written by Morton and Pentico [299] and by Blazewicz, Ecker, Schmidt, and Weglarz [41].

5.2 TWO-MACHINE FLOW-SHOP PROBLEM

The two-machine flow-shop problem with a makespan objective is known as *Johnson's problem*. An optimal sequence for the two-machine problem can be determined by the well-known Johnson's rule [21]. The ideas in Johnson's rule provide a foundation for later heuristics for the m-machine problem.

Suppose there are n jobs with required processing time $t_{i1}, i = 1, 2, \ldots, n$, on machine 1 and $t_{i2}, i = 1, 2, \ldots, n$, on machine 2. An optimal sequence can be characterized by the following rule for ordering pairs of jobs:

Theorem 5.1 (Johnson's rule). Job i precedes job j in an optimal sequence if $\min\{t_{i1}, t_{j2}\} \leq \min\{t_{i2}, t_{j1}\}$.

An optimal sequence is directly constructed with an adaptation of this result by a one-pass scanning procedure. Let J denote the job list and let S denote the schedule; then Johnson's algorithm can be described as follows:

Algorithm: Johnson's Rule

Step 1. Let $U = \{ j | t_{j1} < t_{j2} \}$ and $V = \{ j | t_{j1} \geq t_{j2} \}$.

Step 2. Sort jobs in U with nondecreasing order of t_{j1}.

Step 3. Sort jobs in V with nonincreasing order of t_{j2}.

Step 4. An optimal sequence is the ordered set U followed by the ordered set V.

To illustrate the algorithm, consider an eight-job problem shown in Table 5.1. The solution is constructed as follows:

Step 1. Job set $U = \{2, 3, 6\}$ and $V = \{1, 4, 5, 7, 8\}$;

Step 2. Sort jobs in U as follows:

Job i:	3	2	6
t_{i1}:	1	2	3

Step 3. Sort jobs in V as follows:

Job i:	5	4	7	1	8
t_{i2}:	6	5	2	2	1

Step 4. an optimal sequence is $\{3, 2, 6, 5, 4, 7, 1, 8\}$.

Table 5.1. Example of Eight-Job Problem

Job i:	1	2	3	4	5	6	7	8
t_{i1}:	5	2	1	7	6	3	7	5
t_{i2}:	2	6	2	5	6	7	2	1

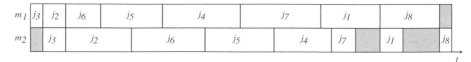

Figure 5.2. Gantt chart of the optimal schedule.

The Gantt chart is shown in Figure 5.2.

5.3 HEURISTICS FOR GENERAL *m*-MACHINE PROBLEMS

Over the past three decades, extensive research has been done on the pure flow-shop problem: There is no easy algorithm which can provide an optimal solution. Integer programming and branch-and-bound techniques can be used to find the optimal solution [227]. However, they are not very effective on large problems or even medium-sized problems. The flow-shop problem has been proved to be NP-complete [78, 356]. Hence, many heuristics have been developed to provide a good and quick solution. Some of the well-known heuristics will be discussed below.

5.3.1 Palmer's Heuristic Algorithm

Palmer proposed a *slope order index* to sequence the jobs on the machines based on the processing time [330]. The idea is to give priority to jobs so that jobs with processing times that tend to increase from machine to machine will receive higher priority, while jobs with processing times that tend to decrease from machine to machine will receive lower priority. The slope index s_i for job i is calculated as

$$s_i = \sum_{j=1}^{m} (2j - m - 1)t_{ij}, \qquad i = 1, 2, \ldots, n \qquad (5.6)$$

Then a permutation schedule is constructed by sequencing the jobs in nonincreasing order of s_i such as

$$s_{i_1} \geq s_{i_2} \geq \cdots \geq s_{i_n} \qquad (5.7)$$

5.3.2 Gupta's Heuristic Algorithm

Gupta suggested another heuristic which is similar to Palmer's heuristic except that he defined his slope index in a different manner, taking into account some interesting facts about optimality of Johnson's rule for the three-machine problem [195]. The slope index s_i for job i is calculated as

$$s_i = \frac{e_i}{\min_{1 \le k \le m-1} \{t_{ik} + t_{i,k+1}\}} \tag{5.8}$$

where

$$e_i = \begin{cases} 1 & \text{if } t_{i1} < t_{im} \\ -1 & \text{if } t_{i1} \ge t_{im} \end{cases} \tag{5.9}$$

Thereafter the jobs are sequenced according to the slope index (5.8).

5.3.3   CDS Heuristic Algorithm

The Campbell, Dudek, and Smith (CDS) heuristic is basically an extension of Johnson's algorithm [53]. Its efficiency relies on two properties:

- Use Johnson's rule in a heuristic fashion.
- Generally create several schedules from which a best schedule can be chosen.

This algorithm first generates a set of $m - 1$ two machine problems by aggregating the m machines into two groups systematically, then applies Johnson's two-machine algorithms to find the $m - 1$ schedules, and finally selects the best one among the schedules. At stage 1, consider the two-machine problem formed with machines 1 and m. At stage 2, consider the artificial two-machine problem: Artificial machine 1 is formed with machine group $\{1, 2\}$, and artificial machine 2 is formed with machine group $\{m, m - 1\}$. At stage k, consider the artificial two-machine problem: artificial machine 1 with machine group $\{1, 2, \dots, k\}$ and artificial machine 2 with machine group $\{m, \dots, m - k + 1\}$. The aggregated processing times are given in Table 5.2. Aggregated processing times for the stage k are defined as follows:

Table 5.2.   The Set of Artificial Two-Machine Problems

Stage i	Artificial Two-Machine Problem		Aggregated Processing Time	
	Group 1	Group 2	t'_{i1}	t'_{i2}
1	1	m	t_{i1}	t_{im}
2	1, 2	$m, m - 1$	$t_{i1} + t_{i2}$	$t_{im} + t_{i,m-1}$
3	1, 2, 3	$m, m - 1, m - 2$	$t_{i1} + t_{i2} + t_{i3}$	$t_{im} + t_{i,m-1} + t_{i,m-2}$
...				
$m - 1$	$1, 2, \dots, m - 1$	$m, m - 1, \dots, 2$	$t_{i1} + t_{i2} + \cdots + t_{i,m-1}$	$t_{im} + t_{i,m-1} + \cdots + t_{i,2}$

$$t'_{i1} = \sum_{j=1}^{k} t_{ij} \quad \text{and} \quad t'_{i2} = \sum_{j=1}^{k} t_{i,m-j+1} \tag{5.10}$$

The CDS heuristic has been found to be a very good and robust heuristic, and it has been used in many studies as a standard of comparison.

5.3.4 RA Heuristic Algorithm

Dannenbring developed a procedure called *rapid access* (RA), which attempts to combine the advantages of Palmer's slope index and the CDS method [91]. Its purpose is to provide a good solution as quickly and easily as possible. Instead of solving $m - 1$ artificial two-machine problems, it solves only one artificial problem using Johnson's rule, in which the processing times are determined from a weighting scheme as follows:

$$t'_{i1} = \sum_{j=1}^{m} w_{j1} t_{ij} \quad \text{and} \quad t'_{i2} = \sum_{j=1}^{m} w_{j2} t_{ij} \tag{5.11}$$

where weights are defined as follows:

$$W_1 = \{w_{j1} \mid j = 1, 2, \ldots, m\} = \{m, m-1, \ldots, 2, 1\}$$
$$W_2 = \{w_{j2} \mid j = 1, 2, \ldots, m\} = \{1, 2, \ldots, m-1, m\}$$

5.3.5 NEH Heuristic Algorithm

The Nawaz, Enscore, and Ham (NEH) heuristic algorithm is based on the assumption that a job with a high total processing time on all the machines should be given higher priority than a job with a low total processing time [315]. The NEH algorithm does not transform the original m-machine problem into one artificial two-machine problem. It builds the final sequence in a constructive way, adding a new job at each step and finding the best partial solution.

Algorithm: NEH

Step 1. Order the n jobs by decreasing the sums of processing times on the machines.

Step 2. Take the first two jobs and schedule them in order to minimize the partial makespan as if there were only these two jobs.

Step 3. For $k = 3$ to n do: insert the k-job at the place, among the k possible ones, which minimizes the partial makespan.

5.4   GEN, TSUJIMURA, AND KUBOTA'S APPROACH

Genetic algorithms have been successfully applied to solve flow-shop problems [57, 233, 256, 350, 366, 405]. In this section, we describe Gen, Tsujimura, and Kubota's approach for the flow-shop problem in detail and show how it works.

5.4.1   Representation

Because the flow-shop problem is essentially a permutation schedule problem, we can use the permutation of jobs as the representation scheme of chromosomes, which is the natural representation for a sequencing problem. For example, let the kth chromosome be

$$\boldsymbol{v}_k = [3\ 2\ 4\ 1]$$

This means that the job sequence is j_3, j_2, j_4, j_1.

5.4.2   Evaluation Function

A simple way to determine the fitness for each chromosome is to use the inverse of makespan. Let $c^k_{\max}$ denote the makespan for the kth chromosome, the fitness is then calculated as follows:

$$eval(\boldsymbol{v}_k) = 1/c^k_{\max} \tag{5.12}$$

5.4.3   Crossover and Mutation

There are many well-known crossover operations available for permutation representation, such as PMX (partially mapped crossover), OX (order crossover), and CX (cycle crossover) (see Chapter 3). Here, Goldberg's PMX was used. An example is given as follows:

Mutation is designed to perform *random exchange*; that is, it selects two genes randomly in a chromosome and exchanges their positions. An example is given as follows:

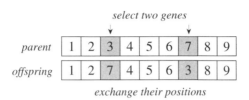

5.4.4   Examples

Example 5.1. Consider a simple example of a five-job two-machine problem given by Baker [21]. The processing time for each job is given in Table 5.3. The parameters are set as follows: *pop_size* = 20, *max_gen* = 20, p_c = 0.3, and p_m = 0.1. We run the genetic algorithm 10 times and obtain an optimal schedule in every trail. The genetic algorithm also finds two new job orderings with the optimal makespan. The results are presented in Table 5.4, and the Gantt charts are shown in Figure 5.3.

Table 5.3. Processing Time

Job j:	1	2	3	4	5
t_{j1}:	3	5	1	6	7
t_{j2}:	6	2	2	6	5

Table 5.4. Comparison of Genetic Algorithm with Johnson's Results

Method	Schedule	Makespan	Method	Schedule	Makespan
Johnson	3-1-4-5-2	24	GA	1-3-4-5-2	24
GA	3-1-4-5-2	24	GA	1-4-3-5-2	24

GA: genetic algorithm

Figure 5.3. Gantt chart for optimal schedules.

Table 5.5. Processing Time

Job j:	1	2	3	4	5
t_{j1}:	31	19	23	13	33
t_{j2}:	41	55	42	22	5
t_{j3}:	25	3	27	14	57
t_{j4}:	30	34	6	13	19

Table 5.6. The Results for Genetic Algorithm Runs

Total Run	Best One	Worst One	Average	Frequency for Getting Best One	Average Generation for Getting Best One
12	213	216	213.5	0.833	37.75

Example 5.2. Consider another example of five-job four-machine problem given by Ho and Chang [218]. The processing time for each job is given in Table 5.5. The parameters are set as follows: $pop\_size = 20$, $max\_gen = 150$, $p_c = 0.3$, and $p_m = 0.1$. We run the genetic algorithm 12 times. The results are summarized in Table 5.6. The comparison with well-known heuristic algorithms is given in Table 5.7, and the Gantt chart for the best one is shown in Figure 5.4.

Table 5.7. The Comparison with Heuristics

Method	Solution	Makespan
Genetic algorithm	4-2-5-1-3	213
Ho	4-2-5-1-3	213
CDS	4-2-1-5-3	246
Dannenbring	5-1-2-3-4	256
Gupta	2-1-5-3-4	251
Palmer	5-2-4-1-3	245
Random	1-2-3-4-5	286

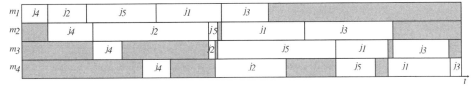

Figure 5.4. Gantt chart for optimal schedules.

The schedule obtained with the genetic algorithm method is the same as one obtained by Ho and is better than other heuristics.

5.5   REEVES' APPROACH

Reeves proposed an implementation of the genetic algorithm method for solving flow-shop sequencing problem [350]. He tested his genetic algorithm on Taillard's benchmarks [393] and concluded that simulated annealing algorithms and genetic algorithms produce comparable results for the flow-shop sequencing problem for most sizes and types of problems, but genetic algorithms will perform relatively better for large problems and will reach a near-optimal solution more quickly. This is encouraging, because Ogbu and Smith regard simulated annealing as outperforming all other heuristics [319].

5.5.1   Initial Population

Reeves hybridized the conventional heuristic in the initial generation to produce a good seed chromosome. One chromosome is generated with the NEH heuristic algorithm, while the remaining chromosomes are generated randomly. He compared this hybrid approach with the one without a seeded population and commented that the one with the seeded population appeared to arrive at its final solution more quickly, with no observed diminution in solution quality.

Chen, Vempati, and Aljaber have proposed a similar hybrid method to generate the initial population [57]. For n-job m-machine flow-shop problem, $m-1$ schedules were generated with CDS and one schedule was generated with the RA heuristic; the remaining ones were generated by randomly mutating the already generated schedules.

5.5.2   Genetic Operators

Reeves adopted one-cut-point crossover in his implementation. One-cut-point crossover picks one-cut-point randomly, and then it takes the pre-cut section of the first parent and fills up the offspring by taking in order each legitimate gene from the second parent as shown in Figure 5.5.

Two types of mutation were tried. The first, an *exchange* mutation, was a simple exchange of two genes of the chromosome, chosen at random. The second, a *shift* mutation, was a shift of one gene (chosen randomly) to the right or left a random number of places.

A few experiments were made in which *shift* seemed to be better than *exchange*, so this was used as shown in Figure 5.6.

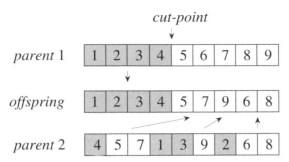

Figure 5.5. One-point crossover.

5.5.3 Selection

In essence, if the population consists of many chromosomes whose fitness values are relatively close, the resolution between "good" and "bad" ones is not of a high order. In this case, relative fitness measure is not very good at distinguishing between chromosomes. Reeves used a simple ranking method as a selection mechanism. The selection of parents was then made in accordance with the following probability distribution:

$$p_k = \frac{2k}{M(M+1)} \tag{5.13}$$

where k refers to the kth chromosome in ascending order (i.e., descending order of makespan) and M refers to the fittest one. This implies that the median value has a chance of $1/M$ of being selected, while the fittest one (the Mth chromosome) has a chance of $2/(M+1)$, roughly twice that of the median.

A pseudo-code description of Reeves' implementation is given below:

Step 0. System parameters

$$M = 30, \quad p_c = 1.0, \quad p_m^{\text{init}} = 0.8, \quad \theta = 0.99, \quad p_m = p_m^{\text{init}}, \quad \text{and} \quad D = 0.95$$

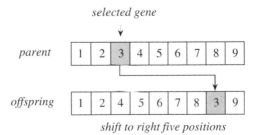

Figure 5.6. Shift mutation.

Step 1. Initial population

    generate(NEH_sequence);
    evaluate(NEH_sequence);
    $pop\_no \leftarrow 1$;
    repeat
      $pop\_no \leftarrow pop\_no + 1$;
      generate(random_sequence);
      evaluate(random_sequence);
    until $pop\_no = M$
    sort(population);
    calculate(population_statistics).

Step 2. Crossover

    if $(random\_no < p_c)$ **then**
      select(parent_1) using fitness rank distribution;
      select(parent_2) using uniform distribution;
      choose(crossover_point);
      crossover;
    end

Step 3. Mutation

    if $(random\_no < p_m)$ **then**
      mutation;

Step 6. Evaluation and selection

    evaluate(new_sequence);
    select(old_sequence) from unfit_members;
    delete(old_sequence) from population;
    insert(new_sequence) into population;
    update(population_statistics);

Step 4. Update data

    $p_m \leftarrow \theta p_m$;
    if $(v_{\min}/v_{\mean} > D)$ **then**
    $p_m := p_m^{\text{init}}$;
    if $cpu\_time > max\_cpu\_time$ **then**
      stop;
    else
      go to step 2.
    end

5.5.4   Numerical Example

Taillard [393] has produced a set of test problems which he found to be particularly difficult, in the sense that the best solutions he could find by a very lengthy tabu search (TS) procedure were still substantially inferior to their lower bounds. The problems range from small instances with 20 jobs and 5 machines to large instances with 500 jobs and 20 machines. There were 10 instances for each size of problem, and all the processing times were generated randomly from a $U(1, 100)$ distribution.

Reeves tested his method with these problems. The results are presented in Table 5.8, where the solution quality is measured by the mean percentage difference from Taillard's upper bounds, averaged over the ten instances within each group of problems.

It can be seen that simulated annealing algorithms and genetic algorithms outperform neighborhood search algorithms on all except the smallest problems. In comparing simulated annealing algorithms and genetic algorithms directly, there is again little to choose between them overall in that their distances from the upper bounds were very similar. On the whole, simulated annealing algorithms performed very slightly better than genetic algorithms, but the figures show that genetic algorithms did better for the larger problems. It is also worth

Table 5.8. Comparison of Solution Quality Among Genetic Algorithms, Simulated Annealing Algorithms, and Neighborhood Search Algorithms

Problem Size	Method		
	Genetic Algorithm	Simulated Annealing Algorithm	Neighborhood Search Algorithm
20/5	1.61	1.27	1.46
20/10	2.29	1.71	2.02
20/20	1.95	0.86	1.10
50/5	0.45	0.78	0.79
50/10	2.28	1.98	3.21
50/20	3.44	2.86	3.90
100/5	0.23	0.56	0.76
100/10	1.25	1.33	2.69
100/20	2.95	2.32	3.98
200/10	0.50	0.83	3.81
200/20	1.35	1.74	6.07
500/20	0.22	0.85	9.07
Average	1.50	1.42	3.24

Reprinted from *Computers and Operations Research*, vol. 22, C. Reeves, "A genetic algorithm for flow shop sequencing," p. 11, 1995 with kind permission.

pointing out that for the last set, genetic algorithms actually found nine solutions (out of ten) which improved on Taillard's upper bounds.

5.6   ISHIBUCHI, YAMAMOTO, MURATA, AND TANAKA'S APPROACH

Ishibuchi, Yamamoto, Murata, and Tanaka have proposed fuzzy flow-shop scheduling problems and have applied genetic algorithm and neighborhood search algorithms to solve fuzzy flow-shop scheduling problems [233]. They showed that the hybrid genetic algorithm with neighborhood search performs well.

5.6.1   Fuzzy Flow-Shop Problem

Ishibuchi et al. formulated two fuzzy flow-shop problems with the concept of fuzzy due dates. One fuzzy flow-shop scheduling problem is to maximize the minimum grade of satisfaction over given jobs, and the other is to maximize the total grade of satisfaction. The membership function of a fuzzy due date assigned to each job represents the grade of satisfaction of a decision maker for the completion time of that job, which is represented as a trapezoid membership function as shown in Figure 5.7, where c_j is the completion time of job j and $\mu_j(c_j)$ is the membership function of the fuzzy due date of job j. The membership function can be written as

$$\mu_j(c_j) = \begin{cases} 0 & \text{if} \quad c_j \le e_j^U \\ 1 - (e_j^U - c_j)/(e_j^U - e_j^L) & \text{if} \quad e_j^L < c_j < e_j^U \\ 1 & \text{if} \quad e_j^U \le c_j \le d_j^L \\ 1 - (c_j - d_j^L)/(d_j^U - d_j^L) & \text{if} \quad d_j^L < c_j < d_j^U \\ 0 & \text{if} \quad d_j^U \le c_j \end{cases} \qquad (5.14)$$

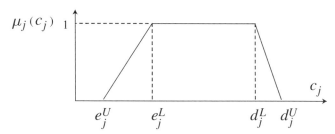

Figure 5.7. Trapezoid membership function.

Then the maximization problem of the minimum grade of satisfaction is formulated as

$$\max f_{\min} = \min\{\mu_j(c_j): j = 1, 2, \ldots, n\} \tag{5.15}$$

and the maximization problem of the total grade of satisfaction is formulated as

$$\max f_{\text{sum}} = \sum_{j=1}^{n} \mu_j(c_j) \tag{5.16}$$

5.6.2 Hybrid Genetic Algorithm

Genetic Operations. Various genetic operators are applicable for sequencing problems. Position-based crossover and shift mutation were used because they showed good performance in their preliminary computation simulations. Position-based crossover was discussed by Syswerda [392]. The crossover picks some positions randomly, and then it takes the genes under the positions from the first parent and fills up the offspring by taking in order each legitimate gene from the second parent as shown in Figure 5.8.

Fitness Calculation. Let Ψ_t denote the population in the tth generation, and let x_t^i denote the ith chromosome in Ψ_t; that is,

$$\Psi_t = \{x_t^i \quad i = 1, 2, \ldots, pop\_size\}$$

The objective function $f(x)$ is employed as a fitness value, and selection probability $p(x_t^i)$ of the ith chromosome in the tth generation ψ_t is calculated as follows:

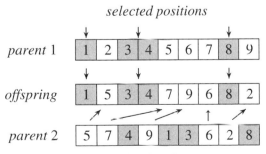

Figure 5.8. Position-based crossover.

$$p(x_t^i) = \frac{[f(x_t^i) - f_M(\Psi_t)]^2}{\sum_{x_t^i \in \Psi_t} [f(x_t^i) - f_M(\Psi_t)]^2} \tag{5.17}$$

where

$$f_M(\Psi_t) = \min\{f(x_t^i): x_t^i \in \Psi_t\} \tag{5.18}$$

Multistart Descent Algorithm. Ishibuchi et al. incorporated a multistart descent algorithm into a genetic algorithm to improve the performance of the conventional version of the latter. A multistart descent algorithm from random initial solution can be described as follows:

Procedure: Multistart Descent

Step 1. Initialization. Randomly generate an initial solution x.

Step 2. Neighborhood search. Examine the solutions in the neighborhood of the current solution x in random order. Let $y*$ be the first solution that improves the current one. If there is no solution in the neighborhood that improves the current one, then let $y*$ be ϕ, where ϕ shows that $y*$ is empty.

Step 3. Termination test. If a prespecified stopping condition is satisfied, then stop this algorithm; otherwise, return to step 1 if $y*$ is empty. If $y*$ is not empty, let $x := y*$ and return to step 2. The total number of evaluations of solutions is employed as a stopping condition.

Hybrid Genetic Algorithm. The overall procedure of Ishibuchi, Yamamoto, Murata, and Tanaka's approach is summarized as follows:

Procedure: Ishibuchi, Yamamoto, Murata, and Tanaka's Approach

Step 1. Initialization.

Step 2. Descent search. Apply the multistart descent algorithm to the current population, where chromosomes in the current population are used as the initial solution of the descent search procedure. Let the solutions obtained through multistart descent be the current population.

Step 3. Selection.

Step 4. Position-based crossover.

Step 5. Shift mutation.

Step 6. Elitist strategy. Randomly remove one solution from the current population and add the best solution of the previous population to the current population.

Step 7. Termination test. If a prespecified stopping condition is satisfied, stop the algorithm; otherwise, return to step 2.

5.6.3   Numerical Example

Ishibuchi et al. compared the hybrid genetic algorithm with other methods on randomly generated problems whose size ranged from 10 jobs to 50 jobs [233]. The comparative results are presented in Table 5.9. It is clear that the genetic descent algorithm outperforms the others.

Table 5.9. GAs and Other Methods: Comparative Results[a]

Problem Size	Total Number of Examined Chromosomes	GA	DA	TS	SA	GDA
10 jobs	10,000	0.49	0.50	0.50	0.50	0.50
	100,000	0.50	0.50	0.50	0.50	0.50
20 jobs	10,000	0.34	0.40	0.39	0.39	0.41
	100,000	0.40	0.47	0.49	0.47	0.49
50 jobs	10,000	0.00	0.04	0.01	0.10	0.04
	100,000	0.11	0.18	0.15	0.26	0.24

[a]GA, genetic algorithm; DA, multistart descent algorithm; TS, tabu search; SA, simulated annealing algorithm; GDA, genetic descent algorithm.

Reprinted from *Fuzzy Sets and Systems*, vol. 67, H. Ishibuchi, N. Yamamoto, T. Murata, and H. Tanaka, "Genetic algorithms and neighborhood search algorithms for fuzzy flowshop scheduling problems," p. 67, 1994, with kind permission.

6

JOB-SHOP SCHEDULING PROBLEMS

6.1 INTRODUCTION

Machine scheduling problems arise in diverse areas such as flexible manufacturing systems, production planning, computer design, logistics, communication, and so on. A common feature of many of these problems is that no efficient solution algorithm is known yet for solving it to optimality in polynomial time. The classical job-shop scheduling problem is one of the best-known machine scheduling problems. Informally, the problem can be described as follows: There are a set of jobs and a set of machines. Each job consists of a chain of operations, each of which needs to be processed during an uninterrupted time period of a given length on a given machine. Each machine can process at most one operation at a time. A schedule is an allocation of the operations to time intervals on the machines. The problem is to find a schedule of minimum length. During the past three decades, the problem has captured the interest of a significant number of researchers, and a huge amount of literature has been published. Among these are the books by Muth and Thompson [307], Conway et al. [84], Baker [21], French [144], Rinnooy Kan [356], Coffman [78], Blazewicz et al. [41], and, most recently, Morton and Pentico [299].

The job-shop scheduling problem (JSP) is among the hardest combinatorial optimization problems [148]. Because of its inherent intractability, heuristic procedures are an attractive alternative. Most conventional heuristic procedures use a *priority rule*—that is, a rule for choosing an operation from a specified subset of as yet unscheduled operations. In recent years, a resurgence of interest in using a local search to solve job-shop problems has occurred because of the development of probabilistic local search methods such as simulated annealing (SA) [409], tabu search (TS) [107], and genetic algorithms (GAs) [15, 86, 100,

Table 6.1. Recent Papers on Genetic Algorithms and Job-Shop Problems

Authors	Description
Davis (1985) [100]	Preference-list-based representation
Nakano and Yamada (1991) [313]	Job-pair-relation-based representation
Falkenauer and Bouffoix (1991) [122]	Preference-list-based representation
Bagchi et al. (1991) [15]	Problem-specific representation
Yamada and Nakano (1992) [423]	Completion-time-based representation
Tamaki and Nishikawa (1992) [397]	Disjunctive-graph-based representation
Paredis (1992) [332]	Job-pair-relation-based representation
Holsapple et al. (1993) [221]	Job-based representation
Fang et al. (1993) [124]	Operation-based representation
Gen et al. (1994) [29]	Operation-based representation
Bean (1994) [29]	Random key representation
Dorndorf and Pesch (1995) [112]	Priority-rule-based and machine-based representation
Croce et al. (1995) [86]	Preference-list-based representation
Kobayashi et al. (1995) [249]	Preference-list-based representation
Norman and Bean (1995) [317]	Random key representation

112, 122, 124, 160, 165, 221, 313, 332, 363, 378, 397, 423, 460]. Recent papers on applying genetic algorithms to solve classical job-shop scheduling problem are listed in Table 6.1.

In this chapter, we give a review of recent literature on solving the classical job-shop scheduling problem using genetic algorithms and illustrate in detail some typical implementations of genetic algorithms.

6.2 CLASSICAL JOB-SHOP MODEL

The classical job-shop scheduling problem can be stated as follows: there are m different machines and n different jobs to be scheduled. Each job is composed of a set of operations and the operation order on machines is prespecified. Each operation is characterized by the required machine and the fixed processing time. There are several constraints on jobs and machines:

- A job does not visit the same machines twice.
- There are no precedence constraints among operations of different jobs.
- Operation cannot be interrupted.
- Each machine can process only one job at a time.
- Neither release times nor due dates are specified.

The problem is to determine the operation sequences on the machines in order to minimize the makespan—that is, the time required to complete all jobs.

Table 6.2. Example of a Three-Job Three-Machine Problem

	Processing Time				Machine Sequence		
	Operations				Operations		
Job	1	2	3	Job	1	2	3
j_1	3	3	2	j_1	m_1	m_2	m_3
j_2	1	5	3	j_2	m_1	m_3	m_2
j_2	3	2	3	j_3	m_2	m_1	m_3

An example of a three-job three-machine problem is presented in Table 6.2. We use Gantt charts to illustrate the feasible schedules. There are two kinds of Gantt charts: machine Gantt chart and job Gantt chart. We use an index *jom* to name each operation, where *j* indicates the job, *o* indicates the operation of job *j*, and *m* indicates the machine to perform this operation. For example, index 321 means that the second operation of job 3 will be processed on machine 1. Figure 6.1 shows this schedule from the perspective of what time the various jobs are on each machine, while Figure 6.2 shows the schedule from the perspective of when the operations of a job are processed.

In principle, there are an infinite number of feasible schedules for a job-shop problem, because superfluous idle time can be inserted between two operations.

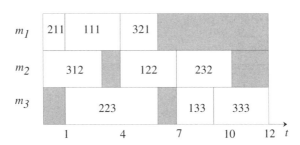

Figure 6.1. Machine Gantt chart.

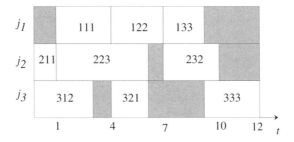

Figure 6.2. Job Gantt chart.

We may shift the operation to the left as compact as possible. A shift in a schedule is called *local left-shift* if some operations can be started earlier in time without altering the operation sequence. A shift is called *global left-shift* if some operation can be started earlier in time without delaying any other operation even though the shift has changed the operation sequence. Based on these two concepts, three kinds of schedules can be distinguished as follows:

Definition 6.1 (Semiactive Schedule). A schedule is semiactive if no local left-shift exists.

Definition 6.2 (Active Schedule). A schedule is active if no global left-shift exists.

Definition 6.3 (Nondelay Schedule). A schedule is nondelay if no machine is kept idle at a time when it could begin processing some operations.

The relationship among active, semiactive, and nondelay schedules is shown by the Venn diagram in Figure 6.3. Optimal schedule is within the set of active schedules. The nondelay schedules are smaller than active schedules, but there is no guarantee that the former will contain an optimum [21].

6.2.1 IP Model

The integer programming (IP) formulation is discussed by Baker [21] and more completely by Greenberg [184]. This formulation relies on the indicator variable to specify operation sequence. Cheng, Gen, and Tsujimura proposed a new integer program formulation in order to give a better representation of a precedence constraint using a system of linear inequalities than previous formulations [71].

Two kinds of constraints need to be considered for the classical job-shop scheduling problem:

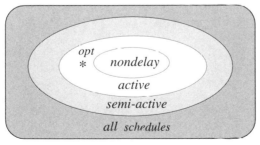

Figure 6.3. Venn diagram of schedule relationship. Reprinted from *Introduction to Sequencing and Scheduling*, K. R. Baker, p. 186, 1974, with kind permission.

- Operation precedence constraint for a given job
- Operation un-overlapping constraint for a given machine

As usual, let c_{jk} denote the completion time of job j on machine k and let t_{jk} denote the processing time of job j on machine k. First we consider the operation precedence constraint for a given job-shop schedule. For a job i, if the processing on machine h precedes that on machine k, we need the following constraint:

$$c_{ik} - t_{ik} \geq c_{ih}$$

If, on the other hand, the processing on machine k comes first, then we need the following constraint:

$$c_{ih} - t_{ih} \geq c_{ik}$$

Since one or the other of these must hold, they are called *disjunctive constraints*. It is useful to define an indicator coefficient a_{ihk} as follows:

$$a_{ihk} = \begin{cases} 1, & \text{processing on machine } h \text{ precedes that on machine } k \text{ for job } i \\ 0, & \text{otherwise} \end{cases}$$

Then we can rewrite the above constraints as follows:

$$c_{ik} - t_{ik} + M(1 - a_{ihk}) \geq c_{ih}, \qquad i = 1, 2, \ldots, n, \quad h, k = 1, 2, \ldots, m$$

where M is a large positive number. Note that the inequality very cleverly capture the precedent constraints. If processing on machine h comes first, $M(1 - a_{ihk})$ is zero, giving the inequality as desired. On the other hand, $M(1 - a_{ihk})$ becomes very large, making the inequality true also. Now we consider the operation un-overlapping constraint for a given machine. For two jobs i and j, both need to be processed on machine k. If job i comes before job j, we need the following constraint:

$$c_{jk} - c_{ik} \geq t_{jk}$$

If, on the other hand, job j comes first, then we need the following constraint:

$$c_{ik} - c_{jk} \geq t_{ik}$$

We define an indicator variable x_{ijk} as follows:

$$x_{ijk} = \begin{cases} 1, & \text{if job } i \text{ precedes job } j \text{ on machine } k \\ 0, & \text{otherwise} \end{cases}$$

Then we can rewrite the above constraints as follows:

$$c_{jk} - c_{ik} + M(1 - x_{ijk}) \geq t_{jk}, \qquad i,j = 1, 2, \ldots, n, \quad k = 1, 2, \ldots, m$$

The job-shop scheduling problem with a makespan objective can be formulated as follows:

$$\min \quad \max_{1 \leq k \leq m} \left\{ \max_{1 \leq i \leq n} \{c_{ik}\} \right\} \tag{6.1}$$

$$\text{s.t.} \quad c_{ik} - t_{ik} + M(1 - a_{ihk}) \geq c_{ih}, \qquad i = 1, 2, \ldots, n, \quad h, k = 1, 2, \ldots, m \tag{6.2}$$

$$c_{jk} - c_{ik} + M(1 - x_{ijk}) \geq t_{jk}, \qquad i,j = 1, 2, \ldots, n, \quad k = 1, 2, \ldots, m \tag{6.3}$$

$$c_{ik} \geq 0, \qquad i = 1, 2, \ldots, n, \quad k = 1, 2, \ldots, m \tag{6.4}$$

$$x_{ijk} = 0 \text{ or } 1, \qquad i,j = 1, 2, \ldots, n, \quad k = 1, 2, \ldots, m \tag{6.5}$$

The objective is to minimize the makespan. Constraint (6.2) ensures that the processing sequence of operations for each job corresponds to the prescribed order. Constraint (6.3) ensures that each machine can process only one job at a time.

If we consider the mean flowtime problem, the objective can be rewritten as

$$\min \quad \sum_{i=1}^{n} \max_{1 \leq k \leq m} \{c_{ik}\} \tag{6.6}$$

6.2.2   LP Model

The linear programming (LP) formulation is discussed in Adams et al. [3]. Let $N = \{0, 1, 2, \ldots, n\}$ denote the set of operations where 0 and n are considered as the dummy operations "start" and "finish," let $M = \{1, 2, \ldots, m\}$ denote the set of machines, let A denote the set of pairs of operations constrained by precedence relations for each job, and let E_k denote the set of pairs of operations to be performed on machine k and which therefore cannot overlap in time. For each operation i, its processing time d_i is fixed, and the start time of the operation t_i is a variable that has to be determined during the optimization. Hence, the job-shop scheduling problem can be formulated as follows:

$$\min \quad t_n \tag{6.7}$$

$$\text{s.t.} \quad t_j - t_i \geq d_i, \quad (i,j) \in A \tag{6.8}$$

$$t_j - t_i \geq d_i \quad \text{or} \quad t_i - t_j \geq d_j, \quad (i,j) \in E_k, \quad k \in M \tag{6.9}$$

$$t_i \geq 0 \tag{6.10}$$

The objective is to minimize the makespan. Constraint (6.8) ensures that the processing sequence of operations for each job corresponds to the prescribed order. Constraint (6.9) ensures that each machine can process only one job at a time. A feasible solution to this problem is called a *schedule*.

6.2.3 Graph Model

The job-shop scheduling problem can be represented with a *disjunctive graph* [23, 360]. The disjunctive graph $G = (N, A, E)$ is defined as follows: N contains nodes representing all operations, A contains arcs connecting consecutive operations of the same job, and E contains disjunctive arcs connecting operations to be processed by the same machine. A disjunctive arc can be settled by either of its two possible orientations. The construction of a schedule will settle the orientations of all disjunctive arcs so as to determine the sequences of operations on same machines. Once a sequence is determined for a machine, the disjunctive arcs connecting operations to be processed by the machine will be replaced by the usual (oriented) precedence arrow, or *conjunctive* arc. The set of disjunctive arcs E can be decomposed into cliques E_k, $E = E_1 \cup E_2 \cup E_3 \cdots \cup E_m$, one for each machine. Figure 6.4 illustrates the disjunctive graph for a three-job three-machine instance, where each job consists of three operations. The nodes of $N = \{0, 1, 2, 3, 4, 5, 6, 7, 8, 9, 10\}$ correspond

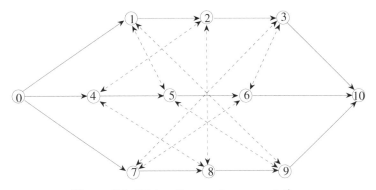

Figure 6.4. Disjunctive graph representation.

to operations, where nodes 0 and n are the dummy operations "start" and "finish." The conjunctive arcs of $A = \{(1,2),(2,3),(4,5),(5,6),(7,8),(8,9)\}$ correspond to precedence constraints on operations of same jobs. The disjunctive arcs (dashed lines) of $E_1 = \{(1,5),(5,9),(9,1)\}$ correspond to operations to be performed on machine 1, disjunctive arcs of $E_2 = \{(4,2),(2,8),(8,4)\}$ correspond to operations to be performed on machine 2, and disjunctive arcs of $E_3 = \{(7,3),(3,6),(6,7)\}$ correspond to operations to be performed on machine 3.

The job-shop scheduling problem is to find the order of the operations on each machine—that is, to settle the orientation of the disjunctive arcs such that the resulting graph is acyclic (there are no precedence conflicts between operations) and that the length of the maximum weight path between the start and end nodes is minimal. The length of a maximum weight (or longest) path determines the makespan.

6.3   CONVENTIONAL HEURISTICS

Job-shop scheduling is a very important everyday practical problem. Since job-shop scheduling is among the hardest combinatorial optimization problems, it is therefore natural to look for approximation methods that produce an acceptable schedule in useful time. The heuristic procedures for a job-shop problem can be roughly classified into two classes:

- One-pass heuristic
- Multipass heuristic

A one-pass heuristic simply builds up a single complete solution by fixing one operation in the schedule at a time based on *priority dispatching rules*. There are many rules for choosing an operation from a specified subset to be scheduled next. This heuristic is fast, and it usually finds solutions that are not too difficult. In addition, a one-pass heuristic may be used repeatedly to build a more sophisticated multipass heuristic in order to obtain better schedules at some extra computational cost.

6.3.1   Priority Dispatching Heuristics

Priority rules are probably the most frequently applied heuristics for solving scheduling problems because of their ease of implementation and their low time complexity. The algorithms of Giffler and Thompson can be considered as the common basis of all priority-rule-based heuristics [389]. Giffler and Thompson have proposed two algorithms to generate schedule: *active schedule* and *nondelay schedule* generation procedures [167]. A nondelay schedule has the property that no machine remains idle if a job is available for processing. An active schedule has the property that no operation can be started earlier without delaying another job. Active schedules form a much larger set and include non-delay schedules as a subset. The generation procedure of Giffler and Thompson is a tree-structured approach. The nodes in the tree correspond to partial schedules, the arcs represent the possible choices, and the leaves of the tree are the set of enumerated schedules. For a given partial schedule, the algorithm essentially identifies all processing conflicts (i.e., operations competing for the same machine), and an enumeration procedure is used to resolve these conflicts in all possible ways at each stage. By contrast, heuristics resolve these conflicts with priority dispatching rules; that is, they specify a priority rule for selecting one operation among the conflicting operations.

Generation procedures operate with a set of schedulable operations at each stage. Schedulable operations are unscheduled operations with immediately scheduled predecessors; this set can be simply determined from the precedence structure. The number of stages for a one-pass procedure is equal to the number of operations $m \times n$. At each stage, one operation is selected to be added into the *partial schedule*. Conflicts among operations are solved by priority dispatching rules. Following the notations of Baker [21], we have

PS_t = a partial schedule containing t scheduled operations

S_t = the set of schedulable operations at stage t, corresponding to a given PS_t

σ_i = the earliest time at which operation $i \in S_t$ could be started

ϕ_i = the earliest time at which operation $i \in S_t$ could be completed

For a given active partial schedule, the potential start time σ_i is determined by the completion time of the direct predecessor of operation i and the latest completion time on the machine required by operation i. The larger of these two quantities is σ_i. The potential finishing time ϕ_i is simply $\sigma_i + t_i$, where t_i is the processing time of operation i. The procedure to generate an active schedule is as follows:

Algorithm: Priority Dispatching Heuristic (Active Schedule Generation)

Step 1. Let $t = 0$ and begin with PS_t as the null partial schedule. Initially S_t includes all operations with no predecessors.

Step 2. Determine $\phi_t^* = \min_{i \in S_t} \{\phi_i\}$ and the machine m^* on which ϕ_t^* could be realized.

Step 3. For each operation $i \in S_t$ that requires machine m^* and for which $\sigma_i < \phi_t^*$, calculate a priority index according to a specific priority rule. Find the operations with the smallest index and add this operation to PS_t as early as possible, thus creating a new partial schedule PS_{t+1}.

Step 4. For PS_{t+1}, update the data set as follows:
1. Remove operations i from S_t.
2. Form S_{t+1} by adding the direct successor of operation j to S_t.
3. Increment t by one.

Step 5. Return to step 2 until a complete schedule is generated.

If we replace step 2 and step 3 of the algorithm with those given in the following algorithm, the procedure can generate a nondelay schedule, which is shown below.

Algorithm: Priority Dispatching Heuristic (Nondelay Schedule Generation)

Step 1. Let $t = 0$ and begin with PS_t as the null partial schedule. Initially S_t includes all operations with no predecessors.

Step 2. Determine $\sigma_t^* = \min_{i \in S_t} \{\sigma_i\}$ and the machine m^* on which σ_t^* could be realized.

Step 3. For each operation $i \in S_t$ that requires machine m^* and for which $\sigma_i = \sigma_t^*$, calculate a priority index according to a specific priority rule. Find the operations with the smallest index and add this operation to PS_t as early as possible, thus creating new partial schedule PS_{t+1}.

Step 4. For PS_{t+1}, update the data set as follows:
1. Remove operations i from S_t.
2. Form S_{t+1} by adding the direct successor of operation j to S_t.
3. Increment t by one.

Step 5. Return to step 2 until a complete schedule is generated.

The remaining problem is to identify an effective priority rule. For an extensive summary and discussion see Panwalkar and Iskander [331], Haupt [210], and Blackstone et al. [38]. Table 6.3 consists of some of the priority rules commonly used in practice.

6.3.2 Randomized Dispatching Heuristic

While one-pass heuristics limit themselves to constructing a single solution, multipass heuristics (also called search heuristics) try to obtain much better solutions by generating many of them, usually at the expense of a much higher computation time. Techniques such as the branch-and-bound method and dynamic programming can guarantee an optimal solution, but they are not practical for large-scale problems.

Randomized heuristics are an early attempt to provide more accurate solutions [21]. The idea of randomized dispatch is to start with a family of dispatch-

Table 6.3. A List of Job-Shope Dispatch Rules

Rule	Description
SPT (shortest processing time)	Select an operation with the shortest processing time.
LPT (longest processing time)	Select an operation with the longest processing time.
MWR (most work remaining)	Select an operation for the job with the most total processing time remaining.
LWR (least work remaining)	Select an operation for the job with the least total processing time remaining.
MOR (most operations remaining)	Select an operation for the job with the greatest number of operations remaining.
LOR (least operations remaining)	Select an operation for the job with the smallest number of operations remaining.
EDD (earliest due date)	Select a job with the earliest due date.
FCFS (first come, first served)	Select the first operation in the queue of jobs for the same machines.
RANDOM (random)	Select an operation at random.

ing rules. At each selection of an operation to run, choose the dispatching rule randomly, repeated throughout an entire schedule generation. Repeat the entire process several times and choose the best result. The procedure to generate an active schedule is given as follows:

Algorithm: Randomized Dispatching Heuristic (Active Schedule Generation)

Step 0. Let the best schedule BS as a null schedule.

Step 1. Let $t = 0$ and begin with PS_t as the null partial schedule. Initially S_t includes all operations with no predecessors.

Step 2. Determine $\phi_t^* = \min_{i \in S_t}\{\phi_i\}$ and the machine m^* on which ϕ_t^* could be realized.

Step 3. Select a dispatching rule randomly from the family of rules. For each operation $i \in S_t$ that requires machine m^* and for which $\sigma_i < \phi_t^*$, calculate a priority index according to the specified rule. Find the operations with the smallest index and add this operation to PS_t as early as possible, thus creating new partial schedule PS_{t+1}.

Step 4. For PS_{t+1}, update the data set as follows:

   1. Remove operations i from S_t.
   2. Form S_{t+1} by adding the direct successor of operation j to S_t.
   3. Increment t by one.

Step 5. Return to step 2 until a complete schedule is generated.

Step 6. If the generated schedule in the above step is better than the best one found so far, save it as *BS*. Return to step 1 until iteration equals the predetermined times.

Various researchers have tried to improve on the randomization approach. One change is to have a learning process so that more successful dispatching rules will have higher chances of being selected in the future. Morton and Pentico proposed a guided random approach [299]. The term *guided* means that an excellent heuristic is needed first, to "explore" the problem and provide good guidance as to where to search.

6.3.3 Shifting Bottleneck Heuristic

At the present time, the *shifting bottleneck heuristic* from Adams, Balas, and Zawack [3] is probably the most powerful procedure among all heuristics for the job-shop scheduling problem. It sequences the machines one by one, successively, taking each time the machine identified as a bottleneck among the machines not yet sequenced. Every time after a new machine is sequenced, all previously established sequences are locally reoptimized. Both the bottleneck identification and the local reoptimization procedures are based on repeatedly solving a certain one-machine scheduling problem that is a relaxation of the original problem. The method of solving the one-machine problems is not new, although Adams, Balas, and Zawack have speeded up considerably the time required for generating these problems. Instead, the main contribution of their approach is the way to use this relaxation to decide upon the order in which the machines should be sequenced. This is based on the classic idea of giving priority to bottleneck machines. Let M_0 be the set of machines already sequenced ($M_0 = \varnothing$ at the start). A brief statement of the shifting bottleneck procedure is as follows:

Algorithm: Shifting Bottleneck Heuristic

Step 1. Identify a bottleneck machine m among the machines $k \in M \backslash M_0$ and sequence it optimally. Set $M_0 \leftarrow M_0 \cup \{m\}$.

Step 2. Reoptimize the sequence of each critical machine $k \in M_0$ in turn, while keeping the other sequences fixed. Then if $M_0 = M$, stop; otherwise go to step 1.

The details of this implementation can be found in Adams, Balas, and Zawack [3] or in Applegate and Cook [9]. According to the computational experience of Adams, Balas, and Zawack, the typical improvement of their procedure is somewhere between 4% and 10%. Storer, Wu, and Vaccari reported that the Applegate and Cook implementation of the shifting bottleneck procedure was unable to solve two 10×50 problems [389]. Recently, Dauzere-Peres and Lasserre gave a modified version of the shifting bottleneck heuristic [92].

6.4   GENETIC ALGORITHMS FOR JOB-SHOP SCHEDULING PROBLEMS

6.4.1   Representation

Because of the existence of the precedence constraints of operations, JSP is not as easy as the traveling salesmen problem (TSP) to determine a nature representation. There is not a good representation with a system of inequalities for the precedence constraints. Therefore, the penalty approach is not easily applied to handle such types of constraint. Orvosh and Davis [322] have shown that for many combinatorial optimization problems, it is relatively easy to repair an infeasible or illegal chromosome, and the repair strategy does indeed surpass other strategies such as the rejecting strategy or the penalizing strategy. Most genetic algorithm/JSP researchers prefer to use the repairing strategy to handle the infeasibility and illegality. A very important issue in building a genetic algorithm for the job-shop problem is to devise an appropriate representation of solutions together with problem-specific genetic operations so that all chromosomes generated in either the initial phase or the evolutionary process will produce feasible schedules. This is a crucial phase that affects all the subsequent steps of genetic algorithms. During the past few years, the following nine representations for job-shop scheduling problem have been proposed:

- Operation-based representation
- Job-based representation
- Preference-list-based representation
- Job-pair-relation-based representation
- Priority-rule-based representation
- Disjunctive-graph-based representation
- Completion-time-based representation
- Machine-based representation
- Random key representation

These representations can be classified into the following two basic encoding approaches:

- Direct approach
- Indirect approach

In the direct approach, a schedule (the solution of JSP) is encoded into a chromosome, and genetic algorithms are used to evolve those chromosomes to determine a better schedule. The representations such as operation-based representation, job-based representation, job-pair-relation-based representation, completion-time-based representation, and random key representation belong to this class.

In the indirect approach, such as priority-rule-based representation, a sequence of dispatching rules for job assignment (i.e., a schedule) is encoded into a chromosome, and genetic algorithms are used to evolve those chromosomes to determine a better sequence of dispatching rules. A schedule then is constructed with the sequence of dispatching rules. Preference-list-based representation, priority-rule-based representation, disjunctive-graph-based representation, and machine-based representation belong to this class.

We shall discuss them in turn using the simple example given in Table 6.2.

Operation-Based Representation. This representation encodes a schedule as a sequence of operations, and each gene stands for one operation. There are two possible ways to name each operation. One natural way is to use natural number to name each operation, like the permutation representation for TSP. Unfortunately because of the existence of the precedence constraints, not all the permutations of natural numbers define feasible schedules. Gen, Tsujimura, and Kubota proposed an alternative: They name all operations for a job with the same symbol and then interpret them according to the order of occurrence in the sequence for a given chromosome [105, 256]. For an n-job m-machine problem, a chromosome contains $n \times m$ genes. Each job appears in the chromosome exactly m times, and each repeating (each gene) does not indicate a concrete operation of a job but refers to an operation which is context-dependent. It is easy to see that any permutation of the chromosome always yields a feasible schedule.

Consider the three job three-machine problem given in Table 6.2. Suppose a chromosome is given as [3 2 2 1 1 2 3 1 3], where 1 stands for job j_1, 2 for job j_2, and 3 for job j_3. Because each job has three operations, it occurs exactly three times in the chromosome. For example, there are three 2's in the chromosome, which stands for the three operations of job j_2. The first 2 corresponds to the first operation of job j_2 which will be processed on machine 1, the second 2 corresponds to the second operation of job j_2 which will be processed on machine 3, and the third 2 corresponds to the third operation of job j_2 which will be processed on machine 2. We can see that all operations for job j_2 are named with the same symbol 2 and then interpreted according to their orders of occurrence in the sequence of this chromosome. The corresponding relations of the operations of jobs and processing machines are shown in Figure 6.5. According to these relations, we can obtain a corresponding machine list

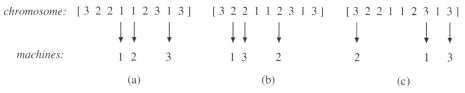

Figure 6.5. Operations of jobs and corresponding machines: (a) for job 1, (b) for job 2 and (c) for job 3.

chromosome: [3 2 2 1 1 2 3 1 3]

machines: [2 1 3 1 2 2 1 3 3]

Figure 6.6. The processing order of jobs on machine 1 (indicated by shading).

as [2 1 3 1 2 2 1 3 3], which is shown in Figure 6.6. From this figure we can see that the job processing order for machine 1 is 2–1–3 (indicated by shading), for machine 2 is 3–1–2, and for machine 3 is 2–1–3. We then summarize them to form a feasible schedule as shown in Figure 6.7.

Fang, Ross, and Corne also proposed a type of operation-based representation for a job-shop schedule problem [124]. For a $j \times m$ problem, the chromosome is a string containing a $j \times m$ chunk that is large enough to hold the largest job number j. A chunk is atomic as far as a genetic algorithm is concerned. It is indubitable that one chunk corresponds to one operation, but the description about "chunk" given in this paper is not clear enough for us to be able to code an operation into a chunk. When constructing an actual schedule, their method maintains a circular list of uncompleted jobs and lists of unscheduled operations for each such job. A chunk provides information on how to find out the operation to be scheduled next among these lists. A chunk is used as the index of a job in the circular list of uncompleted jobs, and the first untackled operation of the job will be scheduled into the earliest place where it can fit in.

Job-Based Representation. This representation consists of a list of n jobs and a schedule is constructed according to the sequence of jobs. For a given sequence of jobs, all operations of the first job in the list are scheduled first, and then the operations of the second job in the list are considered. The first operation of the job under treatment is allocated in the best available processing time for the corresponding machine the operation requires; and then the second operation is allocated, and so on, until all operations of the job are scheduled. The process is repeated with each of the jobs in the list considered in the appropriate sequence.

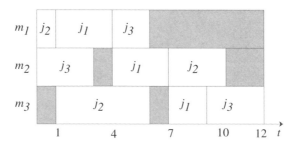

Figure 6.7. One feasible schedule.

Any permutation of jobs corresponds to a feasible schedule. Holsapple, Jacob, Pakath, and Zaveri have used this representation to deal with static scheduling problem in flexible manufacturing context [221].

We use the same example of a three-job three-machine problem given in Table 6.2. Suppose a chromosome is given as [2 3 1]. The first job to be processed is job j_2. The operation precedence constraint for j_2 is $[m_1 \ m_3 \ m_2]$ and the corresponding processing time for each machine is [1 5 3]. First, job j_2 is scheduled as shown in Figure 6.8(a). Then job j_3 is processed. Its operation precedence among machines is $[m_2 \ m_1 \ m_3]$ and the corresponding processing time for each machine is [3 2 3]. Each of its operations is scheduled in the best available processing time for the corresponding machine the operation requires, as shown in Figure 6.8(b). Finally, job j_1 is scheduled as shown in Figure 6.8(c).

Preference-List-Based Representation. This representation was first proposed by Davis for a kind of scheduling problem [100]. Falkenauer and Bouffouix used it for dealing with a job-shop scheduling problem with release times and due dates [122]. Croce, Tadei, and Volta applied it for classical job-shop scheduling problem [86].

For an n-job m-machine job-shop scheduling problem, a chromosome consisting of m subchromosomes is formed, each for one machine. Each subchromosome is a string of symbols with length n, and each symbol identifies an operation that has to be processed on the relevant machine. Subchromosomes do not describe the operation sequence on the machine, because they are *preference lists*; each machine has its own preference list. The actual schedule is deduced from the chromosome through a simulation, which analyzes the state of the waiting queues in front of the machine and, if necessary, use the preference lists to determine the schedule; that is, the operation which appears first in the preference list will be chosen.

Now we show how to deduce an actual schedule for a given chromosome. Consider the same example given in Table 6.2. Suppose a chromosome is [(2 3 1) (1 3 2) (2 1 3)]. The first gene (2 3 1) is the preference list for machine m_1, (1 3 2) is the list for machine m_2, and (2 1 3) is the list for machine m_3. From these preference lists we can see that the first preferential operations are job j_2 on machine m_1, j_1 on m_2, and j_2 on m_3. According to the given operation precedence constraints, only j_2 on m_1 is schedulable, so it is first scheduled on m_1 as shown in Figure 6.9(a). The next schedulable operation is j_2 on m_3 as shown in Figure 6.9(b). Now the current preferential operations are j_3 on m_1 and j_1 on m_2 and m_3. Because all of them are not schedulable at current time, we look for the second preferential operation in each list. They are j_1 on m_1 and j_3 on m_2 and m_3. The schedulable operations are j_1 on m_1 and j_3 on m_2. We schedule them as shown in Figure 6.9(c). The next schedulable operations are j_3 on m_1 and j_1 on m_2 as shown in Figure 6.9(d). After that, the schedulable operations are j_2 on m_2 and j_1 on m_3 as shown in Figure 6.9(e). The last operation to be scheduled is j_3 on m_3 as shown in Figure 6.9(f). Now we complete schedule deduction from a chromosome and obtain a feasible schedule with

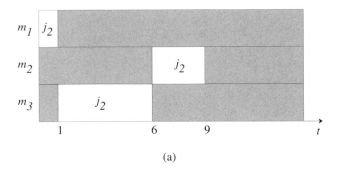

(a)

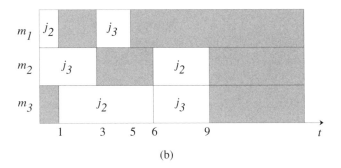

(b)

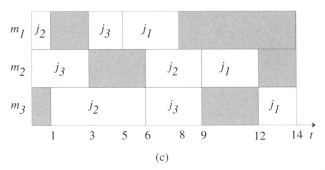

(c)

Figure 6.8. Scheduling jobs: (a) for the first job j_2, (b) for the second job j_3, and (c) for the third job j_1.

makespan 12. With the decoding procedure, all possible chromosomes always produce feasible schedules.

Croce et al. have argued that the deduction procedure generates only nondelay schedules, so we cannot be sure that the optimal solution is encoded [21]. These authors presented a rather complex lookahead evaluation procedure to help the deduction procedure generate an active schedule. But the argument is not true. When generating a nondelay schedule, we must first identify a critical machine which can start earliest and then select an operation which can be processed earliest on the critical machine. Because in the process of deduction the next operation is selected according to the preference lists, we may not obtain a nondelay schedule. The deduction procedure can guarantee to generate an active schedule. For example, the schedule given in Figure 6.9(f) is just an active schedule but not a nondelay schedule because machine m_3 remains idle at time 6 when it could start on the third operation of job j_3.

Kobayashi, Ono, and Yamamura also adopted this kind of representation in their studies [249]. The difference with the above method is that they use Giffler and Thompson heuristics to decode a chromosome into a schedule.

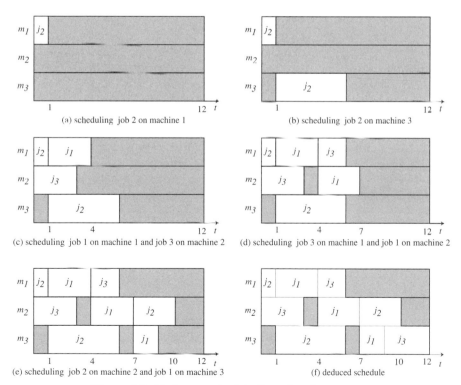

Figure 6.9. Deduce a schedule from the chromosome.

Table 6.4. Example of a Three-Job Three-Machine Scheduling Problem

Operation Precedence				Feasible Schedule			
Job	Machine Sequence			Machine	Job Sequence		
j_1	m_1	m_2	m_3	m_1	j_2	j_1	j_3
j_2	m_1	m_3	m_2	m_2	j_3	j_1	j_2
j_3	m_2	m_1	m_3	m_3	j_2	j_1	j_3

Job-Pair-Relation-Based Representation. Nakano and Yamada use a binary matrix to encode a schedule, and the matrix is determined according to the precedence relation of a pair of jobs on corresponding machines [313].

Consider a three-job three-machine scheduling problem. The operation precedence constraints of jobs and one feasible schedule of the problem are presented in Table 6.4.

A binary variable is defined to indicate the precedence relation for a pair of jobs as follows:

$$x_{ijm} = \begin{cases} 1, & \text{if job } i \text{ is processed prior to job } j \text{ on machine } m \\ 0, & \text{otherwise} \end{cases}$$

Let us examine the precedence relation of a job pair (j_1, j_2) on machines (m_1, m_2, m_3). According to the given schedule, we have $(x_{121}\ x_{122}\ x_{123}) = (0\ 1\ 0)$. For the job pair (j_1, j_3), we have $(x_{131}\ x_{132}\ x_{133}) = (1\ 0\ 1)$. For the job pair (j_2, j_3), the precedence relation on machines (m_1, m_3, m_2) are $(x_{231}\ x_{233}\ x_{232}) = (1\ 1\ 0)$. Note that the sequence of variable x_{ijm} for a pair of jobs should keep consistent with the sequence of operations of the first job i. For example, for the job pair (j_2, j_3), the sequence of operations of job j_2 is (m_1, m_3, m_2), so the relative variables are sequenced as $(x_{231}\ x_{233}\ x_{232})$ rather than $(x_{231}\ x_{232}\ x_{233})$. By summarizing these results, we get a binary matrix representation for the given feasible schedule as follows:

$$\begin{array}{l} (j_1, j_2) \text{ on } (m_1, m_2, m_3): \\ (j_1, j_3) \text{ on } (m_1, m_2, m_3): \\ (j_2, j_3) \text{ on } (m_1, m_3, m_2): \end{array} \begin{pmatrix} x_{121} & x_{122} & x_{123} \\ x_{131} & x_{132} & x_{133} \\ x_{231} & x_{233} & x_{232} \end{pmatrix} = \begin{pmatrix} 0 & 1 & 0 \\ 1 & 0 & 1 \\ 1 & 1 & 0 \end{pmatrix}$$

The representation is perhaps the most complex one among these representations and is highly redundant [124]. Besides its complexity, with this representation the chromosomes produced either by initial procedure or by genetic operations are illegal in general.

Priority-Rule-Based Representation. Dorndorf and Pesch proposed a priority-rule-based genetic algorithm [112], where a chromosome is encoded as a sequence of dispatching rules for job assignment and a schedule is constructed with priority dispatching heuristic based on the sequence of dispatching rules. Genetic algorithms are used here to evolve those chromosomes that will help produce a better sequence of dispatching rules.

Priority dispatching rules are probably the most frequently applied heuristics for solving scheduling problems in practice because of their ease of implementation and their low time complexity. The algorithms of Giffler and Thompson can be considered as the common basis of all priority-rule-based heuristics [167, 389]. The problem is to identify an effective priority rule. For an extensive summary and discussion on priority dispatching rules see Panwalkar and Iskander [331], Haupt [210], and Blackstone et al. [38].

For an n-job m-machine problem, a chromosome is a string of $n \times m$ entry $(p_1, p_2, \ldots, p_{nm})$. An entry p_i represents one rule of the set of prespecified priority dispatching rules. The entry in the ith position says that a conflict in the ith iteration of the Giffler and Thompson algorithm should be resolved using priority rule p_i. More precisely, an operation from the conflict set has to be selected by rule p_i; ties are broken by a random choice. Let

PS_t = a partial schedule containing t scheduled operations

S_t = the set of schedulable operations at iteration t, corresponding to PS_t

σ_i = the earliest time at which operation $i \in S_t$ could be started

ϕ_i = the earliest time at which operation $i \in S_t$ could be completed

C_t = the set of conflicting operations in iteration t

The procedure to deduce a schedule from a given chromosome $(p_1, p_2, \ldots, p_{nm})$ is as follows:

Procedure: Deduce a Schedule for Priority-Rule-Based Encoding

Step 1. Let $t = 1$ and begin with PS_t as the null partial schedule and let S_t include all operations with no predecessors.

Step 2. Determine $\phi_t^* = \min_{i \in S_t} \{\phi_i\}$ and the machine m^* on which ϕ_t^* could be realized. If more than one such machine exists, the tie is broken by a random choice.

Step 3. Form a conflicting set C_t which includes all operations $i \in S_t$ with $\sigma_i < \phi_t^*$ that requires machine $m*$. Select one operation from C_t by the priority rule p_t and add this operation to PS_t as early as possible, thus creating a new partial schedule PS_{t+1}. If more than one operation exists according to the priority rule p_t, the tie is broken by a random choice.

Step 4. Update PS_{t+1} by removing the selected operation from S_t and adding the direct successor of the operation to S_t. Increment t by one.

Table 6.5. Selected Priority Rules

Number	Rule	Description
1	SPT	Select an operation with the shortest processing time.
2	LPT	Select an operation with the longest processing time.
3	MWR	Select an operation for the job with the most total processing time remaining.
4	LWR	Select an operation for the job with the least total processing time remaining.

Step 5. Return to step 2 until a complete schedule is generated.

Let us see an example. We will use the four priority rules listed in Table 6.5. Consider the following chromosome [1 2 2 1 4 4 2 1 3], where 1 stands for rule SPT, 2 for rule LPT, 3 for rule MWR, and 4 for rule LWR. At the initial step, we have

$$S_1 = \{o_{111}, o_{211}, o_{312}\}$$
$$\phi_1^* = \min\{3, \ 1, \ 3\} = 1$$
$$m^* = 1$$
$$C_1 = \{o_{111}, o_{211}\}$$

Now operations o_{111} and o_{211} compete for machine m_1. Because the first gene in the given chromosome is 1 (which means SPT priority rule), operation 211 is scheduled on machine m_1 as shown in Figure 6.10(a). After updating the data, we have

$$S_2 = \{o_{111}, o_{223}, o_{312}\}$$
$$\phi_2^* = \min\{4, 6, 3\} = 3$$
$$m^* = 2$$
$$C_2 = \{o_{312}\}$$

Operation o_{312} is scheduled on machine m_2 as shown in Figure 6.10(b). After updating the data, we have

$$S_3 = \{o_{111}, o_{223}, o_{321}\}$$
$$\phi_3^* = \min\{4, 6, 3\} = 3$$
$$m^* = 1$$
$$C_3 = \{o_{111}, o_{321}\}$$

Now operations o_{111} and o_{321} compete for machine m_1. Because the third gene

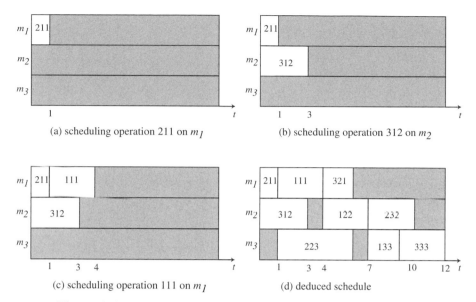

Figure 6.10. Deduce a schedule from a priority-rule-based encoding.

in the given chromosome is 2 (which means LPT priority rule), operation o_{111} is scheduled on machine m_1 as shown in Figure 6.10(c). Repeat these steps until a complete schedule is deduced from the given chromosome as shown in Figure 6.10(d).

Disjunctive-Graph-Based Representation. Tamaki and Nishikawa proposed a disjunctive graph-based representation [397], which also can be viewed as one kind of job-pair-relation-based representation. The job-shop scheduling problem can be represented with a *disjunctive graph* [23, 360]. The disjunctive graph $G = (N, A, E)$ is defined as follows: N contains nodes representing all operations, A contains arcs connecting consecutive operations of the same job, and E contains disjunctive arcs connecting operations to be processed by the same machine. The disjunctive constraints are represented by an edge in E. A disjunctive arc can be settled by either of its two possible orientations. The construction of a schedule will settle the orientations of all disjunctive arcs so as to determine the sequences of operations on same machines. Once a sequence is determined for a machine, the disjunctive arcs will be replaced by the usual *conjunctive* arcs. Figure 6.11 illustrates the disjunctive graph for a three-job three-machine example. A chromosome consists of a binary string corresponding to an order list of disjunctive arcs in E as shown in Figure 6.12, where e_{ij} stands for the disjunctive arc between nodes i and j and is defined as follows:

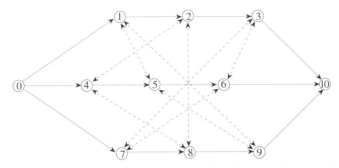

Figure 6.11. Disjunctive graph of a three-job three-machine problem.

$$e_{ij} = \begin{cases} 1, & \text{settle the orientation of the disjunctive arc from node } j \text{ to node } i \\ 0, & \text{settle the orientation of the disjunctive arc from node } i \text{ to node } j \end{cases}$$

The job-shop scheduling problem is to find an order of the operations on each machine—that is, to settle the orientation of the disjunctive arcs such that the resulting graph is acyclic to guarantee that there are no precedence conflicts between operations. It is easy to see that an arbtirary chromosome may yield a cyclic graph, which means that the schedule is infeasible. Thus this chromosome is not used to represent a schedule but is used only as a decision preference. They used a critical-path-based procedure to deduce a schedule. During the process of deduction, when a conflict of two nodes (operations) occurs on a machine, the corresponding bit of the chromosome is used to settle the processing order of the two operations—that is, to settle the orientation of the disjunctive arc between the two nodes.

Completion-Time-Based Representation. Yamada and Nakano proposed a completion-time-based representation [423]. A chromosome is an ordered list of completion times of operations. For the same example given in Table 6.2, the chromosome can be represented as follows:

$$[c_{111} \; c_{122} \; c_{133} \; c_{211} \; c_{223} \; c_{232} \; c_{312} \; c_{321} \; c_{333}]$$

where c_{jir} denotes the completion time for operation i of job j on machine r. It is easy to see that such representation is not suitable for most genetic operators

order list of disjunctive arcs: $\quad e_{15} \; e_{19} \; e_{59} \; e_{24} \; e_{28} \; e_{48} \; e_{36} \; e_{37} \; e_{67}$

chromosome: $\quad [\, 0 \quad 0 \quad 1 \quad 1 \quad 0 \quad 0 \quad 0 \quad 1 \quad 1 \,]$

Figure 6.12. Disjunctive-graph-based representation.

because it will yield an illegal schedule. Yamada and Nakano designed a special crossover operator for it.

Machine-Based Representation. Dorndorf and Pesch proposed a machine-based genetic algorithm [112] where a chromosome is encoded as a sequence of machines and a schedule is constructed with a shifting bottleneck heuristic based on the sequence.

The *shifting bottleneck heuristic*, proposed by Adams, Balas, and Zawack [3], is probably the most powerful procedure among all heuristics for the job-shop scheduling problem. It sequences the machines one by one, successively, taking each time the machine identified as a bottleneck among the machines not yet sequenced. Whenever a new machine is sequenced, all previously established sequences are locally reoptimized. Both the bottleneck identification and the local reoptimization procedures are based on repeatedly solving a certain one-machine scheduling problem that is a relaxation of the original problem. The main contribution of this approach is the way to use this relaxation to decide upon the order in which the machines should be sequenced.

The *shifting bottleneck heuristic* is based on the classic idea of giving priority to bottleneck machines. Different measures of bottleneck quality of machines will yield different sequences of bottleneck machines. The quality of the schedules obtained by the shifting bottleneck heuristic depends heavily on the sequence of bottleneck machines. Adams, Balas, and Zawack also proposed an enumerative version of the shifting bottleneck heuristic to consider different sequences of machines.

Instead of an enumerative tree search, Dorndorf and Pesch proposed a genetic strategy to determine the best machine sequence for the shifting bottleneck heuristic. A chromosome is a list of ordered machines. Genetic algorithms are used here to evolve those chromosomes that will find a better sequence of machines for the shifting bottleneck heuristic. The difference between the shifting bottleneck heuristic and the genetic algorithm is that the bottleneck is no longer a decision criterion for the choice of the next machine, which is controlled by a given chromosome.

Let M_0 be the set of machines already sequenced and let a given chromosome be $[m_1, m_2, \ldots, m_m]$. The procedure of deducing a schedule from the chromosome works as follows:

Procedure: Deduce a Schedule for Machine-Based Encoding

Step 1. Let $M_0 \leftarrow \{\phi\}$ and $i \leftarrow 1$ and let the chromosome be $[m_1, m_2, \ldots, m_m]$.

Step 2. Sequence machine m_i optimally. Update the set $M_0 \leftarrow M_0 \cup \{m_i\}$.

Step 3. Reoptimize the sequence of each critical machine $m_i \in M_0$ in turn, while keeping the other sequences fixed.

Step 4. Let $i \leftarrow i + 1$. Then if $i > m$, stop; otherwise go to step 2.

The details of step 3 can be found in Adams, Balas, and Zawack [3] or in Applegate and Cook [9].

Random Key Representation. Random key representation was first introduced by Bean [29]. With this technique, genetic operations can produce feasible offspring without creating additional overhead for a wide variety of sequencing and optimization problems. Norman and Bean successfully generalized the approach to the job-shop scheduling problem [317, 318].

Random key representation encodes a solution with *random number*. These values are used as sort *keys* to decode the solution. For an n-job m-machine scheduling problem, each gene (a random key) consists of two parts: an integer in set $\{1, 2, \ldots, m\}$ and a fraction generated randomly from $(0, 1)$. The integer part of any random key is interpreted as the machine assignment for that job. Sorting the fractional parts provides the job sequence on each machine. Consider the same example given in Table 6.2. Suppose that a chromosome is

$$[1.34 \; 1.09 \; 1.88 \; 2.66 \; 2.91 \; 2.01 \; 3.23 \; 3.21 \; 3.44]$$

Sort the keys for machine 1 in ascending order in the job sequence $2 \rightarrow 1 \rightarrow 3$, for machine 2 in the job sequence $3 \rightarrow 1 \rightarrow 2$, and for machine 3 in the job sequence $2 \rightarrow 1 \rightarrow 3$. Let o_{jm} denote job j on machine m. The chromosome can be translated into a unique list of ordered operations as follows:

$$[o_{21} \; o_{11} \; o_{31} \; o_{32} \; o_{12} \; o_{22} \; o_{23} \; o_{13} \; o_{33}]$$

It is easy to see that the job sequences given above may violate the precedence constraints. Accompanying this coding, a pseudocode is presented by Norman and Bean for handling precedence constraints [317].

6.4.2   Discussion

Various encoding techniques were summarized in the previous section. In this section, we present a comparative discussion of the following aspects:

- Lamarckian property of the chromosome
- Complexity of the decoder
- The property of coding space and mapping
- Memory requirements

Lamarckian Property. The simple genetic algorithms developed by John Holland [220] were inspired by Darwin's theory of natural selection, which performs a robust search but slower than other methods. Kennedy [246, 415] introduced *Lamarckian evolution* into simple genetic algorithms to improve their efficiency, in which offspring organism first pass through Darwin's biological evolution and then pass through Lamarckian's intelligence evolution to inject some "intelligence" into the offspring organism before returning it to be evaluated.

The Lamarckian property for a coding technique concerns the issue whether a chromosome can pass on its "merits" to future populations through a common genetic operation. Let us see an example. A tour of nine-city TSP can be coded into the following chromosomes:

Random key: [0.34 0.09 0.88 0.66 0.91 0.01 0.23 0.21 0.44]

Permutation: [6 2 8 7 1 9 4 3 5]

Suppose a subtour is formed with the second and sixth cut-points. For permutation representation, the subtour is [8 7 1 9]. An offspring receives the same subtour from its parent. For random key representation, the subtour is [0.88 0.66 0.91 0.01]. Usually, it may refer to a different subtour in offspring but the same subtour of $8 \rightarrow 7 \rightarrow 1 \rightarrow 9$ as its parents. What subtour it refers to will depend on the values of other genes in the offspring. In fact, the offspring receives nothing (about good subtour) from its parents except for a mere random change.

In general, we hope that a coding technique has the Lamarckian property. In fact, most codings have this property; that is, an offspring can inherit goodness from its parents. The opposite extreme is *no Lamarckian*; that is, an offspring inherits nothing from its parents. The third type is called *half-Lamarckian*. In this case, part of the segments inherited from parents refers to the same things as the parents while the remaining part refers to different things.

Let us consider the codings given in the above section. Random key representation and completion-time-based representation have no Lamarckian, job-based representation and priority-rule-based representation have Lamarckian property; and the remaining representations have half-Lamarckian property. In general, for an n-job m-machine problem, a chromosome contains $n \times m$ genes and each gene corresponds to an operation. The 1-to-1 corresponding relation is realized mainly by the following three ways: (1) job sequence, (2) preference list, and (3) priority rule. For the first two cases, the corresponding relation is context-dependent; which means that a gene with the same value may be interpreted into same things, or may not.

Complexity of Decoder. In principle, there are an infinite number of feasible schedules for a job-shop scheduling problem. Generally, three kinds of schedules can be distinguished as follows: *semiactive schedule*, *active schedule*, and *nondelay schedule* [21]. An optimal schedule is within the set of active schedules. The nondelay schedules are smaller than the active schedules, but there is no guarantee that it will contain an optimum. Thus we hope that a schedule decoded from a chromosome would be an active one. All of the coding techniques proposed for the job-shop scheduling problem can generate an active schedule by use of the decoder. The degree of complexity of the decoder can be classified into the following four levels:

Level 0: no decoder. Completion-time based representation belongs to this class. In this case, all burdens are put on genetic operators.

Level 1: simple mapping relation. Operation-based representation, job-pair-relation-based representation, and job-based representation belong to the class.

Level 2: simple heuristic. Preference-list-based representation and priority-rule-based representation belong to the class.

Level 3: complex heuristic. Preference-list based coding used by Kobayashi et al., disjunctive-graph-based representation, and machine-based representation belong to this class.

The Property of Coding Space and Mapping. As we can see for any coding technique, through a decoding procedure, a chromosome always corresponds to a legal, feasible, and active schedule and the mapping relation is one-to-one. But the coding space of these methods can be classified into two classes: One contains only feasible solution space, whereas the other includes illegal solution space. This can be tested by applying the simple one-cut-point crossover operator to an encoding and then checking if the yielded offspring is illegal. Priority-rule-based representation, job-based representation, and disjunctive-graph-based representation belong to the first class, and the remaining representations belong to the second class.

In addition, the spaces of job-based and machine-based representations correspond only to the partial space of whole solution space.

Memory Requirements. For an n-job m-machine problem, if we define the standard length for a chromosome as with $n \times m$ genes, the codings for the job-shop problem can be classified into three classes:

1. The coding length is less than standard length. Job-based length, disjunctive-graph-based length, and machine-based length belong to this class. Note that the coding space for job-based length corresponds to only part of the solution space, but not the whole solution space. Thus there is no guarantee for finding the optima with the coding techniques.

2. The coding length is larger than standard length. Only the job-pair-relation-based length belongs to the class. The representation is highly redundant.

3. The coding length is equal to standard length. The remaining representations belong to this class.

Finally, we would like to point out that to give a fair judgment on these coding techniques, it is necessary to perform further comparative studies of these techniques under standard experimental conditions using benchmark problems to reveal the advantages and disadvantages for each of them. We leave the topic about how to design a standard experiment in Section 6.9.

6.4.3   Hybrid Genetic Search

Problems from combinatorial optimization are well within the scope of genetic algorithms. Compared to conventional heuristics, genetic algorithms are not well suited for fine-tuning structures which are very close to optimal solutions. Therefore it is essential to incorporate conventional heuristics (such as local search) into genetic algorithms to construct a more competitive genetic algorithm. Various methods of hybridization have been proposed for the job-shop scheduling problem, which can be classified into two basic approaches [101, 353]:

- Adapt genetic operators
- Incorporate conventional heuristics into genetic algorithms

The first approach attempts to invent new genetic operators inspired from conventional heuristics, such as (a) designing a new crossover operator based on the Giffler and Thompson algorithm [423] or (b) designing a new mutation operator based on neighborhood search mechanism [71]. The second approach involves hybridizing conventional heuristics into genetic algorithms where possible. This can be done in a variety of ways, including the following:

1. Incorporate heuristics into initialization to generate a well-adapted initial population. In this way, a hybrid genetic algorithm with elitism can guarantee to do no worse than the conventional heuristic does.
2. Incorporate heuristics into an evaluation function to decode chromosomes to schedules.
3. Incorporate a local search heuristic as an add-on to the basic loop of the genetic algorithm, working together with mutation and crossover operators, to perform quick and localized optimization in order to improve offspring before returning it to be evaluated.

With the hybrid approach, genetic algorithms are used to perform global exploration among population while heuristic methods are used to perform local exploitation around chromosomes. Because of the complementary properties of genetic algorithms and conventional heuristics, the hybrid approach often outperforms either method operating alone.

Giffler and Thompson Algorithm-Based Crossover.   Yamada and Nakano proposed a crossover operator based on the Giffler and Thompson algorithm [423]. The generation procedure of the Giffler and Thompson algorithm is a tree search approach. At each step, it essentially identifies all processing conflicts (the operations competing for the same machine), and an enumeration procedure is used to resolve these conflicts in all possible ways. Yamada and Nakano's procedure of crossover operator is essentially a kind of one-pass procedure but not a tree search approach. When generating offspring, at each step, it identifies all processing conflicts in a manner similar to that of the Giffler and Thompson algorithm, and then it chooses one operation from the conflict set of operations according to one of their parents schedules. Let

o_{ji} = the ith operation of job j

S = the set of schedulable operations for a given partial schedule

ϕ_{ji} = the earliest completion time of the ith operation of job j in S

G_r = the set of conflicting operations in S on machine r

The procedure to generate offspring from two parents is as follows:

Procedure: Giffler and Thompson Algorithm-Based Crossover

Step 1. Let S include all operations with no predecessors initially.

Step 2. Determine $\phi^* = \min\{\phi_{ji}|o_{ji} \in S\}$ and the machine r^* on which ϕ^* could be realized.

Step 3. Let G_{r^*} include all operations $o_{ji} \in S$ that requires machine r^*.

Step 4. Choose one of the operations from G_{r^*} as follows:

1. Generate a random number $\epsilon \in [0, 1)$ and compare it with mutation rate p_m if $(\epsilon < p_m)$, and then choose an arbitrary operation from G_{r^*} as o_{ji}^*.

2. Otherwise select one parent with an equal probability, say parent p_s; find and operation o_{ji}^* which was scheduled earliest in p_s among all the operations in G_{r^*},

3. Schedule o_{ji}^* in offspring according to ϕ_{ji},

Step 5. Update S as follows:

1. Remove operation o_{ji}^* from S.

2. Add the direct successor of operation o_{ji}^* to S.

Step 6. Return to step 2 until a complete schedule is generated.

In step 4.1, conflict is resolved by choosing an operation randomly, while in step 4.2, conflict is resolved by giving priority to the operation which was scheduled earliest in one of its parents p_s among all conflicting operations in G_{r^*}. The parent p_s is selected randomly with equal probability. Thus Yamada and Nakano's approach is essentially based on priority dispatching heuristics and not on a pure Giffler and Thompson approach. At each step, one operation is selected to be added into the partial schedule of offspring, and conflicts among

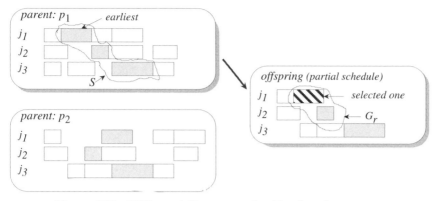

Figure 6.13. Giffler and Thompson algorithm-based crossover.

operations are resolved by specifying priority to operations according to their parents' schedules.

Figure 6.13 demonstrates the selection of operation in step 4.2. For the offspring, operations o_{11}, o_{21}, o_{31}, and o_{32} are scheduled. The schedulable operations are $S = \{o_{12}, o_{22}, o_{33}\}$. Assume that the set of conflicting operations are $S = \{o_{12}, o_{22}\}$. Parent p_1 is selected. Because operation o_{12} was scheduled earliest in parent p_1, it is scheduled into the offspring.

Neighborhood-Search-Based Mutation. In conventional genetic algorithms, mutation is a background operator, which is just used to produce small perturbation on chromosomes in order to maintain the diversity of population. When designing a hybrid genetic algorithm, a fundamental principle is to *hybridize where possible*. Here mutation is designed with the neighbor search technique. It is not a background operator and is used to perform intensive search in order to find an improved offspring.

A lot of definitions may be considered for the neighborhood of a schedule. For operation-based representation, the neighborhood for a given chromosome can be considered as the set of chromosomes (schedules) transformable from a given chromosome by exchanging the positions of λ genes (randomly selected and nonidentical genes). A chromosome (schedule) is said to be λ-*optimum* if it is better than any others in the neighborhood according to some measure. Let us see an example of the four-job four-machine problem. Suppose the genes on positions 4, 8, and 12 are randomly selected. They are (4 3 2) and their possible permutations are (3 4 2), (3 2 4), (2 3 4), (2 4 3), and (4 2 3). The permutations of the genes together with remaining genes of the chromosome form the neighbor chromosomes shown in Figure 6.14. Then we evaluate all neighbor chromosomes, and the best one is used as the offspring of mutation. The overall procedure of mutation is given below.

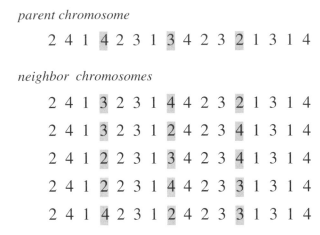

Figure 6.14. Neighbor schedules.

Procedure: Neighbor-Search-Based Mutation

begin
  $i \leftarrow 0$;
  while $(i \leq pop\_size \times p_m)$ **do**
    choose an unmutated chromosome randomly;
    pick up λ nonidentical genes randomly from it;
    make its neighbors based on all permutations of the genes;
    evaluate all neighbor schedules;
    select the best neighbor as offspring;
    $i \leftarrow i + 1$;
  end
end

Combining Genetic Algorithm with Local Search. A common form of hybrid genetic algorithm is the combination of local search with a genetic algorithm. A genetic algorithm is good at global search but slow to converge, while local search is good at fine-tuning but often falls into local optima. The hybrid approach complements the properties of genetic algorithm and local search heuristic methods. A genetic algorithm is used to perform global search to escape from local optima, while local search is used to conduct fine-tuning.

Local search in this context can be thought of as being analogous to a kind of learning that occurs during the lifetime of an individual string. With a simple genetic algorithm, the selection of chromosomes is based on the instantaneous fitness at their birth; while with hybrid methods, the selections are based on the fitness at the end of the individual life, the life being performed by local search. The improved offspring will pass on its learned traits (local optimization) to future offspring through common crossover. This phenomenon is called *Lamarckian evolution.* Thus this hybrid approach can be viewed as the combination of *Darwin's evolution* with *Lamarckian's evolution.*

Procedure: Genetic Algorithms + Local Search

begin
  $t \leftarrow 0$;
  initialize $P(t)$;
  improve $P(t)$ with local search;
  evaluate $P(t)$;
  while (not termination condition) **do**
    recombine $P(t)$ to yield $C(t)$;
    improve $C(t)$ with local search;
    evaluate C(t);
    select $P(t + 1)$ from $P(t)$ and $C(t)$;
    $t \leftarrow t + 1$;
  end
end

Combining Genetic Algorithm with Beam Search. The hybrid approach proposed by Holsapple, Jacob, Pakath, and Zaveri is to combine genetic algorithms with the beam search technique [221]. Roughly, beam search is used to convert a chromosome (job-based representation) to a schedule. Beam search is a heuristic refinement of breadth-first search that relies on the notion of *beam width* to restrict the number of nodes that we branch from at each stage (or level) of the search tree. At each level of the search process, all nodes are evaluated using a predefined evaluation function. Then the beam width w is used to pick w best nodes (in terms of their evaluations) to branch out from (i.e., expand) at the current level to generate the offspring nodes at the next level in the tree. The remaining nodes at the current level are pruned out permanently. If there are only fewer than w nodes available to begin with, then all of these are chosen for expansion. From each of the chosen w nodes, all possible successor nodes are generated. The process continues until no further expansion is possible. Figure 6.15(a) illustrates beam search with a simple example.

Filtered beam search is a refinement of the beam search method. It uses another construct called the *filter width* to further prune the state space. For each of the w nodes selected for expansion at a particular level, the filter width f determines the maximum number of successor nodes that could be generated. That is, we could generate as many successor nodes as possible without exceeding f. For a given value of f, the choice of which f successors (from the set of possible successor nodes) to generate is random. Both w and f are user-specified quantities. The search process is given in Figure 6.15(b).

Holsapple et al. adopted a job-based representation. Genetic algorithm is used to manipulate the job sequence, and filtered beam search is used to generate the "best" schedule for a given job sequence (or a chromosome). We can

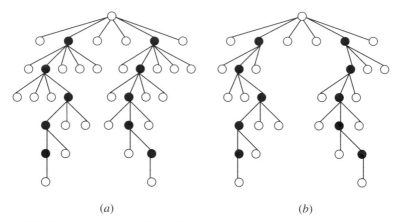

Figure 6.15. Beam search and filtered beam search. (*a*) beam search with $w = 2$ and (*b*) filtered beam search with $w = 2$ and $f = 3$.

consider the beam search as part of the evaluation function—that is, converting a job sequence (the genotype of chromosome) to a schedule (the phenotype of chromosome)—but it does more than converting. They considered a type of static scheduling problem in flexible manufacturing contexts. The difference between this problem and the classical job-shop problem is that any operation of jobs can be processed on any machine. In such a case, a chromosome of the job-based representation corresponds to many feasible schedules, so they use beam search to locate the "best" one among the feasible schedules for a given chromosome. For a classical job-shop scheduling problem, the processing machine for each operation is predetermined, so a job-based chromosome corresponds to only one feasible schedule as illustrated in last section. In this case, we do not need any tree search techniques such as beam search to convert a chromosome into a feasible schedule. The overall procedure of their genetic algorithms is given below.

Procedure: Genetic Algorithms + Beam Search

begin
　　$t \leftarrow 0$;
　　initialize $P(t)$ (job sequences);
　　perform beam search to generate the best schedule for each chromosome
　　　　in $P(t)$;
　　evaluate schedules to determine fitness for $P(t)$;
　　while (not termination condition) **do**
　　　　recombine $P(t)$ to yield $C(t)$;
　　　　perform beam search to generate the best schedule for each chromosome
　　　　　　in $C(t)$;
　　　　evaluate schedules to determine fitness for $C(t)$;

select next population $P(t + 1)$ from $P(t)$ and $C(t)$;
$t \leftarrow t + 1$;
  end
end

6.5   GEN, TSUJIMURA, AND KUBOTA'S APPROACH

Gen, Tsujimura, and Kubota proposed an implementation of GA for solving a job-shop scheduling problem [165]. They proposed the operation-based representation and designed a *partial schedule exchange* crossover operator for it, which considers partial schedules to be the natural building blocks, and they intend to use such crossover to maintain building blocks in offspring in much the same manner as Holland described [220]. The partial schedule is identified with the same job in the first and last positions of the partial schedule. For example, we have two parent chromosomes p_1 and p_2 as shown in Figure 6.16. The partial schedules are randomly picked up as follows:

(a) Choose one position in parent p_1 randomly. Suppose that it is the 6th position at which job 4 is located.

(b) Find out the next-nearest job 4 in the same parent p_1, which is in position 9. Now we get *partial schedule*$_1$ as (4 1 2 4).

(c) The next partial schedule is not randomly generated from parent p_2. It must begin and end with job 4, as with the first partial schedule. Note that the two jobs 4 are the first and second job 4 in parent p_1. Thus *partial schedule*$_2$ is formed with the genes between the first and second job for job 4 in parent p_2 as (4 1 3 1 1 3 4).

Exchange the partial schedules as shown in Figure 6.17. Usually, the partial schedules contain a different number of genes, so the offspring generated after exchanging may be illegal. The missed and exceeded genes for offspring can be determined as shown in Figure 6.18.

parent chromosomes:

partial schedule$_1$

p_1 [3 2 1 2 3 **4 1 2 4** 4 1 3 4 1 2 3]

p_2 [**4 1 3 1 1 3 4** 1 2 3 4 2 2 2 3 4]

partial schedule$_2$

Figure 6.16. Select partial schedules.

exchange partial schedules:

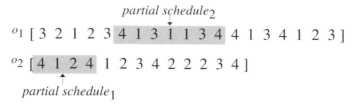

Figure 6.17. Exchange partial schedules.

The next step is to legalize offspring by deleting exceeded genes and adding missed genes. For offspring o_1, its exceeded genes are (3 1 1 3). Offspring o_1 received *partial schedule*$_2$ from parent p_2. Because the job 3 and job 1 in *partial schedule*$_2$ are the first job 3 and first job 1 in the parent p_2, we must delete the jobs 3 and 1 before the partial schedule in o_1 so as to keep their order the same as in parent p_2. The reason is that genes in the operation-based representation refer to operations which are context-dependent. If the genes of *partial schedule*$_2$ in offspring o_1 keep the same order as those in parent p_2, these genes will refer to the same operations in both cases; otherwise they will refer to different operations. We hope that offspring o_1 inherits the same partial schedule identified by *partial schedule*$_2$ in parent p_2. We must let the genes of *partial schedule*$_2$ have the same order in both o_1 and p_2. For offspring o_1, its missed gene is job 2. Because there is no job 2 in *partial schedule*$_2$, we insert job 2 into anywhere (except the inside of *partial schedule*$_2$), which does not change the order of the genes in *partial schedule*$_2$. The process of offspring legalization is shown in Figure 6.19. For offspring o_2, its exceeded gene is job 2. We can delete any job 2 in offspring o_2 except the inside of *partial schedule*$_1$. Its missed genes are (3 1 1 3). We must insert one job 1 before *partial schedule*$_1$ and one job 1 after *partial schedule*$_1$ in offspring o_2 so as to keep the job 1 of

missed and exceeded genes for o_1:

partial schedule$_1$ 4 1 2 4	$\xrightarrow{\text{missed gene}}$	2
partial schedule$_2$ 4 1 3 1 1 3 4	$\xrightarrow{\text{exceeded genes}}$	3 1 1 3

missed and exceeded genes for o_2:

partial schedule$_2$ 4 1 3 1 1 3 4	$\xrightarrow{\text{missed genes}}$	3 1 1 3
partial schedule$_1$ 4 1 2 4	$\xrightarrow{\text{exceeded gene}}$	2

Figure 6.18. Missed and exceeded genes.

legalize offspring o_1:

(1) *delete the exceeded genes*   3 1 1 3

o_1 [⬚ 2 ⬚ 2 ⬚ 4 1 3 1 1 3 4 4 ⬚ 3 4 1 2 3]

↓

o_1 [2 2 4 1 3 1 1 3 4 4 3 4 1 2 3]

(2) *insert the missed genes*   2

↓

o_1 [2 2 4 1 3 1 1 3 4 2 4 3 4 1 2 3]

Figure 6.19. Legalizing offspring o_1.

partial schedule$_1$ being the second job 1 in offspring o_2. Because there is no job 3 in *partial schedule$_1$*, we insert job 3 into anywhere (except the inside of *partial schedule$_1$*). This is shown in Figure 6.20.

Gen, Tsujimura, and Kubota used a job-pair exchange mutation; that is, they randomly picked up two nonidentical jobs and then exchanged their positions as shown in Figure 6.21. For operation-based representation, the mapping relation between chromosome and schedule is many-to-one, and an offspring obtained by exchanging two nearest jobs may yield the same schedule as its parents do. Thus the greater the separation between two nonidentical jobs, the better.

Furthermore, Tsujimura, Gen, and Kubota extended their work into a fuzzy job-shop scheduling problem where the processing time of each job was given as a triangular fuzzy number in order to handle the uncertainty of processing time [406].

legalize offspring o_2:

(1) *delete the exceeded genes*   2

o_2 [4 1 2 4 1 ⬚ 3 4 2 2 2 3 4]

↓

o_2 [4 1 2 4 1 3 4 2 2 2 3 4]

(2) *insert the missed genes*   3 1 1 3

↓   ↓ ↓ ↓

o_2 [1 4 1 2 4 1 3 3 1 3 4 2 2 2 3 4]

Figure 6.20. Legalizing offspring o_2.

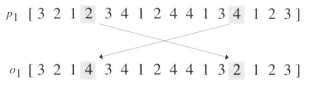

p_1 [3 2 1 **2** 3 4 1 2 4 4 1 3 **4** 1 2 3]

o_1 [3 2 1 **4** 3 4 1 2 4 4 1 3 **2** 1 2 3]

Figure 6.21. Mutation.

6.6   CHENG, GEN, AND TSUJIMURA'S APPROACH

Cheng, Gen, and Tsujimura further modified Gen, Tsujimura, and Kubota's method in order to enhance its efficiency. They have made three major modifications:

1. A new decoding procedure was devised to guarantee generation of an active schedule.
2. A simplified crossover operation was proposed.
3. Mutation was designed with the neighbor search technique and was used to perform an intensive search in order to find an improved offspring, which has been given in Section 6.4.3.

In Gen, Tsujimura, and Kubota's method, when decoding a chromosome to a schedule, they consider only two things: the operation order in the given chromosome and the precedence constraints of operations of jobs. They also specify the order of the operations on each machine in the schedule. The decoding procedure can only guarantee the generation of a *semiactive schedule* but not an *active schedule*, that is, there may exist *permissible left-shift* (or *global left-shift*) within the schedule. Because an optimal schedule is an active schedule, the decoding procedure will inhibit the efficiency of the proposed algorithm for a large-scale job-shop problem.

Let us see an example. We still use the same three-job three-machine problem presented in Table 6.2. Suppose we have a chromosome as [2 1 1 1 2 2 3 3 3]. By the decoding procedure, we can obtain a corresponding machine list as [1 1 2 3 3 2 2 1 3]. According to the machine list, we can make a schedule as shown in Table 6.6, which specifies the order of the operations on

Table 6.6. Schedule

Machine	Job Sequence		
m_1	j_2	j_1	j_3
m_2	j_1	j_2	j_3
m_3	j_1	j_2	j_3

each machine. Figure 6.22(a) shows the Gantt chart of the semiactive schedule obtained by the decoding procedure for the given chromosome, and the schedule has a makespan of 25. There are two permissible left-shifts in the semiactive schedule:

1. Job j_3 can start at time 0 on machine m_2.
2. Job j_2 can start at time 1 on machine m_3.

By performing these left-shifts, we get an active schedule shown in Figure 6.22(b), which has a makespan of 12.

Cheng, Gen, and Tsujimura modified the decoding procedure to guarantee generation of an active schedule from a given chromosome. For the operation-based representation, each gene uniquely indicates an operation. The modified decoding procedure first translates the chromosome to a list of ordered operations. A schedule is generated by a one-pass heuristic based on the list. The first operation in the list is scheduled first, and then the second operation in the list is considered, and so on. Each operation under treatment is allocated in the best available processing time for the corresponding machine the operation requires. The process is repeated until all operations in the list are scheduled into the appropriate places. A schedule generated by this modified procedure

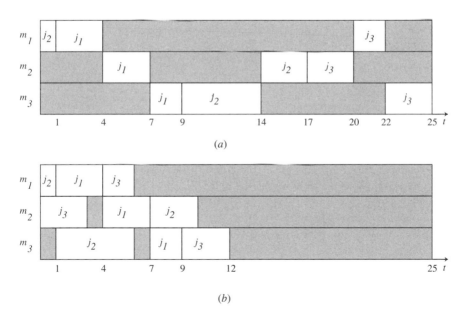

Figure 6.22. The effect of left-shifts in altering a semiactive schedule; (*a*) a semiactive schedule obtained by decoding procedure, and (*b*) an active schedule by performing left-shifts on the semiactive schedule.

can be guaranteed to be an active schedule. A chromosome here can be considered to assign a priority to each operation. Let us label the position from left to right of the list as from lowest to highest. An operation with lower position in the list has higher priority and will be scheduled first prior to those with higher position than it.

Let o_{jim} denote the ith operation of job j on machine m. The chromosome considered above ([2 1 1 1 2 2 3 3 3]) can be translated into a unique list of ordered operations of $[o_{211}\ o_{111}\ o_{122}\ o_{133}\ o_{223}\ o_{232}\ o_{312}\ o_{321}\ o_{333}]$. Operation o_{211} has the highest priority and is first scheduled, then o_{111} is scheduled next, and so on. The resulting active schedule is shown in Figure 6.22(b).

The crossover operator given in the last section is rather complex because it tries to keep the orders of operations of a partial schedule in offspring the same as they are in the parent. The modified approach uses chromosomes to determine the priority for each operation, and a schedule is generated by a one-pass priority dispatching heuristic. Therefore it is not necessary to keep the same orders for the operations of partial schedule in both parents and offspring. The partial schedule exchange crossover can then be simplified as follows:

1. First, we pick up two partial schedules from parents p_1 and p_2, respectively. Two partial schedules contain the same number of genes. Exchange the partial schedules to generate offspring. Figure 6.23(a) illustrates an offspring o generated by this method.

2. Then find the missed and exceeded genes for offspring o by comparing two partial schedules as shown in Figure 6.23(b).

3. Legalize offspring o by deleting the exceeded genes and inserting the missed genes in a random way as shown in Figure 6.23(c).

6.7   FALKENAUER AND BOUFFOUIX'S APPROACH

Falkenauer and Bouffouix proposed an implementation of the genetic algorithm for the job-shop scheduling problem with release times and due dates [122]. They adopted the preference-list-based representation. Croce, Tadei, and Volta further modified Falkenauer and Bouffouix's approach and dealt with the classical job-shop scheduling problem [86].

For an n-job m-machine classical job-shop scheduling problem, a chromosome encoded as preference lists is formed from m subchromosomes and each subchromosome consisting of n genes is a *preference list* of operations for one machine. The actual schedule is deduced from the chromosome through a simulation based on the preference lists of operations.

With this chromosome, the aim of genetic search is to find the best permutation of the operations in each list. Since, during the deduction of schedules from chromosomes, the priorities of operations are determined according to the relative position for each operation (gene) on the preference lists, it is desired

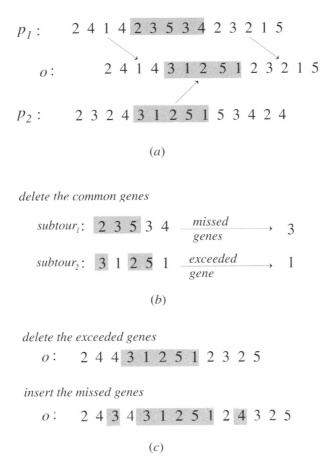

Figure 6.23. Crossover: (*a*) exchanging partial schedules, (*b*) finding the missed and exceeded genes, and (*c*) legalizing offspring.

to preserve the relative positions between genes as much as possible when conducting a crossover operation. Falkenauer and Bouffouix proposed a modified version of order crossover (OX). OX was first discussed by Davis [99]; it tends to transmit the relative positions of genes rather than the absolute ones. In the OX the chromosome is considered to be circular since the operator is devised for the TSP. In the job-shop problem the chromosome cannot be considered circular. For this reason they developed a variant of the OX called linear order crossover (LOX), where the chromosome is considered linear instead of circular. Crossover LOX is applied to each subchromosome independently. For the ease of explanation, we use the subchromosome as parent. The LOX works as follows:

1. Select sublists from parents randomly as shown in Figure 6.24(a).

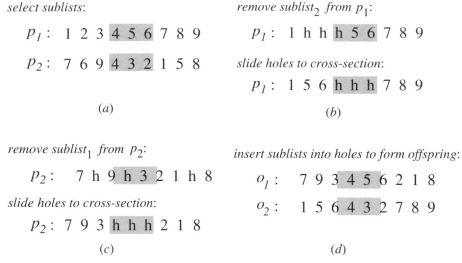

Figure 6.24. Linear order crossover.

2. Remove *sublist*$_2$ from parent p_1, leaving some "holes" (marked with h); then slide the holes from the extremities toward the center until they reach the cross section as shown in Figure 6.24(b). Similarly, remove *sublist*$_1$ from parent p_2 and slide the holes to cross section as shown in Figure 6.24(c).
3. Insert *sublist*$_1$ into the holes of parent p_2 to form the offspring o_1 and insert *sublist*$_2$ into the holes of parent p_1 to form the offspring o_2 as shown in Figure 6.24(d).

The crossover operator can preserve as much as possible both the relative positions between genes and the absolute positions relative to the extremities of parents. The extremities correspond to the high- and low-priority operations.

Falkenauer and Bouffouix used inversion as mutation operation, and Croce, Tadei, and Volta used the swap of two genes as mutation operation.

6.8   DORNDORF AND PESCH'S APPROACH

Dorndorf and Pesch proposed two kinds of implementation of genetic algorithms for the job-shop scheduling problem: One is the priority-rule-based representation, while the other is the machine-based representation [112]. The common feature of these algorithms is to hybridize genetic algorithms with conventional heuristics.

For the priority-rule-based representation, they incorporated the well-known Giffler and Thompson algorithm into genetic algorithm, where the genetic algorithm is used to evolve a sequence of priority dispatching rules and the Giffler and Thompson algorithm is used to deduce a schedule from the encoding of priority dispatching rules. Because the Giffler and Thompson algorithm is a tree-structured approach, they, in fact, used the priority dispatching heuristic, a kind of one-pass heuristic, not a pure Giffler and Thompson algorithm [21].

For the machine-based representation, they incorporated the well-known shifting bottleneck heuristic into a genetic algorithm, where the genetic algorithm is used to evolve a sequence of machines and the shifting bottleneck heuristic is used to deduce a schedule from the encoding of machine sequence. In fact, when deducing a schedule from a given chromosome, they used the basic idea of the pure shifting bottleneck heuristic—that is, to repeatedly solve one machine scheduling problem until all machines are treated. At each iteration, the shifting bottleneck heuristic identifies one bottleneck machine and schedules it and then identifies the next bottleneck machine. In their genetic algorithm, the sequence of machines is determined by a given chromosome, so there is no need to identify one bottleneck machine at each iteration.

Because these two encodings can be viewed as a permutation representation such as the traveling salesman problem, many conventional genetic operators can be applied here. They used the scaling window technique to calculate fitness values of chromosomes in order to increase selective pressure [206]. They also used elitist strategy in selection so that the best one in each generation always survived [171].

6.9 COMPUTATIONAL RESULTS AND DISCUSSION

Fisher and Thompson have proposed three well-known benchmarks for the job-shop scheduling problem in 1963 [126]; since then, researchers in operations research have tested their new algorithms on these problems. Most GA/JSP researchers used these benchmarks to test the performance of their genetic algorithms. Table 6.7 summarizes the available results of some genetic algorithms on the three benchmarks. The makespan of the best solution found by each of them and the makespan of an optimal solution for the corresponding problem are given in the table, where Dorndorf1 stands for the hybrid approach of the genetic algorithm with the Giffler and Thompson heuristic while Dorndorf2 stands for the hybrid approach of the genetic algorithm with the shifting bottleneck heuristic proposed by Dorndorf and Pesch. The 6×6 problem is a relatively easy one, and all of the genetic algorithms have obtained the optimal solution with a makespan of 55. For 10×10 and 20×5 problems, most genetic algorithms can obtain relatively good solutions. It is difficult to judge which approach is superior to others only on this table. All of them are tested on the same benchmarks, but not on the same experimental condition. The same experimental conditions means that

- Same total number of examined chromosomes

Table 6.7. Fisher and Thompson's Benchmark Problems

Authors	6×6	10×10	20×5
Optimal	55	930	1165
Nakano and Yamada (1991)	55	965	1215
Yamada and Nakano (1992)	55	930	1184
Paredis (1992)	—	1006	—
Gen et al. (1994)	55	962	1175
Fang et al. (1993)	—	949	1189
Dorndorf1[a] and Pesch (1995)	55	960	1249
Dorndorf2 and Pesch (1995)	55	938	1178
Croce et al. (1995)	55	946	1178
Cheng et al. (1995)	55	948	1196

[a]See text for definitions of Dorndorf1 and Dorndorf2.

- Same total number of random runs
- Same soft and hard computational means

For the total number of examined chromosomes, one can use either the same genetic algorithm parameters or different parameters (according to one's preference) while keeping the total number constant. Under the same experimental conditions, the performance of the proposed genetic algorithms can be assessed from the following five aspects:

- The best solution
- The deviation of the best one from the optimal solution
- The frequency for obtaining the best solution
- Solution distribution and statistical analysis
- Memory and computation time requirement

It is worthwhile to perform additional comparative studies of these genetic algorithms under standard experimental conditions using benchmark problems and other large-scale problems to reveal the advantages and disadvantages for each of them.

Table 6.8 presents some available results on parameter settings for the 10×10 problem. From the table we can see that there are two different preferences for genetic algorithm parameter settings. Nakano and Yamada prefer to use a large population size and a small iteration step, whereas Gen and Cheng prefer to begin with a relatively small population and employ a long evolutionary process.

Another basic issue is the comparison of genetic algorithms with conventional heuristics. For example, for the 10×10 problem, Yamada and Nakano have examined 2100 chromosomes on one run [424]. This means that they have run the Giffler and Thompson heuristic almost 2100 times. An optimal schedule

Table 6.8. Genetic Algorithm Parameters (10 × 10)

Authors	pop_size	max_gen	p_c	p_m
Nakano and Yamada (1991)	1000	150	—	—
Yamamda and Nakano (1992)	2000	100	—	—
Gen et al. (1994)	60	5000	0.4	0.3
Dorndorf1[a] and Pesch (1995)	200	—	0.65	0.001
Dorndorf2 and Pesch (1995)	40	—	0.65	—
Croce et al. (1994)	300	2971	1	0.03
Cheng et al. (1995)	40	2000	0.4	0.4

[a]See text for definitions of Dorndorf1 and Dorndorf2.

with a makespan of 930 was obtained five times in 300 trials of their genetic algorithms. For the same problem, Dorndorf and Pesch have examined 250,000 chromosomes with their first hybrid genetic algorithm, which means that they have run the priority dispatching heuristics almost 250,000 times. They have examined 800 chromosomes with their second hybrid genetic algorithms, which means that they have run the shifting bottleneck heuristic almost 800 times. The question is that if we run the same amount of trials with the conventional heuristics independently (if possible), could we obtain a better solution?

Until now, most of the genetic algorithm/JSP researchers directed their attention to the problem of measuring makespan. Because genetic algorithms provide a flexible framework for evolutionary computation and can handle any kind of objective function and any kind of constraint, we are able to deal with very complex problems with nonregular measures or multiple objectives, which are the potential areas worthy of further research.

The studies on genetic algorithms and JSP provide a rich experience for the constrained combinatorial optimization problems. All the techniques developed for the job-shop scheduling problem may be useful for other scheduling problems in modern flexible manufacturing systems (such as open-shop scheduling, mixed-shop scheduling, dynamic job-shop scheduling, etc.) and other combinatorial optimization problems.

7

MACHINE SCHEDULING PROBLEMS

7.1 INTRODUCTION

The machine scheduling problem is a rich and promising field of research with applications in manufacturing, logistics, computer architecture, and so on. Since the publication of the survey papers by Graham et al. [182] and Graves [183], the machine scheduling literature has grown exponentially. The main elements of the machine scheduling problem are

- Machine configuration
- Job characteristics
- Objective function

The machine configuration can be broadly classified into single and multiple machine problems. Gupta and Kyparisis surveyed the single-machine scheduling problem [197], and Cheng and Sin surveyed the parallel-machine scheduling problem [74]. The job characteristics include, but are not limited to, precedence relations among jobs, job release dates, job due dates, and preemption. Sen and Gupta surveyed research involving due dates [376], and Cheng and Gupta surveyed the research where job due dates are treated as decision variables [73]. The objective function can be broadly classified into single and multiple objective problems. Dileepan and Sen surveyed the bicriterion scheduling problem for the single machine case [110]. Objective function can be further classified as regular or nonregular measures. A nonregular performance measure may increase as the job completion time decrease. Baker and Scudder surveyed the research on nonregular performance measures [22]. Koulamas surveyed the total tardiness problem [251].

Machine scheduling problems are combinatorial problems, and their complexity can usually be determined by reducing other problems with known com-

234

plexity to them. Lenstra, Rinnooy Kan, and Brucker surveyed the complexity results for this problem [270].

7.1.1 Single-Machine Sequencing Problem

The basic assumption of the single-machine problem can be described as follows:

- A set of n independent, single-operation jobs is available for processing at time zero.
- Setup times for jobs are independent of job sequence and can be included in processing times.
- Jobs are known in advance.
- One machine is continuously available and is never kept idle while work is waiting.
- No job preempt is permitted.

The problem is to determine the processing sequence of jobs so as to optimize some performance measures.

Let $(j_1, j_2, \ldots, j_n)$ denote the jobs and let p_i denote the processing time, d_i the due date, w_i the weight, c_i the completion time, and r_i the ready time for job j_i, respectively.

Flowtime F_i is defined as $F_i = c_i - r_i$, the amount of time job j_i spends in the system. Flowtime measures the response of the system to individual demands for service and represents the interval a job waits between its arrival and its departure.

Lateness L_i is defined as $L_i = c_i - d_i$, the amount of time by which the completion time of job j_i exceeds it due date. Lateness measures the conformity of the schedule to a given due date. It may have either a positive value or a negative value. Negative lateness means that a job is completed before its due date. In many situation, penalties and costs are associated with positive lateness, but no benefits are associated with negative lateness. Therefore, only the positive lateness needs to be considered, which is called *tardiness*.

Tardiness T_i is defined as $T_i = \max\{0, L_i\}$; it represents the lateness of job j_i if it fails to meet its due date, or zero otherwise.

Let Π denote the set of feasible schedules without idle times between jobs. For a given schedule $\sigma \in \Pi$, let $f(\sigma)$ denote the corresponding objective function value and $r_i = 0$.

Mean Flowtime Problem. The mean flowtime problem is defined as follows:

$$\min_{\sigma \in \Pi} f(\sigma) = \frac{1}{n} \sum_{i=1}^{n} F_i = \frac{1}{n} \sum_{i=1}^{n} c_i \tag{7.1}$$

This problem is solved by sequencing jobs in nondecreasing order of processing times; that is, $p_{i_1} \leq p_{i_2} \leq \cdots \leq p_{i_n}$. This is known as *shortest processing time sequencing* (SPT rule).

Weighted Flowtime Problem. The weighted flowtime problem is defined as follows:

$$\min_{\sigma \in \Pi} f(\sigma) = \sum_{i=1}^{n} w_i F_i = \sum_{i=1}^{n} w_i c_i \tag{7.2}$$

This problem is solved by sequencing jobs in nondecreasing order of p_i/w_i; that is, $p_{i_1}/w_{i_1} \leq p_{i_2}/w_{i_2} \leq \cdots \leq p_{i_n}/w_{i_n}$, which is called *weighted shortest processing time* (WSPT).

Mean Tardiness Problem. The mean tardiness problem is defined as follows:

$$\min_{\sigma \in \Pi} f(\sigma) = \frac{1}{n} \sum_{i=1}^{n} T_i \tag{7.3}$$

Maximum Flowtime Problem. The maximum flowtime problem is defined as follows:

$$\min_{\sigma \in \Pi} f(\sigma) = \max_{1 \leq i \leq n} \{F_i\} = \max_{1 \leq i \leq n} \{c_i\} \tag{7.4}$$

This problem is known as the *makespan problem.*

Maximum Lateness Problem. The maximum lateness problem is defined as follows:

$$\min_{\sigma \in \Pi} f(\sigma) = \max_{1 \leq i \leq n} \{L_i\} \tag{7.5}$$

This problem is solved by sequencing jobs with the earliest due date (EDD) rule; that is, $d_{i_1} \leq d_{i_2} \leq \cdots \leq d_{i_n}$.

Maximum Tardiness Problem. The maximum tardiness problem is defined as follows:

$$\min_{\sigma \in \Pi} f(\sigma) = \max_{1 \leq i \leq n} \{T_i\} \tag{7.6}$$

This problem is solved by the EDD rule.

Minimum Tardy Job Problem. The minimum tardy job problem is defined as follows:

$$\min_{\sigma \in \Pi} f(\sigma) = \sum_{i=1}^{n} \delta(T_i) \tag{7.7}$$

where

$$\delta(T_i) = \begin{cases} 1 & \text{if } T_i > 0 \\ 0 & \text{otherwise} \end{cases}$$

All of the measures discussed above are the function of job completion times, so the general form is given as follows:

$$\min_{\sigma \in \Pi} f(\sigma) = f(c_1, c_2, \ldots, c_n)$$

Furthermore, they belong to an important class of performance measures called *regular measure*. A performance measure is regular if

- the scheduling objective is to minimize $f(\sigma)$ and
- $f(\sigma)$ can increase only if at least one of the completion times in the schedule increases.

Completion Time Variance Problem. The completion time variance (CTV) problem is defined as follows [286]:

$$\min_{\sigma \in \Pi} f(\sigma) = \frac{1}{n} \sum_{j=1}^{n} (c_j - \bar{c})^2 \tag{7.8}$$

where $\bar{c}$ is the average flow time given as follows:

$$\bar{c} = \frac{1}{n} \sum_{j=1}^{n} c_j$$

In the optimal CTV schedule, the longest job is processed first [371].

The weighted version of completion time variance measure (WCTV) problem examined by Gupta et al. is defined as follows [196]:

$$\min_{\sigma \in \Pi} f(\sigma) = \frac{1}{n} \sum_{j=1}^{n} w_j (c_j - \overline{c})^2 \qquad (7.9)$$

Cheng and Cai have proved that the CTV problem is NP-hard and that no polynomial or pseudopolynomial algorithm exists to find the optimal solution for the CTV problem [72].

Waiting Time Variance Problem. If we let w_i denote the waiting time of job j_i, then we have $w_i = c_i - p_i$. The waiting time variance (WTV) problem is defined as follows:

$$\min_{\sigma \in \Pi} f(\sigma) = \frac{1}{n} \sum_{j=1}^{n} (w_j - \overline{w})^2 \qquad (7.10)$$

where $\overline{w}$ is the average waiting time given as follows:

$$\overline{w} = \frac{1}{n} \sum_{j=1}^{n} w_j$$

The optimal WTV schedule is V-shaped, meaning that the jobs preceding and succeeding the shortest job are in LPT and SPT orders, respectively [116].

Total Absolute Differences in Completion Time. This problem is to minimize the total sum of absolute differences in completion time (TADC problem) as follows:

$$\min_{\sigma \in \Pi} f(\sigma) = \sum_{i=1}^{n} \sum_{j=1}^{n} |c_j - c_i| \qquad (7.11)$$

$$= \sum_{i=1}^{n} (i-1)(n-i+1)p_{[i]} \qquad (7.12)$$

where $[i]$ denotes the ith job to be processed.

Because of the difficulty of the CTV problem, Kanet proposed TADC as a measure of variation and proposed an $O(n \log n)$ algorithm for the problem [240]. The crucial difference between TADC and CTV is the difference between the sum of absolute difference and the sum of squared difference. According to Hardy, Littlewood, and Polya, TADC is minimized by matching the positional weights $(i-1)(n-i+1)$ in nonincreasing order with the processing times $p_{[i]}$ in nondecreasing order [208, 240].

Total Absolute Differences in Waiting Time. This problem is to minimize the total sum of absolute differences in waiting time (TADW problem) as follows:

$$\min_{\sigma \in \Pi} f(\sigma) = \sum_{i=1}^{n} \sum_{j=1}^{n} |w_j - w_i| \tag{7.13}$$

$$= \sum_{i=1}^{n} i(n-i)p_{[i]} \tag{7.14}$$

The relationship between TADW and WTV is similar to the relationship between TADC and CTV. TADW is minimized by matching the positional weights $i(n-i)$ in nonincreasing order with the processing times $p_{[i]}$ in nondecreasing order [16].

These measures given above are nonregular measures.

7.1.2   Earliness and Tardiness Scheduling Problems

The study of earliness and tardiness penalties in scheduling models is a relatively recent area of inquiry [22]. For many years, scheduling research focused on regular measures that are nondecreasing in job completion times. Most of the literature deals with such regular measures as mean flowtime, mean lateness, percentage of jobs tardy, and mean tardiness. The mean tardiness criterion, in particular, has been a standard way of measuring conformance to due dates, although it ignores the consequences of jobs completing early. However, this emphasis has changed with the current interest in just-in-time (JIT) production, which espouses the notion that earliness as well as tardiness should be discouraged. The concept of penalizing both earliness and tardiness has spawned a new and rapidly developing line of research in the scheduling field. Because the use of both earliness and tardiness penalties gives rise to a *nonregular performance measure*, it has led to new methodological issues in the design of solution procedures.

To describe a generic E/T model, let n be the number of jobs to be scheduled. Job j_i is described by a processing time p_i and a due date d_i. Jobs are assumed to be simultaneously available. As a result of scheduling decisions, job j_i will be assigned a completion time c_i. Let E_i and T_i represent the earliness and tardiness of job j_i, respectively. These quantities are defined as follows:

$$E_i = \max\{0, d_i - c_i\} = (d_i - c_i)^+$$
$$T_i = \max\{0, c_i - d_i\} = (c_i - d_i)^+$$

Associated with each job is a unit earliness penalty $\alpha_i > 0$ and a unit tardiness penalty $\beta_i > 0$. For a given schedule $\sigma \in \Pi$, let $f(\sigma)$ be the corresponding objective function. The basic E/T scheduling problem can be written as follows:

$$\min_{\sigma \in \Pi} f(\sigma) = \sum_{i=1}^{n} [\alpha_i(d_i - c_i)^+ + \beta_i(c_i - d_i)^+] \tag{7.15}$$

$$= \sum_{i=1}^{n} (\alpha_i E_i + \beta_i T_i) \tag{7.16}$$

However, the penalties can be measured in different ways. Most of the variety in the E/T literature stems from the generality of assumptions made about due dates and penalty costs. In some formulations of the E/T problem the due date is given, while in others the problem is to optimize the due date and the job sequence simultaneously. The simplest model considers a common due date for all jobs, and a more general model allows distinct due dates. In a similar vein, some models prescribe common penalties, while others permit differences among jobs or differences between the earliness and tardiness penalties. Even the simplest model leads to an optimization problem that is NP-hard [73].

Absolute Deviation Problem. An important special case in the family of E/T problems involves minimizing the sum of absolute deviations of the job completion times from a common due date. In particular, the objective function can be written as follows:

$$\min_{\sigma \in \Pi} f(\sigma) = \sum_{i=1}^{n} |c_i - d| \tag{7.17}$$

with the understanding that $d_i = d$. This model penalizes earliness and tardiness at the same rate for all jobs. It is desirable to construct the schedule so that the due date is in the middle of the jobs. If d is too tight, then it will not be possible to fit enough jobs in front of d, because of the restriction that no job can start before time zero. Thus, for a given problem, the common due date d may be too tight; this gives rise to the *restricted* problem, but otherwise it leads to the *unrestricted* problem.

There exists an optimal solution to the unrestricted problem with the following properties:

- There is no inserted idle time in the schedule.
- The optimal schedule is *V*-shaped.
- One job is completed precisely at the due date.
- In an optimal schedule, the kth job is completed at time d, where k is the smallest integer greater than or equal to $n/2$.

The analysis of this problem is due to Kanet [240], Sundararaghavan and Ahmed [390], Hall [200], Bagchi, Chang, and Sullivan [17], and Emmons [119].

Bagchi, Chang, and Sullivan considered the more generic case where earliness and tardiness are penalized at different rates:

$$\min_{\sigma \in \Pi} f(\sigma) = \sum_{i=1}^{n} (\alpha E_i + \beta T_i) \tag{7.18}$$

There is a restricted as well as unrestricted version of the problem. In the unrestricted version, an optimal solution has the same properties as the absolute deviation problem.

Weighted Absolute Deviation Problem. The weighted earliness and tardiness problem examined by Hall and Posner is defined as follows [202]: n jobs with known integer weights $w_i, i = 1, 2, \ldots, n$; a common due date d, where d is unrestrictively large. The objective is to minimize the sum of weighted earliness and tardiness of the jobs; that is,

$$\min_{\sigma \in \Pi} f(\sigma) = \sum_{i-1}^{n} w_i |c_i - d| \tag{7.19}$$

The model derives from the notion that different job should be penalized at different rates. This problem is NP complete in the ordinary sense.

Let the early job set be $E = \{ j_i | c_i \leq d \}$ and tardy job set $T = \{ j_i | c_i > d \}$. There exists several optimality conditions for the problem.

- There exists an optimal solution where some job is completed at due date d.

- The optimal schedule is V-shaped; that is, jobs in E have nondecreasing w_i/p_i ratios, whereas jobs in T have nonincreasing w_i/p_i ratios.

- Given an optimal schedule σ^*, we have

$$\sum_{i \in E} w_i \geq \sum_{i \in T} w_i$$

- Given an optimal schedule σ^*, we have

$$\sum_{i \in E} p_i \geq \sum_{i \in T} p_i$$

Squared Deviation Problem. In some cases, large deviations from the due date are highly undesirable, and it might be more appropriate to use squared deviations from the common due date as the performance measure as follows:

$$\min_{\sigma \in \Pi} f(\sigma) = \sum_{i=1}^{n} (c_i - d)^2 = \sum_{i=1}^{n} (E_i^2 + T_i^2) \tag{7.20}$$

This is the quadratic analog of total absolute deviation. Elion and Chowdhury have proven that the optimal schedule is V-shaped [116].

Bagchi, Chang, and Sullivan also examine the general case in which earliness and tardiness penalties are different [17];

$$\min_{\sigma \in \Pi} f(\sigma) = \sum_{i=1}^{n} (\alpha E_i^2 + \beta T_i^2) \tag{7.21}$$

7.1.3 Parallel Machine Scheduling Problem

Multiple machine scheduling is the study of constructing schedules of machine processing for a set of jobs in order to ensure the execution of all jobs in a reasonable amount of time. Three issues are to be dealt with:

- What machines to be allocated to which jobs
- How to order the jobs in an appropriate processing sequence
- How to rationalize the reasonableness of the schedule

In other words, the major concern of multiple machine scheduling theory is how to provide a perfect match, or near-perfect match, of machines to jobs and subsequently determine the processing sequence of the jobs on each machine in order to achieve some prescribed goal.

The parallel machine scheduling problem can be viewed as a class of problem which is relaxed from the multiple machine scheduling problem. In the parallel machine system, all machines are identical and a job can be processed by any one of the free machine. Each finished job will free a machine and leave the system.

A job is characterized by the following factors:

- **Job processing time:** unit processing time or various processing time.
- **Due date requirement:** without due date, or common due date, or general due date.
- **Preemptive job:** preempting is permitted or not.
- **Precedence constraints:** jobs are independent, or dependent, or with tree precedence constraint.

- **Job ready time:** some ready time for all jobs or arbitrary ready time.

The optimality criteria can be split into three distinct groups [37]:

- Completion-time-based measures
- Due-date-based measures
- Idleness-penality-based measures

General assumptions are made as follows [144]:

- Each job has only one operation.
- No job can be processed on more than one machine simultaneously.
- Any machine can process any job.
- No machine may process more than one job at a time.
- Machines will never break down and are available throughout the scheduling period.
- Job processing times are independent of the schedule.
- Machine setup time is negligible.
- Transportation time between machines is negligible.
- Work-in-process inventory is allowed and its associated costs are negligible.
- Number of jobs is fixed.
- Number of machines is fixed.
- Processing time of j_i on m_k is given and fixed for all i and k.
- Ready time of j_i for all i is known.

Minimum Makespan Problem. Consider the following multiple-machine scheduling problem: a set of simultaneously available and independent nonpreemptive jobs $\{j_1, j_2, \ldots, j_n\}$ on a set of identical and unrelated parallel machines $\{m_1, m_2, \ldots, m_m\}$ $(m < n)$. For a given schedule $\sigma \in \Pi$, let c_j be the completion time of job j under schedule σ for $j = 1, 2, \ldots, n$, and $f(\sigma)$ denote the corresponding objective function value. The minimum makespan problem is defined as follows:

$$\min_{\sigma \in \Pi} f(\sigma) = \max\{c_j | j = 1, \ldots, n\} \qquad (7.22)$$

Garey and Johnson have proved that the problem is NP-hard in the strong sense when the number of machines is arbitrary [147]. The problem is solvable in pseudopolynomial time when the number of machines is fixed and thus NP-hard only in the ordinary sense.

Minimum Weighted Flowtime Problem. Let w_i denote the weight and r_i the ready time associated with job j_i. The problem is to find an optimal nonpre-emptive job schedule to minimize the weighted flowtime as follows:

$$\min_{\sigma \in \Pi} f(\sigma) = \sum_{j=1}^{n} w_j(c_j - r_j) \qquad (7.23)$$

Bruno et al. proved that even a two-machine system for finding the weighted sum of flowtime with an unequally weighted set of jobs is NP-hard [49].

Minmax Weighted Absolute Lateness Problem. Given an unrestricted due date d, which is common to all jobs, as follows:

$$d \geq \sum_{j=1}^{n} p_j$$

the problem is to find an optimal nonpreemptive job schedule to minimize the maximum weighted absolute lateness as follows:

$$\min_{\sigma \in \Pi} f(\sigma) = \max \{ w_j |c_j - d|; j = 1, \ldots, n \} \qquad (7.24)$$

Li and Cheng have shown that the minmax scheduling problem is NP-hard even for a single-machine system and have proposed heuristic procedures to solve it [271].

Minsum Weighted Absolute Lateness Problem. The problem is to find an opti-mal job schedule to minimize the sum of weighted absolute lateness as follows:

$$\min_{\sigma \in \Pi} f(\sigma) = \sum_{j=1}^{n} w_j |c_j - d| \qquad (7.25)$$

A minsum problem attempts to minimize the sum of weighted absolute devi-ation of job completion time about the due date (i.e., to reduce a customers' aggregate disappointment), while a minmax problem attempts to minimize the maximum weighted absolute deviation of job completion time about the due date (i.e., to reduce a customer's maximum disappointment).

7.2   CLEVELAND AND SMITH'S APPROACH

Cleveland and Smith first investigated the use of genetic algorithms to solve a one-machine sequence problem [76]. In fact, they considered the use of genetic algorithms as a means of scheduling the release of jobs into a manufactur-ing facility, called the *sector release scheduling problem*. The sector release scheduling problem can be seen as both a sequencing and a timing problem. In the former case, the problem is restricted to determine the order in which jobs are released. In the latter case, the problem is defined as one of determining absolute release times for each order.

The job sequencing problem is that of finding an ordering of jobs to be released into the sector so that the overall cost of processing those jobs is as low as possible. The overall cost is usually a combination of inventory cost, machine setup costs, and earliness and tardiness costs.

7.2.1 Genetic Operators

This problem has much in common with the traveling salesman problem (TSP). Many GA-based approaches to the TSP are applicable. The main difference between the TSP and job sequencing is that a job's *absolute position* in the sequence, in addition to its *relative position*, is important for the job sequencing problem. Cleveland and Smith have examined many recombination operators proposed for TSP, such as Goldberg's PMX [171], Grefenstette's subtour-swap and subtour-chunk [188], and subtour-replace and weighted chunking. In their experiments, the subtour-chunking operator always produced the best result.

The subtour-chunking operator works by alternately selecting segments (chunks) from the two parent chromosomes and incorporating those into the offspring. The chunks are placed in the offspring in approximately the same position that they occupied in the parent chromosome. Conflicts are resolved by trimming the chunks and by sliding them to the right and to the left to make them fit. An example is given below:

1. Choose a chunk from parent p_1. Suppose it is (6 8 7). Place it into the child in the same position as in the parent.

2. Choose a chunk from parent p_2. Suppose it is (7 10 4 6). Trim the chunk down to (10 4) to avoid conflicts and place the chunk to the left of its original place since the previous chunk is acting as an obstacle.

3. Choose a chunk from parent p_1 again. Suppose it is (9 4 3 2). Trim the chunk down to (9) and place it in the first position of the offspring as the best approximation to its desired position.

4. Choose a chunk from parent p_2 again. Suppose it is (3 5 8). Trim it down to (3 5) and place it in the second position to avoid the obstacle in the first position. Note that the 5 is wasted since the position it should occupy is already filled.

5. Choose a chunk from parent p_1 again. Suppose it is (5 10). Trim the chunk down to (5) and place it in the ninth position.

6. Choose a chunk from parent p_2 again. Suppose it is (9 2 1). Trim it down to (2) and place it in the last position.

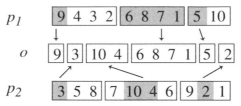

Figure 7.1. Crossover: subtour-chunking.

7. At this time all the chunks have been used or wasted and the operator must resort to random selection to place the remaining pieces. No mutation was used in their implementation.

7.2.2 Selection

Cleveland and Smith adopted *stochastic universal sampling*, proposed by Baker [20], as their selection mechanism. This method uses a single wheel spin. The wheel is constructed as a standard roulette wheel and is spun with a number of equally spaced markers equal to the population size.

Let $eval(v_i)$, $i = 1, 2, \ldots, pop\_size$ denote the fitness value of chromosome v_i. The expected value e_i is calculated as follows:

$$e_i = pop\_size \times p_i = \frac{pop\_size \times eval(v_i)}{\sum_{i=1}^{pop\_size} eval(v_i)}, \qquad i = 1, 2, \ldots, pop\_size$$

The procedure of stochastic universal sampling can be described as follows:

Procedure: Stochastic Universal Sampling
begin
   sum ← 0;
   ptr ← *rand*( );
   for i ← 1 **to** *pop_size* **do**
   sum ← *sum* + e_i;
   while (*sum* > *ptr*) **do**
     select individual i;
     ptr ← *ptr* + 1;
   end
   end
end

where *rand*( ) returns a random real number uniformly distributed within the range [0,1).

7.3   GUPTA, GUPTA, AND KUMAR'S APPROACH

M. Gupta, Y. Gupta, and Kumar have applied a genetic algorithm to the n-job single-machine scheduling problem with an objective to minimize the flow-time variance [196]. They adopted a permutation representation in their work because permutation of n jobs is a natural representation for a one-machine scheduling problem. Their implementation of genetic algorithms includes PMX crossover, swap mutation, linear scaled fitness, simple diversification strategy, and combined selection mechanism of crowding and elitist strategies.

7.3.1   Evaluation Function

Because the CTV and WCTV problems are minimization problems, the following transformation is used in mapping the original objective function to a fitness function:

$$f(v_i) = \begin{cases} c_{\max} - g(v_i), & \text{when } g(v_i) < c_{\max} \\ 0, & \text{otherwise} \end{cases} \tag{7.26}$$

where $g(v_i)$ denotes the original objective function and $f(v_i)$ denotes the fitness function for chromosome v_i, respectively.

The linear scaling method, proposed by Goldberg [171], was used to help in differentiating the chromosomes with average fitness value and the best fitness value.

$$f' = af + b \tag{7.27}$$

where f is the raw fitness, f' is the scaled fitness, and a and b are constants.

7.3.2   Replacement Strategy

A combination of *crowding* and *elitist* was used as replacement strategy [272]. By using genetic operations, a pool of offspring is generated to create a new population. If all of the offspring outperform every existing chromosome in the old population, then all the offspring replace the existing chromosomes in the new population. If some of them fare better, they replace an equal number of existing chromosomes that are lowest in the order of performance in the old population. For the remaining offspring an offspring is selected from the old population using a specified probability value. The replacement strategy ensures that the best-performing chromosomes of the previous generation are stored in the current population.

7.3.3 Convergence Policy

An ideal genetic algorithm should maintain a high degree of diversity within the population as it iterates from one generation to the next. Otherwise, the population may converge prematurely before the desired solution is found. Gupta et al. adapted Grefenstette's entropic measure for monitoring the population divergence [188]. For each job i, a measure of the entropy H_i in the current population is calculated as follows:

$$H_i = \frac{1}{\log n} \sum_{j=1}^{n} \frac{n_{ij}}{2pop\_size} \log \left(\frac{n_{ij}}{2pop\_size} \right) \qquad (7.28)$$

$$H = \frac{1}{n} \sum_{i=1}^{n} H_i \qquad (7.29)$$

where n_{ij} is the number of edges connecting job i and j in the population, n is the number of jobs. As the population converges, H approaches 0.

7.3.4 Overall Procedure

The overall procedure of Gupta et al.'s approach is summarized as follows:

Procedure: Gupta et al.'s Approach

Step 1. Initialization

  1. Set population size pop_size, mutation rate p_m, crossover rate p_c, and the maximum number of generations $maxgen$. Let $gen = 1$.
  2. Generate an initial population randomly and call it oldpop.
  3. Calculate the objective function values using (7.8) or (7.9), and calculate fitness values using (7.27). Sort the population in increasing order of objective function value.
  4. Check to see if the diversity of oldpop is at an acceptable level. If it is not acceptable, execute step 4.

Step 2. Recombination

  1. Perform PMX crossover and swap mutation.
  2. Calculate the objective function values and fitness values for every offspring.
  3. Sort the selection pool in increasing order of objective function value.

Step 3. Replacement. Compare the chromosomes of sorted oldpop and the selection pool for their fitness value and create a newpop using the replacement policy:

1. If all the offspring outperform every existing chromosome in the old population, then all the offspring replace all the existing chromosomes in the new population.

2. If only some offspring fare better than the existing population, they replace an equal number of the existing chromosomes which are lowest in the order of performance in the old population.

3. For the remaining offspring, a selection is made with a specified probability.

Step 4. Diversification. Apply the mutation process to diversify the population.

1. Calculate the diversity parameter H for the current population using (7.28) and (7.29).

2. Compare the diversity with the given acceptable level. If the diversity is below an acceptable level, execute the mutation process repeatedly until the diversity of the population is equal to an acceptable level.

Step 5. New generation

If $gen < maxgen$, let $gen \leftarrow gen + 1$ and return to step 2; otherwise stop.

The proposed algorithm was tested on several problems with sizes ranging from 10 jobs to 20 jobs. The basic parameter settings are given in Table 7.1.

7.4   LEE AND KIM'S APPROACH

Lee and Kim have developed a parallel genetic algorithm for the job scheduling problem for a single machine [266]. The objective of the scheduling is to minimize the total generally weighted earliness and tardiness penalties from a common due date. A binary representation scheme was employed for coding

Table 7.1. Basic Setting for Genetic Algorithm Parameters

Parameters	Range
pop_size	(75, 100)
max_gen	(100, 400)
p_c	(0.6, 0.9)
p_m	(0.05, 0.15)

job schedules into chromosomes. The genetic algorithm was parallelized by keeping the population in separate subgroups.

The general weighted earliness and tardiness (GWET) problem is defined as follows:

$$\min_{\sigma \in \Pi} f(\sigma) = \sum_{i=1}^{n} \{\alpha_i(d - c_i)^+ + \beta_i(c_i - d)^+\}$$

There are no restrictions on the weights α_i and β_i. This problem is known as NP-hard [201, 202, 265].

7.4.1   Representation

An optimal schedule for the GWET problem is V-shaped around the due date [18, 22]; that is, the jobs that are completed on or before the due date are processed in nonincreasing order of p_i/α_i, and the jobs that are started on or after the due date are processed in nondecreasing order of p_i/β_i. Because the optimal schedule is V-shaped, Lee and Kim proposed a binary representation which can guarantee that all chromosomes are V-shaped schedules.

For an n-job problem, a chromosome contains n binary bits, each of which indicates one of two sets (the tardy job set T and the nontardy job set E) a corresponding job belongs to. For example, if the ith bit is 0, the corresponding job i belongs to set E; otherwise, job i belongs to set T. A schedule is constructed by sequencing jobs in set E in nonincreasing order of p_i/α_i and jobs in set T in nondecreasing order of p_i/β_i. Let us consider a simple example with nine jobs given in Table 7.2. Suppose a chromosome is given as

$$[1 \quad 0 \quad 0 \quad 1 \quad 0 \quad 0 \quad 1 \quad 1 \quad 0]$$

Because first bit, fourth bit, seventh bit, and eighth bit are 1, the tardy set is formed as $T = \{j_1, j_4, j_7, j_8\}$ and the nontardy set is formed as $E = \{j_2, j_3, j_5, j_6, j_9\}$. Now sequence the jobs in set E in nonincreasing order of p_i/α_i

Table 7.2. Nine-Job Example

	j_1	j_2	j_3	j_4	j_5	j_6	j_7	j_8	j_9
p_i	2	2	4	5	1	4	3	2	1
α_i	1	2	6	4	4	8	10	5	9
β_i	8	8	8	2	9	2	2	1	2
$p_i\beta_i/\alpha_i$	16	8	5.3	2.5	2.25	1	0.6	0.4	0.2
p_i/α_i	2	1	0.6	1.25	0.25	0.5	0.3	0.4	0.1
p_i/β_i	0.25	0.25	0.5	2.5	0.11	2	1.5	2	0.5

and jobs in set T in nondecreasing order of p_i/β_i to form a V-shaped schedule as follows:

$$[j_2 \quad j_3 \quad j_6 \quad j_5 \quad j_9 \quad j_1 \quad j_7 \quad j_8 \quad j_4]$$

There are two main components for this coding representation:

1. Separate the jobs into either set E or set T by using a binary string.

2. Construct a V-shaped schedule by sequencing jobs in set E in nonincreasing order of p_i/α_i and jobs in set T in nondecreasing order of p_i/β_i.

7.4.2   Parallel Subpopulations

The parallel genetic algorithm consists of a group of local subpopulations. Each subgroup independently generates offspring using genetic operators. Each subpopulation is characterized by the first job of its schedules. In other words, each group of strings has the same job as the first one to be processed.

The problem is how to determine the number of subgroups m. The random V-shaped schedule generator given by Lee and Kim will try to place some jobs with nonincreasing order of p_i/α_i in the time interval $[0, d]$ and place the remaining jobs with nondecreasing order of p_i/β_i in the time interval $[d, \sum_{i=1}^{n} p_i]$. A schedule beginning with some jobs could not form a V-shaped schedule because in this case the total processing time of the jobs in the set T will exceed d $\sum_{i=1}^{n} p_i$. Lee and Kim devised a lemma to identify the critical job k. Then the subgroups are formed as follows: First sequence the jobs with the order $p_1/\alpha_1, p_2/\alpha_2, \ldots, p_n/\alpha_n$. The strings in subpopulation 1 start with the job that has the highest p_i/α_i value. In subpopulation 2, the first job in each string has the second highest p_i/α_i value. In subpopulation 3, a job with the third or fourth highest p_i/α_i value is first processed. In subpopulation m, jobs start with the $(2^{m-2}+1)$th, $(2^{m-2}+2)$th, $\ldots, (2^{m-1})$th highest p_i/α_i values. This is repeated until we encounter the critical job k.

The initial procedure works with the following five main steps:

Procedure: Initialization

Step 1. For each subgroup, first assign 0 to the starting job (i.e., place it into the nontardy set E).

Step 2. All jobs with a higher value of p_i/α_i than that of the starting job is fixed to 1 (i.e., place them into the tardy set T).

Step 3. Other jobs are randomly assigned to the early job set or the tardy job set as far as the total processing time of early jobs does not exceed the due date.

Step 4. When the completion time of a job with bit value 0 first exceeds the due date d, its bit value is altered to 1 and all remaining unassigned jobs' bit values are fixed to 1.

Step 5. When the total completion time of jobs with bit value 1 exceeds $\sum_{i=1}^{n} p_i - d$, all remaining jobs' bit values are fixed to 0.

7.4.3 Crossover and Mutation

Since the chromosome is encoded as a binary string, one-cut-point and two-cut-point crossovers are employed. However, offspring generated by crossover may be infeasible in the sense that the V-shape is not maintained because the total processing time of jobs in set E may either exceed or be far less than the due date. In order to adjust the offspring to a V-shaped schedule, the following mutation operator is employed:

1. If the total processing time of jobs in set E exceeds the due date, alter the jobs $\{i|c_i - p_i \geq d\}$ to set T. Also let job i with $c_i - p_i \geq d < c_i$ straddle the due date.

2. If the total processing time of the job in set T exceeds $\sum_{i=1}^{n} p_i - d$, alter the jobs $\{i|c_i \geq d\}$ to set E.

The mutation moves some appropriate jobs before or after the due date to maintain the V-shape of the schedule.

7.4.4 Evaluation and Selection

To evaluate the penalty of each chromosome, it is decoded into a schedule by sequencing the jobs in set E in nonincreasing order of p_i/α_i and jobs in set T in nondecreasing order of p_i/β_i. Then the linear scaling method is employed to determine fitness for each schedule as follows:

1. The raw fitness value for a schedule is obtained by subtracting the penalty from the maximum penalty in each subpopulation.

2. The fitness value then is computed by linear scaling in which the lowest fitness is zero and the average fitness is equal to the average raw fitness.

Two reproduction methods were tested: roulette wheel selection and deterministic selection (N-best selection).

7.4.5   Parallel Genetic Algorithm

The proposed parallel genetic algorithm is based on the island model and has the following structure:

Algorithm: Parallel Genetic Algorithm

begin
   initialization;
   evaluation;
   while (not done of each subpop) **do**
     reproduction;
     crossover;
     mutation;
     evaluation;
     communication;
     deletion;
   end
end

Communication is a medium for exchanging information among subpopulations in parallel genetic algorithms. The best chromosome in each subpopulation is transformed to other subpopulations to generate improved schedules by mating with the group-specific chromosomes.

Two types of communication procedures are prevalent: all node connection and neighbor node connection [303]. In all node connection, the best chromosome in each subpopulation is transferred to every other subpopulation, while in neighbor node connection only some neighbor subpopulations enjoy the common elite. To avoid premature convergence that is mainly due to superindividuals emerging from other subpopulations, Lee and Kim employed the neighbor node connection scheme. Subpopulations are connected as a ring, and each subpopulation sends a copy of the best string to each adjacent group and also receives copies of the best strings.

7.5   CHENG AND GEN'S APPROACH

Cheng and Gen have applied genetic algorithms to the job scheduling problem in identical parallel machine systems with an objective of minimizing the maximum weighted absolute lateness [68, 443]. This problem was first considered by Li and Cheng as follows [271]: There are a set of jobs associated with known processing times and weights, several parallel and identical machines, and a common due date that is not too early to constrain the scheduling decision. The objective is to find an optimal job schedule so as to minimize the maximum weighted absolute lateness. This objective function is known as one of the nonregular performance measures.

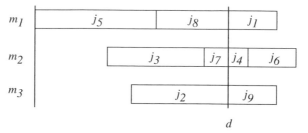

Figure 7.2. A schedule for nine-job three-machine problem.

7.5.1 Representation and Initialization

There are two essential issues to be dealt with for all kinds of multiple machine scheduling problems:

- Partition of jobs to machines
- Sequence jobs for each machine

An extended permutation representation is proposed to encode these two things into a chromosome, where integers represent all possible permutation of jobs (or sequence of jobs) and asterisks * designate the partition of jobs to machines. Let us consider a simple example with nine jobs and three machines. Suppose there is a schedule shown in Figure 7.2. The chromosome can be represented as follows:

$$[5 \quad 8 \quad 1 \quad * \quad 3 \quad 7 \quad 4 \quad 6 \quad * \quad 2 \quad 9]$$

Geneally, for an n-job m-machine problem, a legal chromosome contains n job symbols and $m-1$ partitioning symbols, resulting in a total size of $(n+m-1)$.

Initial population is randomly generated. The overall procedure is shown as follows:

Procedure: Initialization

begin
  $i \leftarrow 0$;
  while $(i \leq pop\_size)$ **do**
    generate a job permutation list randomly;
    put $m-1$ asterisks into the job list randomly;
    $i \leftarrow i+1$;
  end
end

7.5.2 Crossover

As mentioned, the essential issue of the multiple-machine scheduling problem is the combination and permutation for jobs and machines. Both crossover and mutation operators are used to adjust the partition and permutation of jobs.

Let us call a schedule for one machine in a multiple-machine system a *subschedule*. The subschedule can be considered as the natural building blocks for genetic search. A *subschedule preservation* crossover operator is given in order to maintain such building blocks in offspring in much the same manner as Holland described [220]. The proposed crossover takes two parents and creates a single offspring by propagating the overall partitioning structure and a subschedule into offspring from one parent and then completing the offspring with the remaining jobs derived from another parent. It performs as follows:

Procedure: Crossover

Step 1. It obtains asterisk positions (or overall partitioning structure) from one parent.

Step 2. It obtains a randomly selected subschedule from the same parent.

Step 3. It obtains the remaining jobs from the other parent by making a left-to-right scan.

Suppose we have two parents, p_1 and p_2, and that the rightmost subschedule of p_1 is selected to be propagated into offspring o. The operation of crossover is illustrated in Figure 7.3. From Figure 7.3 we can see that the proposed crossover can adjust job partition and job order simultaneously.

7.5.3 Mutation

Two kinds of mutation are used. One is *swapping mutation*, where we select two random positions and then swap their genes. Another is *heuristic mutation*, where we use greedy heuristics to adjust the orders of jobs so as to form V-shaped subschedules for each machine.

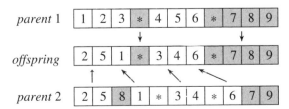

Figure 7.3. Illustration of crossover operator.

Swapping Mutation. The randomly swapped genes may be either job or asterisk. The different combinations of job and asterisk result in four basic types of mutation.

1. If both genes are job, two cases may occur. One case is that two selected jobs are processed by same machine. In such a case the mutation alters the job order for the machine as shown in Figure 7.4(a).
2. Another case is that two jobs are processed by different machines. In such a case the mutation alters both job order and job partition to machines for the chromosome as shown in Figure 7.4(b).
3. If both genes are asterisk, the mutation perform a trivial operation as shown in Figure 7.4(c). This kind of mutation is prohibited in our implementation.
4. If one gene is asterisk and another is job, the mutation alters both job order and job partition to machines for the chromosome as shown in Figure 7.4(d).

The last one is the only genetic operation which can alter the position of asterisks. Without this type of mutation, the position of asterisks will never change throughout the evolutionary process, and therefore genetic search is confined by the initial population (or initial positions of asterisks). Hence this type of mutation plays a vital role in genetic search. The overall procedure is shown below:

Procedure: Mutation

begin
  $i \leftarrow 0$;
  while $(i \leq pop\_size \times p_m)$ **do**
    select a chromosome randomly;
    pick up two genes randomly;
    if (both genes are asterisk) **then**
     pick up a job randomly;
    end
    exchange their positions;
    $i \leftarrow i + 1$;
  end
end

Heuristic Mutation. This mutation uses a greedy strategy to adjust job order to form a V-shaped subschedule for each machine; that is, all jobs before the common due date are sequenced in nonincreasing order of p_i/w_i, whereas all jobs after the common due date are sequenced in nondecreasing order of p_i/w_i.

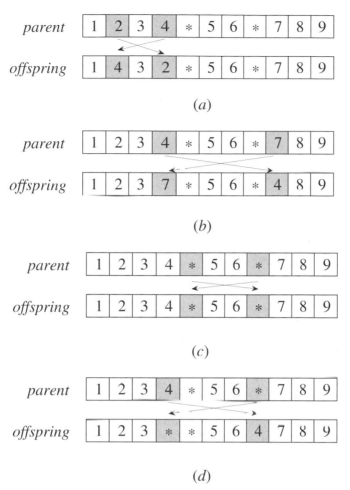

Figure 7.4. Illustration of a mutation operator: (a) swap two jobs within one machine; (b) swap two jobs within different machine; (c) a trivial swap; and (d) swap the position of a job and an asterisk.

Let $d - x$ denote the start time of the earliest job and $d + y$ the completion time of the latest job that have been scheduled on a machine as shown in Figure 7.5. The procedure of heuristic mutation is as follows:

Procedure: Heuristic Mutation

Step 1. Sort jobs in descending order of w_i/p_i. Let the sorted jobs be $(j_1, j_2, \ldots, j_n)$. Set $x \leftarrow 0$ and $y \leftarrow 0$.

Step 2. Put job j_1 in the position before due date d and let $x \leftarrow x + p_1$.

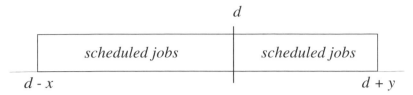

Figure 7.5. The start and completion time for scheduled jobs.

Step 3. for $i \leftarrow 2$ **to** n
    if $(y + p_i < x)$ **then**
        add job j_i to the end of the scheduled job list;
        $y \leftarrow y + p_i$;
    else
        add job j_i to the beginning of the scheduled job list;
        $x \leftarrow x + p_i$;
    end
    end

Let see a five-job example as given in Table 7.3. The sorted jobs are $(j_5, j_3, j_2, j_4, j_1)$, and the schedule constructed by using the heuristic is shown in Figure 7.6.

7.5.4 Determining the Best Due Date

Until now we just considered how to handle job partition and job order with genetic operations but did not discuss how to determine the best due date. For a given job sequence on a machine, the common due date can be optimally determined. Let due date d be a decision variable. For any two jobs i and j, the

Table 7.3. Five-Job Example

	j_1	j_2	j_3	j_4	j_5
p_i	2	2	4	5	1
w_i	1	2	6	4	4
w_i/p_i	0.5	1.0	1.5	0.8	2.0

d

j_4	j_3	j_5	j_2	j_1

Figure 7.6. V-shaped schedule constructed by using a greedy heuristic.

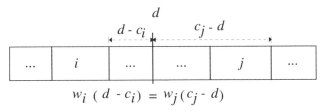

Figure 7.7. Best common due date.

best due date for them can be determined by the following equation:

$$w_i(d - c_i) = w_j(c_j - d) \tag{7.30}$$

That is, the due date for jobs i and j can be calculated as follows:

$$d = \frac{w_i c_i + w_j c_j}{c_i + c_j} \tag{7.31}$$

We can calculate such due dates for all possible pairs of jobs. The best due date for a given job sequence is selected among these due dates which have the largest absolute lateness as shown in equation (7.30).

For the multiple machine case, we first calculate the best due date d_k for machine k using the above method. The common due date for m machines is then determined as follows:

$$d = \max \{d_k | k = 1, 2, \ldots, m\} \tag{7.32}$$

With respect to the common due date d, we need to postpone the start time of each job with

$$\Delta_k = d - d_k \tag{7.33}$$

if the relative job is processed by machine k. The procedure for determining the best common due date is illustrated in Figure 7.8 with a simple example.

7.5.5   Evaluation and Selection

The fitness for each chromosome is calculated as the inverse of its maximum weighted absolute lateness:

$$eval(v_t) = \frac{1}{f(v_t)}, \qquad t = 1, 2, \ldots, pop\_size \tag{7.34}$$

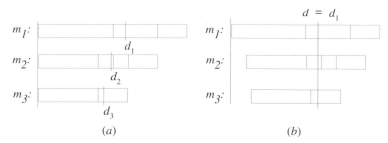

Figure 7.8. Adjustment of start times. (*a*) before adjustment, and (*b*) after adjustment.

where $eval(v_t)$ is the fitness function for the tth chromosome and $f(v_t)$ is the objective function value [see equation (7.24)]. A combined strategy of *roulette wheel* and *elitist* is used as the basic selection mechanism.

7.5.6 Numerical Example

The proposed genetic algorithms were tested on a randomly generated thirty-job five-machine problem. The basic setting of parameters for genetic algorithms is $pop\_size = 40$, $p_c = 0.4$, $p_{m_1} = 0.4$, $p_{m_2} = 0.4$, and $max\_gen = 200$. The results over 20 runs are given in Figure 7.9, where GA stands for the one excluding mutation 2, GA + Heuristic stands for the one including mutation 2, and Heuristic stands for Li and Cheng's greedy heuristic. The results show that the hybrid genetic algorithm performs well for the test problem.

Figure 7.10 shows the frequency of solution distribution by a hybrid genetic algorithm over 50 random runs.

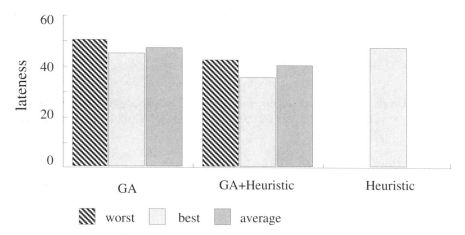

Figure 7.9. Comparison GA with heuristic.

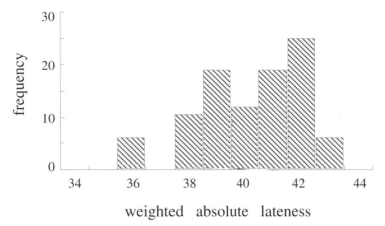

Figure 7.10. Solution distribution.

8

TRANSPORTATION PROBLEMS

8.1 INTRODUCTION

The transportation problem was originally proposed by Hitchcock in 1941 [217]. Since then the research on the problem has received a great deal of attention and various variants of the basic transportation problem have been investigated. Depending on what kind of objective is used, the problem can be characterized as

- Linear problem or nonlinear problem
- Single objective problem or multiple objective problem

Depending on what kind of constraints is under consideration, the problem can be further classified into

- Planar problem or solid problem
- Balanced problem or unbalanced problem

The basic version of the transportation problem is a linear, single objective, balanced, and planar problem. Because the problem possesses a special structure in its constraints, an efficient optimization algorithm has been proposed for it, which is the variation of the simplex method adapted to the particular structure [26, 419].

Vignaux and Michalewicz first discussed the use of a genetic algorithm for solving the linear transportation problem (LTP) [410]. The purpose is not, of course, to compare the genetic methods with a conventional optimization algorithm, because genetic methods will be unable to compete. They used it as an example of a constrained optimization problem, investigated how to handle such constraints with genetic algorithms, and demonstrated the power of genetic algorithms that allow us to use any data structure suitable for a problem together with any set of meaningful genetic operators. Michalewicz, Vignaux, and Hobbs [294] also developed a genetic algorithm, called GENETIC-2, for balanced nonlinear transportation problems, for which the standard transportation methods cannot be used. Although GENETIC-2 was specifically tailored to transportation problems, an important characteristic is that it can handle any type of cost function. It is also possible to modify it to solve many matrix-based constrained optimization problems.

Yang and Gen further extended Michalewicz's work to the bicriteria linear transportation problem (BLTP) [426] and the bicriteria solid transportation problem (BSTP) [155]. Several multiobjective linear programming methods specialized for transportation problems have been developed [8, 77, 355]. A conventional approach for multiobjective cases is to generate nondominated extreme points in decision space or criteria space. Following this methodology, they intended to use genetic algorithms to determine such a set of nondominated points. They used the basic idea of the criteria space approach in the evaluation phase so as to direct the genetic search toward exploiting the nondominated points in the criteria space and demonstrated that the genetic algorithm energized with conventional optimization techniques is a promising way for solving such complex multiobjective optimization problems.

8.2 LINEAR TRANSPORTATION PROBLEM

8.2.1 Formulation of LTP

The linear transportation problem (LTP) involves the shipment of some homogeneous commodity from various origins or sources of supply to a set of destinations, each demanding specified levels of the commodity. The goal is to allocate the supply available at each origin so as to optimize a criterion while satisfying the demand at each destination. The usual objective function is to minimize the total transportation cost or total weighted distance or to maximize the total profit contribution from the allocation [51].

Given m origins and n destinations, the transportation problem can be formulated as the linear programming model:

$$\min \quad z = \sum_{i=1}^{m} \sum_{j=1}^{n} c_{ij} x_{ij} \tag{8.1}$$

$$\text{s.t.} \quad \sum_{j=1}^{n} x_{ij} = a_i, \qquad i = 1, 2, \ldots, m \tag{8.2}$$

$$\sum_{i=1}^{m} x_{ij} = b_j, \qquad j = 1, 2, \ldots, n \tag{8.3}$$

$$x_{ij} \geq 0, \qquad \text{for all } i \text{ and } j \tag{8.4}$$

where x_{ij} is the amount of units shipped from origin i to destination j, c_{ij} is the cost of shipping one unit from origin i to destination j, a_i is the number of units available at origin i, and b_j is the number of units demanded at destination j. Constraint (8.2) is the supply constraint, and constraint (8.3) is the demand constraint.

Feasibility of Transportation Problem. The above formulation assumes that total supply and total demand are equal to one another; that is,

$$\sum_{i=1}^{m} a_i = \sum_{j=1}^{n} b_j \qquad (8.5)$$

Under the assumption of balanced condition, transportation problems always have a feasible solution. For example, it is easy to show that

$$x_{ij} = \frac{a_i b_j}{\sum_i a_i}, \qquad i = 1, 2, \dots, m, \quad j = 1, 2, \dots, n$$

is a feasible solution. Note that for each feasible solution, every component x_{ij} is bounded as follows:

$$0 \le x_{ij} \le \min \{a_i, b_j\}$$

We know that a bounded linear program with a feasible solution has an optimal solution [26].

The problem is usually represented by a *transportation tableau* as shown in Figure 8.1, where the rows represent the origins, the columns represent the destinations, and the cell in row i and column j represents the decision variable x_{ij}. The corresponding cost coefficient c_{ij} is often placed on the upright corner of the cell (i, j).

The transportation problem is graphically illustrated in Figure 8.2. The graph is comprised of the origin and destination nodes O_i, $i = 1, 2, \dots, m$, and D_j, $j = 1, 2, \dots, n$, respectively, and the connecting arcs. It is a *complete bipartite graph* since the nodes can be partitioned into two sets such that all arcs in one set are directed to another set. It is *complete* in the sense that all such possible arcs are present.

Vignaux and Michalewicz proposed two implementations of a genetic algorithm, called GENETIC-1 and GENETIC-2, to solve transportation problem [410]. In GENETIC-2, they used a matrix as a chromosome representation and introduced extra problem-specific knowledge into a genetic algorithm to handle constraint satisfaction.

Destination

to / from	1	2	$\cdots$	n	Supply
1	c_{11} x_{11}	c_{12} x_{12}		c_{1n} x_{1n}	a_1
2	c_{21} x_{21}	c_{22} x_{22}		c_{2n} x_{2n}	a_2
$\cdots$					$\cdots$
m	c_{m1} x_{m1}	c_{m2} x_{m2}		c_{mn} x_{mn}	a_m
Demand	b_1	b_2	$\cdots$	b_n	

Origin

Figure 8.1. Generic transportation tableau.

8.2.2 Representation

A matrix is perhaps the most natural representation of a solution for a transportation problem. The allocation matrix of a transportation problem can be written as follows:

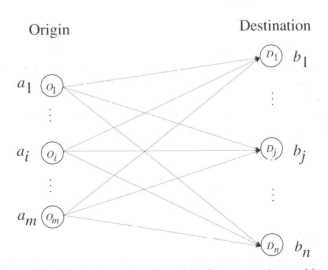

Figure 8.2. Illustrative network model of transportation problem.

$$X_p = \begin{bmatrix} x_{11} & x_{12} & \cdots & x_{1n} \\ x_{21} & x_{22} & \cdots & x_{2n} \\ \cdots & \cdots & \cdots & \cdots \\ x_{m1} & x_{m2} & \cdots & x_{mn} \end{bmatrix}$$

where X_p denotes the pth chromosome and x_{ij} is the corresponding decision variable.

Based on the non-negative condition (8.4) and the balance condition (8.5), the following initialization procedure is proposed to generate the initial population which satisifies all constraints.

Procedure: Initialization

begin
   $\pi \leftarrow \{1, 2, \ldots, mn\}$;
   repeat
      select a random number k from set π;
      calculate corresponding row and column;
      $i \leftarrow \lfloor (k-1)/n + 1 \rfloor$;
      $j \leftarrow (k-1) \bmod n + 1$;
      assign available amount of units to x_{ij};
      $x_{ij} \leftarrow \min \{a_i, b_j\}$;
      update data;
      $a_i \leftarrow a_i - x_{ij}$;
      $b_j \leftarrow b_j - x_{ij}$;
      $\pi \leftarrow \pi/\{k\}$;
   until (π becomes empty)
end

The basic idea of the procedure is to (1) select a random decision variable from the allocation matrix, say x_{ij}; (2) assign x_{ij} available amount of units as much as possible; and (3) update the data of supply and demand to guarantee balance condition.

The objective function (8.1) is used to evaluate each chromosome. Let $eval(X_p)$ denote the fitness function for chromosome X_p, then we have

$$eval(X_p) = \sum_{i=1}^{m} \sum_{j=1}^{n} c_{ij} x_{ij} \tag{8.6}$$

8.2.3 Genetic Operation

Crossover. Assume that two matrices $X_1 = (x_{ij}^1)$ and $X_2 = (x_{ij}^2)$ are selected as parents for the crossover operation. The crossover is performed in three steps:

Step 1. Create two temporary matrices $D = (d_{ij})$ and $R = (r_{ij})$ as follows:

$$d_{ij} = \lfloor (x_{ij}^1 + x_{ij}^2)/2 \rfloor \tag{8.7}$$

$$r_{ij} = (x_{ij}^1 + x_{ij}^2) \bmod 2 \tag{8.8}$$

Matrix D keeps rounded average values from both parents, and matrix R keeps track of whether any rounding is necessary. The relationship between the two matrices is given by the following equations:

$$a_i - \sum_{j=1}^{n} d_{ij} = \frac{1}{2} \sum_{j=1}^{n} r_{ij}, \qquad i = 1, 2, \ldots, m \tag{8.9}$$

$$b_j - \sum_{i=1}^{m} d_{ij} = \frac{1}{2} \sum_{i=1}^{m} r_{ij}, \qquad j = 1, 2, \ldots, n \tag{8.10}$$

These equations describe two interesting properties: (1) The number of 1s in each row and each column is even; that is, the marginal sums of rows and columns $\sum_{j=1}^{n} r_{ij}$ and $\sum_{i=1}^{m} r_{ij}$ are even integers. (2) The values of row marginal sums of matrix R equal twice the difference between row marginal sums of matrix D and corresponding supplies $a_i - \sum_{j=1}^{n} d_{ij}$, and the values of column marginal sums of matrix R equal twice the difference between column marginal sums of matrix D and corresponding demands $b_j - \sum_{i=1}^{m} d_{ij}$. It is easy to verify these properties by using the definitions (8.7) and (8.8). Now we give a proof for equation (8.9):

$$\sum_{j=1}^{n} d_{ij} = \sum_{j=1}^{n} \left\lfloor \frac{1}{2} (x_{ij}^1 + x_{ij}^2) \right\rfloor$$

$$= \sum_{j=1}^{n} \frac{1}{2} ((x_{ij}^1 + x_{ij}^2) - (x_{ij}^1 + x_{ij}^2) \bmod 2)$$

$$= \frac{1}{2} \sum_{j=1}^{n} (x_{ij}^1 + x_{ij}^2) - \frac{1}{2} \sum_{j=1}^{n} ((x_{ij}^1 + x_{ij}^2) \bmod 2)$$

$$- a_i - \frac{1}{2} \sum_{j=1}^{n} r_{ij}$$

Step 2. Divide matrix R into two matrices $R^1 = (r_{ij}^1)$ and $R^2 = (r_{ij}^2)$ such that

$$R = R^1 + R^2 \tag{8.11}$$

$$\sum_{j=1}^{n} r_{ij}^1 = \sum_{j=1}^{n} r_{ij}^2 = \frac{1}{2} \sum_{j=1}^{n} r_{ij}, \qquad i = 1, 2, \ldots, m \tag{8.12}$$

$$\sum_{i=1}^{m} r_{ij}^1 = \sum_{i=1}^{m} r_{ij}^2 = \frac{1}{2} \sum_{i=1}^{m} r_{ij}, \qquad j = 1, 2, \ldots, n \tag{8.13}$$

It is easy to see that there are too many possible ways to divide R into R^1 and R^2 while satisfying above conditions.

Step 3. Then we produce two offspring of X_1' and X_2' as follows:

$$X_1' = D + R^1 \tag{8.14}$$
$$X_2' = D + R^2 \tag{8.15}$$

Let us see a simple example.

Example 8.1. Consider a problem with four sources and five destinations and the following constraints:

$$a_1 = 8, \quad a_2 = 4, \quad a_3 = 12, \quad a_4 = 6$$
$$b_1 = 3, \quad b_2 = 5, \quad b_3 = 10, \quad b_4 = 7, \quad b_5 = 5$$

The following matrices are selected as parents for crossover:

Parents X_1

1	0	0	7	0
0	4	0	0	0
2	1	4	0	5
0	0	6	0	0

Parents X_2

0	0	5	0	3
0	4	0	0	0
0	0	5	7	0
3	1	0	0	2

The temporary matrices D and R are

Matrix D

0	0	2	3	1
0	4	0	0	0
1	0	4	3	2
1	0	3	0	1

Matrix R

1	0	1	1	1
0	0	0	0	0
0	1	1	1	1
1	1	0	0	0

Then divide R into R^1 and R^2 as follows:

Matrix R^1

0	0	1	0	1
0	0	0	0	0
0	1	0	1	0
1	0	0	0	0

Matrix R^2

1	0	0	1	0
0	0	0	0	0
0	0	1	0	1
0	1	0	0	0

Finally, two offspring X_1' and X_2' are

Offspring X_1'

0	0	3	3	2
0	4	0	0	0
1	1	4	4	2
2	0	3	0	1

Offspring X_2'

1	0	2	4	1
0	4	0	0	0
1	0	5	3	3
1	1	3	0	1

Mutation. The mutation is performed in three steps:

Step 1. Make a submatrix from a parent matrix. Randomly select $\{i_1, \ldots, i_p\}$ rows and $\{j_1, \ldots, j_q\}$ columns to create a $(p*q)$ submatrix $Y = (y_{ij})$, where $\{i_1, \ldots, i_p\}$ is a proper subset of $\{1, 2, \ldots, m\}$ and $2 \le p \le m$, $\{j_1, \ldots, j_q\}$ is a proper subset of $\{1, 2, \ldots, n\}$ and $2 \le q \le n$, and y_{ij} takes the value of the element in the crossing position of selected row i and column j in the parent matrix.

Step 2. Reallocate a commodity for the submatrix. The available amount of commodity a_i^y and the demands b_j^y for the submatrix are determined as follows:

$$a_i^y = \sum_{j \in \{j_1, \dots, j_q\}} y_{ij}, \qquad i = i_1, i_2, \dots, i_p$$

$$b_j^y = \sum_{i \in \{i_1, \dots, i_p\}} y_{ij}, \qquad j = j_1, j_2, \dots, j_q$$

We can use the initialization procedure to assign new values to the submatrix such that all constrains a_i^y and b_j^y are satisfied.

Step 3. Replace the appropriate elements of the parent matrix by new elements from the reallocated submatrix Y.

Example 8.2. Consider the same problem given above. Assume that the following matrix X is selected as a parent for mutation:

Parents X

0	0	5	0	3
0	4	0	0	0
0	0	5	7	0
3	1	0	0	2

Select two rows $\{2, 4\}$ and three columns $\{2, 3, 5\}$ randomly. The corresponding submatrix and the reallocated submatrix are

Submatrix Y After Reallocation

4	0	0
1	0	2

2	0	2
3	0	0

Then the offspring after mutation is

Offspring

0	0	5	0	3
0	2	0	0	2
0	0	5	7	0
3	3	0	0	0

8.3 BICRITERIA LINEAR TRANSPORTATION PROBLEM

8.3.1 Formulation of BLTP

The transportation problem with a single objective of minimizing the total cost is well known in the literature [198, 419]. However, in most real-world cases, the complexity of the social and economic environment requires the explicit consideration of criteria other than cost. Examples of additional concerns include: average delivery time of the commodities, reliability of transportation, accessibility to the users, product deterioration, among others; for details see references [88] and [89]. These types of transportation problems are multiobjective in nature and, therefore, can be formulated as multiobjective linear programming problems.

The bicriteria linear transportation problem (BLTP) is a special case of multiobjective transportation problem since the feasible region can be depicted in two dimensions in criteria space. Consider the following two objectives: minimizing total cost and minimizing total deterioration. Assume that there are m origins and n destinations; the BLTP can be formulated as follows:

$$\min \quad z_1 = \sum_{i=1}^{m} \sum_{j=1}^{n} c_{ij}^1 x_{ij} \tag{8.16}$$

$$\min \quad z_2 = \sum_{i=1}^{m} \sum_{j=1}^{n} c_{ij}^2 x_{ij} \tag{8.17}$$

$$\text{s.t.} \quad \sum_{j=1}^{n} x_{ij} = a_i, \qquad i - 1, 2, \ldots, m \tag{8.18}$$

$$\sum_{i=1}^{m} x_{ij} = b_j, \qquad j = 1, 2, \ldots, n \tag{8.19}$$

$$x_{ij} \geq 0, \qquad \text{for all } i \text{ and } j \tag{8.20}$$

where x_{ij} is the amount of units shipped from origin i to destination j, c_{ij}^1 is the cost of shipping one unit from origin i to destination j, c_{ij}^2 is the deterioration of shipping one unit from origin i to destination j, a_i is the number of units available at origin i, and b_j is the number of units demanded at destination j. Constraint (8.18) is the supply constraint, and constraint (8.19) is the demand constraint.

8.3.2 Evaluation

Yang and Gen further extended Michalewicz's work to BLTP [426]. Conventional approaches for BLTP are to generate nondominated extreme points in decision space or criteria space [8, 79]. Following this methodology, they used genetic algorithms to find the set of nondominated points. As we know, genetic algorithms are potentially very powerful in the exploration of feasible solutions. Thus the remaining work is how to increase the selection pressure to force genetic algorithms to exploit the efficient extreme points of the nondominated set. Yang and Gen used the adaptively moving line technique to construct a weighted-sum approach which can force the genetic search to exploit the set of Pareto solutions in the criteria space as described in Chapter 2 for the interval programming problem.

Let E denote the set of nondominated solutions examined so far. Two special points in E interest us: One point contains z_{min}^1 and another contains z_{min}^2. Denote these two points as (z_{min}^1, z_{max}^2) and (z_{max}^1, z_{min}^2), respectively, where

$$z_{min}^1 = \min \{z_1(x) | x \in E\}$$
$$z_{max}^1 = \max \{z_1(x) | x \in E\}$$
$$z_{min}^2 = \min \{z_2(x) | x \in E\}$$
$$z_{max}^2 = \max \{z_2(x) | x \in E\}$$

Then we can devise a new objective function based on these two special points as follows:

$$z = \alpha z_1 + \beta z_2 \tag{8.21}$$

where

$$\alpha = |z_{max}^2 - z_{min}^2|$$
$$\beta = |z_{max}^1 - z_{min}^1|$$

This can be rewritten as follows:

$$z = \sum_{i=1}^{m} \sum_{j=1}^{n} c_{ij} x_{ij} \tag{8.22}$$

where

$$c_{ij} = \alpha c_{ij}^1 + \beta c_{ij}^2$$

Figure 8.3 gives an illustrative explanation of the objective. The line formed with points (z_{min}^1, z_{max}^2) and (z_{max}^1, z_{min}^2) divides the criteria space into two half-spaces: One space Z^+ contains the positive ideal solution and the other space Z^- contains the negative ideal solution. The feasible solution space F is cor-

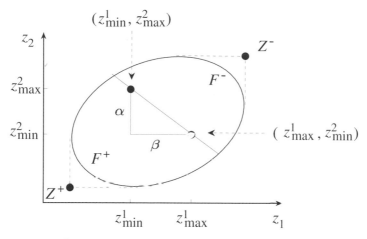

Figure 8.3. Illustrative explanation of objective.

respondingly divided into two parts: One is $F^- = F \cap Z^-$ and another is $F^+ = F \cap Z^+$. It is easy to verify that a solution in F^+ has a lower value of objective function (8.21) or (8.22) than those in F^-.

The fitness value for each chromosome is determined by equation (8.21), so those chromosomes in the half-space F^+ have a relatively larger chance to enter the next generation. At each generation, set E is updated and the two special points may be renewed. It means that along with the evolutionary process, the line formed with points (z_{min}^1, z_{max}^2) and (z_{max}^1, z_{min}^2) will move gradually along the direction from the negative ideal point to the positive ideal point. In other words, this fitness function shows selection pressure that forces the genetic search to exploit the nondominated points in the criteria space.

8.3.3 Overall Procedure

The overall procedure of Yang and Gen's approach is summarized as follows:

Procedure: Yang and Gen's Approach for Solving BLTP

Step 0. Set parameters. Set population size *pop_size*, mutation rate p_m, crossover rate p_c and the maximum number of generations *max_gen*. Let $t \leftarrow \phi$, $E \leftarrow \varnothing$.

Step 1. Initialization. Randomly generate initial solution matrices $X_p = (x_{ij}^p)$, $p = 1, 2, \ldots, pop\_size$ as follows:

begin
  for $p \leftarrow 1$ **to** *pop_size*
  $\pi \leftarrow \{1, 2, \ldots, mn\}$;
  repeat

select a random number k from set π;
calculate corresponding row and column;
$\quad i \leftarrow \lfloor (k - 1)/m + 1 \rfloor$;
$\quad j \leftarrow (k - 1) \bmod m + 1$;
assign available amount of units to x_{ij}^p;
$\quad x_{ij}^p \leftarrow \min \{a_i, b_j\}$;
update data;
$\quad a_i \leftarrow a_i - x_{ij}^p$;
$\quad b_j \leftarrow b_j - x_{ij}^p$;
$\quad \pi \leftarrow \pi \backslash \{k\}$;
until (π becomes empty)
end
end

Step 2. Crossover

1. Generate random numbers r_p in $[0,1]$ ($p = 1, 2, \ldots, pop\_size$). If $r_p < p_c$, then X_p is selected as parents for crossover.

2. For every pair of parents (X_p, X_q), create two temporally matrices:

$$D = (d_{ij}), \qquad R = (r_{ij})$$

where

$$d_{ij} = \lfloor (x_{ij}^p + x_{ij}^q)/2 \rfloor$$
$$r_{ij} = (x_{ij}^p + x_{ij}^q) \bmod 2$$

3. Divide matrix R into two matrices $R^1 = (r_{ij}^1)$ and $R^2 = (r_{ij}^2)$ such that

$$R = R^1 + R^2$$

$$\sum_{j=1}^{n} r_{ij}^1 = \sum_{j=1}^{n} r_{ij}^2 = \frac{1}{2} \sum_{j=1}^{n} r_{ij}, \qquad i = 1, 2, \ldots, m$$

$$\sum_{i=1}^{m} r_{ij}^1 = \sum_{i=1}^{m} r_{ij}^2 = \frac{1}{2} \sum_{i=1}^{m} r_{ij}, \qquad j = 1, 2, \ldots, n$$

4. Then produce two offspring X'_p and X'_q as follows:

$$X'_p = D + R_1$$
$$X'_q = D + R_2$$

Step 3. Mutation

1. Generate random numbers r_p in $[0,1]$ ($p = 1, 2, \ldots, pop\_size$). If $r_p < p_m$, then X_p is selected as parents for mutation.

2. Make submatrix $Y = (y_{ij})_{p \times q}$ by randomly selecting p rows and q columns.

3. Calculate the available amount of commodity a_i^y and the demands b_j^y for the submatrix as follows:

$$a_i^y = \sum_{j \in \{j_1, \ldots, j_q\}} y_{ij}, \qquad i = i_1, i_2, \ldots, i_p$$

$$b_j^y = \sum_{i \in \{i_1, \ldots, i_p\}} y_{ij}, \qquad j = j_1, j_2, \ldots, j_q$$

4. Reallocate the commodity for the submatrix by using the initialization procedure such that all constraints a_i^y and b_j^y are satisfied. Let Y' denote the new submatrix.

5. Replace appropriate elements of matrix X_p by new elements from the submatrix Y' to make offspring X'_p.

Step 4. Update set E

1. Compute objective function values of bicriteria for each chromosome of parents and offspring X_p as follows:

$$z_p = (z_1, z_2)$$

where

$$z_1 = \sum_{i=1}^{m} \sum_{j=1}^{n} c_{ij}^1 x_{ij}$$

$$z_2 = \sum_{i=1}^{m} \sum_{j=1}^{n} c_{ij}^2 x_{ij}$$

2. Update set E by adding new nondominated points into E and deleting dominated points from E.

3. Determine new special points (z_{min}^1, z_{max}^2) and (z_{max}^1, z_{min}^2) in E as follows:

$$z_{min}^1 = \min \{z_1(\pmb{x})|\pmb{x} \in E\}$$
$$z_{max}^1 = \max \{z_1(\pmb{x})|\pmb{x} \in E\}$$
$$z_{min}^2 = \min \{z_2(\pmb{x})|\pmb{x} \in E\}$$
$$z_{max}^2 = \max \{z_2(\pmb{x})|\pmb{x} \in E\}$$

Step 5. Evaluation. Compute fitness values for each chromosome of parents and offspring X_p as follows:

$$eval(X_p) = \alpha z_1 + \beta z_2$$

where

$$\alpha = |z_{max}^2 - z_{min}^2|$$
$$\beta = |z_{max}^1 - z_{min}^1|$$

Step 6. Selection. Select the best *pop_size* chromosomes among parents and offspring to construct next generation.

Step 7. Terminal test. If $t = max\_gen$, stop; otherwise, let $t \leftarrow t + 1$ and go to Step 2.

8.3.4 Numerical Examples

Example 8.3. Consider the following bicriteria linear transportation problem given by Aneja and Nair [8].

$$\min \quad z_1 = x_{11} + 2x_{12} + 7x_{13} + 7x_{14} + x_{21} + 9x_{22} + 3x_{23} + 4x_{24} + 8x_{31}$$
$$+ 9x_{32} + 4x_{33} + 6x_{34}$$
$$\min \quad z_2 = 4x_{11} + 4x_{12} + 3x_{13} + 4x_{14} + 5x_{21} + 8x_{22} + 9x_{23} + 10x_{24} + 6x_{31}$$
$$+ 2x_{32} + 5x_{33} + x_{34}$$

$$\text{s.t.} \quad \sum_{j=1}^{4} x_{1j} = 8, \quad \sum_{j=1}^{4} x_{2j} = 19, \quad \sum_{j=1}^{4} x_{3j} = 17$$

$$\sum_{i=1}^{3} x_{i1} = 11, \quad \sum_{i=1}^{3} x_{i2} = 3, \quad \sum_{i=1}^{3} x_{i3} = 14, \quad \sum_{i=1}^{3} x_{i4} = 16$$

$$x_{ij} \geq 0 \qquad \text{for all } i \text{ and } j$$

Represent this problem with tabular form as follows:

Table for z_1				
	1	2	3	4
1	1	2	7	7	8
2	1	9	3	4	19
3	8	9	4	6	17
	11	3	14	16

Table for z_2				
	1	2	3	4
1	4	4	3	4	8
2	5	8	9	10	19
3	6	2	5	1	17
	11	3	14	16

The population size is fixed at 20 and the number of generations is fixed at 1000. The mutation rate is $p_m = 0.2$ and the crossover rate is $p_c = 0.4$, which are determined according to the experience of Michalewicz that a small proportion of crossover and a large proportion of mutation can work best.

We have run five times and for each running the different random number seeds are used. Almost the same results are obtained and shown in Figure 8.4.

From Figure 8.4, we see that the genetic algorithm approach can find not only efficient extreme points given by [8, 79] but also some new efficient extreme points.

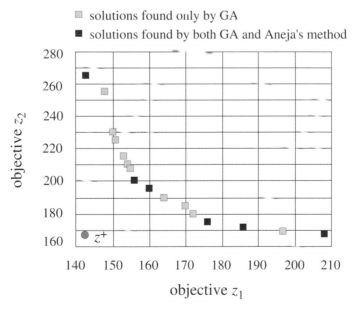

Figure 8.4. The nondominated solutions obtained by a genetic algorithm.

8.4   BICRITERIA SOLID TRANSPORTATION PROBLEM

8.4.1   Formulation of BSTP

The solid transportation problem (STP) is a generalization of the classical transportation problem. The necessity of considering this special type of transportation problem arises when there are heterogeneous conveyances available for
shipment of products. The solid transportation problem is of much use in public distribution systems and is usually considered with more than one objective
[36].

Assume that there are m origins (or sources), n destinations, and K conveyances. At each origin, let a_i be the amount of a homogeneous product which
will be transported to n destinations to satisfy the demand for b_j units of the
product there. Let e_k be the units of products which can be carried by the kth
conveyance. The penalties c_{ijk}^q, $q = 1, 2$, are associated with transportation of a
unit of the product from origin i to destination j by means of the kth conveyance
for the qth objective function. The penalty could represent transportation cost,
delivery time, quantity of goods delivered, underused capacity, and so on. A
variable x_{ijk} represents the unknown quantity to be transported from origin i to
destination j by means of the kth conveyance. The bicriteria solid transportation
problem (BSTP) can be formulated as follows:

$$\min \quad z_q = \sum_{i=1}^{m} \sum_{j=1}^{n} \sum_{k=1}^{K} c_{ijk}^q x_{ijk}, \qquad q = 1, 2 \tag{8.23}$$

$$\text{s.t.} \quad \sum_{j=1}^{n} \sum_{k=1}^{K} x_{ijk} = a_i, \qquad i = 1, 2, \ldots, m \tag{8.24}$$

$$\sum_{i=1}^{m} \sum_{k=1}^{K} x_{ijk} = b_j, \qquad j = 1, 2, \ldots, n \tag{8.25}$$

$$\sum_{i=1}^{m} \sum_{j=1}^{n} x_{ijk} = e_k, \qquad k = 1, 2, \ldots, K \tag{8.26}$$

$$x_{ijk} \geq 0, \qquad \forall \, i, j, k \tag{8.27}$$

where $a_i > 0$ for all i, $b_j > 0$ for all j, $e_k > 0$ for all k, and $c_{ijk}^q \geq 0$ for all i,
j, k, q. The above formulation assumes that total supply and total demand are
equal to one another; that is,

$$\sum_{i=1}^{m} a_i = \sum_{j=1}^{n} b_j = \sum_{k=1}^{K} e_k \qquad \text{(balanced condition)} \tag{8.28}$$

The balanced condition is treated as a necessary and sufficient condition for the existence of a feasible solution to the problem. Gen et al. further extended their work to BSTP [155].

8.4.2   Initialization

For solid transportation problem, a chromosome can be represented as a three-dimensional array and also equivalently represented as a matrix with nmK elements:

$$\lfloor x_{111}x_{211} \cdots x_{m11}, \ldots x_{1n1}x_{2n1} \cdots x_{mn1}, \ldots,$$
$$x_{11K}x_{21K} \cdots x_{m1K}, \ldots, x_{1nK}x_{2nK} \cdots x_{mnK} \rfloor$$

where x_{ijk} are the corresponding decision variables in the chromosome.

We can easily extend the initialization procedure for LTP to STP to adapt for three-dimensional case.

Procedure: Initialization

begin
   $\pi \leftarrow \{1, 2, \ldots, mnK\}$;
   repeat
     select a random number l from set π;
     calculate corresponding subscript indexes;
      $i \leftarrow (l - 1) \bmod m + 1$;
      $j \leftarrow \lfloor (l - 1)/m \rfloor \bmod 1$;
      $k \leftarrow \lfloor (l - 1)/mn \rfloor + 1$;
     assign available amount of units to x_{ijk};
     $x_{ijk} \leftarrow \min \{a_i, b_j, e_k\}$;
     update data;
     $a_i \leftarrow a_i - x_{ijk}$;
     $b_j \leftarrow b_j - x_{ijk}$;
     $e_k \leftarrow e_k - x_{ijk}$;
     $\pi \leftarrow \pi \backslash \{l\}$;
   until (π becomes empty)
end

8.4.3   Genetic Operation

Similarly, the genetic operators for LTP can be adapted to deal with a three-dimensional case as follows.

Crossover. Assume that two parents are $X_1 = (x^1_{ijk})$ and $X_2 = (x^2_{ijk})$. The crossover is performed in three steps:

Step 1. Create two temporary matrices $D = (d_{ijk})$ and $R = (r_{ijk})$ as follows:

$$d_{ijk} = \lfloor (x_{ijk}^1 + x_{ijk}^2)/2 \rfloor \tag{8.29}$$

$$r_{ijk} = (x_{ijk}^1 + x_{ijk}^2) \bmod 2 \tag{8.30}$$

Matrix D keeps rounded average values from both parents, and matrix R keeps the track of whether any rounding is necessary. The relationship between the two matrices is given by the following equations:

$$a_i - \sum_{j=1}^{n} \sum_{k=1}^{K} d_{ijk} = \frac{1}{2} \sum_{j=1}^{n} \sum_{k=1}^{K} r_{ijk}, \qquad i = 1, 2, \ldots, m \tag{8.31}$$

$$b_j - \sum_{i=1}^{m} \sum_{k=1}^{K} d_{ijk} = \frac{1}{2} \sum_{i=1}^{m} \sum_{k=1}^{K} r_{ijk}, \qquad j = 1, 2, \ldots, n \tag{8.32}$$

$$e_k - \sum_{i=1}^{m} \sum_{j=1}^{n} d_{ijk} = \frac{1}{2} \sum_{i=1}^{m} \sum_{j=1}^{n} r_{ijk}, \qquad k = 1, 2, \ldots, K \tag{8.33}$$

Step 2. Divide matrix R into two matrices $R^1 = (r_{ijk}^1)$ and $R^2 = (r_{ijk}^2)$ such that

$$R = R^1 + R^2 \tag{8.34}$$

$$\sum_{j=1}^{n} \sum_{k=1}^{K} r_{ijk}^1 = \sum_{j=1}^{n} \sum_{k=1}^{K} r_{ijk}^2 = \frac{1}{2} \sum_{j=1}^{n} \sum_{k=1}^{K} r_{ijk}, \qquad i = 1, 2, \ldots, m \tag{8.35}$$

$$\sum_{i=1}^{m} \sum_{k=1}^{K} r_{ijk}^1 = \sum_{i=1}^{m} \sum_{k=1}^{K} r_{ijk}^2 = \frac{1}{2} \sum_{i=1}^{m} \sum_{k=1}^{K} r_{ijk}, \qquad j = 1, 2, \ldots, n \tag{8.36}$$

$$\sum_{i=1}^{m} \sum_{j=1}^{n} r_{ijk}^1 = \sum_{i=1}^{m} \sum_{j=1}^{n} r_{ijk}^2 = \frac{1}{2} \sum_{i=1}^{m} \sum_{j=1}^{n} r_{ijk}, \qquad k = 1, 2, \ldots, K \tag{8.37}$$

It is easy to see that there are too many possible ways to divide R into R^1 and R^2 while satisfying the above conditions.

Step 3. Then we can produce two offspring of X'_1 and X'_2 as follows:

$$X'_1 = D + R^1 \tag{8.38}$$
$$X'_2 = D + R^2 \tag{8.39}$$

Mutation. The mutation is performed in three steps:

Step 1. Make a submatrix from a parent matrix. Randomly select $\{i_1, \ldots, i_p\}$, $\{j_1, \ldots, j_q\}$, and $\{k_1, \ldots, k_s\}$ to create a $(p * q * s)$ submatrix $Y = (y_{ijk})$, where $\{i_1, \ldots, i_p\}$ is a proper subset of $\{1, 2, \ldots, m\}$ and $2 \leq p \leq m$, $\{j_1, \ldots, j_q\}$ is a proper subset of $\{1, 2, \ldots, n\}$ and $2 \leq q \leq n$, $\{k_1, \ldots, k_s\}$ is a proper subset of $\{1, 2, \ldots, K\}$ and $2 \leq s \leq K$, and y_{ijk} takes the value of the corresponding element with index ijk in parent matrix.

Step 2. Reallocate the commodity for the submatrix. The available amount of commodity a_i^y, the demands b_j^y, and the capability of conveyances e_k^y for the submatrix is determined as follows:

$$a_i^y = \sum_{j \in \{j_1, \ldots, j_q\}} \sum_{k \in \{k_1, \ldots, k_s\}} y_{ijk}, \qquad i = i_1, i_2, \ldots, i_p$$

$$b_j^y = \sum_{i \in \{i_1, \ldots, i_p\}} \sum_{k \in \{k_1, \ldots, k_s\}} y_{ijk}, \qquad j = j_1, j_2, \ldots, j_q$$

$$e_k^y = \sum_{i \in \{i_1, \ldots, i_p\}} \sum_{j \in \{j_1, \ldots, j_q\}} y_{ijk}, \qquad k = k_1, k_2, \ldots, k_s$$

we can use the initialization procedure to assign new values to a submatrix such that all constraints a_i^y, b_j^y, and e_k^y are satisfied.

Step 3. Replace the appropriate elements of a parent matrix by new elements from the reallocated submatrix Y.

8.4.4   Numerical Examples

Example 8.4. Consider the following bicriteria solid transportation problem given by Bit, Biswal, and Alam [36].

$$\min \quad z_q(X) = \sum_{i=1}^{4} \sum_{j=1}^{4} \sum_{k=1}^{3} c_{ijk}^q x_{ijk}, \qquad q = 1,2$$

$$\text{s.t.} \quad \sum_{j=1}^{4} \sum_{k=1}^{3} x_{1jk} = 24, \qquad \sum_{j=1}^{4} \sum_{k=1}^{3} x_{2jk} = 8$$

$$\sum_{j=1}^{4} \sum_{k=1}^{3} x_{3jk} = 18, \qquad \sum_{j=1}^{4} \sum_{k=1}^{3} x_{4jk} = 10$$

$$\sum_{i=1}^{4} \sum_{k=1}^{3} x_{i1k} = 11, \qquad \sum_{i=1}^{4} \sum_{k=1}^{3} x_{i2k} = 19$$

$$\sum_{i=1}^{4} \sum_{k=1}^{3} x_{i3k} = 21, \qquad \sum_{i=1}^{4} \sum_{k=1}^{3} x_{i4k} = 9$$

$$\sum_{i=1}^{4} \sum_{j=1}^{4} x_{ij1} = 17, \qquad \sum_{i=1}^{4} \sum_{j=1}^{4} x_{ij2} = 31$$

$$\sum_{i=1}^{4} \sum_{j=1}^{4} x_{ij3} = 12$$

$$x_{ijk} \geq, \qquad \forall i,j,k$$

$$C^1 = \begin{array}{c} \\ 1 \\ 2 \\ 3 \\ 4 \end{array} \begin{pmatrix} c_{i11} & c_{i12} & c_{i13} & c_{i21} & c_{i22} & c_{i23} & c_{i31} & c_{i32} & c_{i33} & c_{i41} & c_{i42} & c_{i43} \\ 15 & 18 & 17 & 12 & 22 & 13 & 10 & 4 & 12 & 8 & 11 & 13 \\ 17 & 20 & 19 & 21 & 21 & 22 & 21 & 19 & 18 & 30 & 10 & 23 \\ 14 & 11 & 12 & 25 & 34 & 33 & 20 & 16 & 15 & 21 & 23 & 22 \\ 22 & 18 & 13 & 24 & 35 & 32 & 18 & 21 & 14 & 13 & 23 & 20 \end{pmatrix}$$

$$C^2 = \begin{array}{c} \\ 1 \\ 2 \\ 3 \\ 4 \end{array} \begin{pmatrix} 6 & 7 & 8 & 10 & 6 & 5 & 11 & 3 & 7 & 10 & 9 & 6 \\ 13 & 8 & 11 & 12 & 2 & 9 & 20 & 15 & 13 & 17 & 15 & 13 \\ 5 & 6 & 7 & 11 & 9 & 7 & 10 & 5 & 2 & 15 & 14 & 18 \\ 13 & 6 & 6 & 17 & 11 & 18 & 12 & 16 & 12 & 18 & 14 & 7 \end{pmatrix}$$

The population size is fixed at 20, the mutation rate is $p_m = 0.2$, the crossover rate is $p_c = 0.4$, and the number of generation is fixed at 1000. After 1000 generations, we obtained the following 29 nondominated solutions:

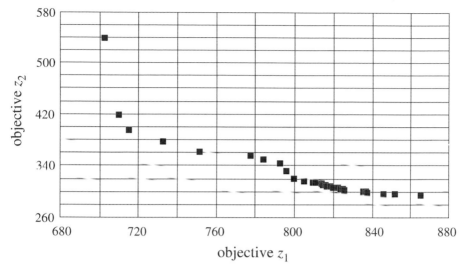

Figure 8.5. The nondominated solutions obtained by genetic algorithms.

$$(703, 537)(710, 418)(715, 394)(733, 376)$$
$$(752, 360)(778, 355)(784, 349)(793, 343)$$
$$(796, 330)(800, 320)(805, 316)(810, 314)$$
$$(811, 313)(814, 312)(815, 310)(816, 309)$$
$$(817, 308)(819, 307)(820, 306)(823, 305)$$
$$(824, 304)(825, 303)(826, 302)(836, 300)$$
$$(837, 299)(838, 298)(846, 296)(852, 295)$$
$$(866, 293)$$

The solutions are depicted in Figure 8.5.

8.5 FUZZY MULTICRITERIA SOLID TRANSPORTATION PROBLEM

8.5.1 Problem Formulation

With the multicriteria approach, besides transportation cost, many other influential factors, such as delivery time, quantity of goods delivered, underused capacity reliability of transportation, accessibility to the users, production deterioration, and so on, can be taken into account when making a transportation plan. In a real-world situation, due to the complexity of the social and economic environment as well as some unpredictable factors such as weather, a common problem is the difficulty in determining the proper values of model parameters. One way of handling such uncertainty in decision making is fuzzy programming [58, 466]. Kaufmann and Gupta first examined fuzzy transportation problems [244]. Recently, some related studies were published by Bit, Biswal, and Alam [35, 36], who dealt with multicriteria transportation problems with fuzzy programming techniques but not the fuzzy transportation problem. Gen, Ida, and Li have considered the following fuzzy multicriteria solid transportation problem [157]:

$$\min \quad \tilde{z}_q = \sum_{i=1}^{m} \sum_{j=1}^{n} \sum_{k=1}^{K} \tilde{c}_{ijk}^{q} x_{ijk}, \qquad q = 1, 2, \ldots, Q \tag{8.40}$$

$$\text{s.t.} \quad \sum_{j=1}^{n} \sum_{k=1}^{K} x_{ijk} = a_i, \qquad i = 1, 2, \ldots, m \tag{8.41}$$

$$\sum_{i=1}^{m} \sum_{k=1}^{K} x_{ijk} = b_j, \qquad j = 1, 2, \ldots, n \tag{8.42}$$

$$\sum_{i=1}^{m} \sum_{j=1}^{n} x_{ijk} = e_k, \qquad k = 1, 2, \ldots, K \tag{8.43}$$

$$x_{ijk} \geq 0, \qquad \forall i, j, k.$$

where $a_i \geq 0$ for all i, $b_j \geq 0$ for all j, and $e_k \geq 0$ for all k. The constraints imply the balanced condition as discussed in the previous section. The only difference is that the coefficients in objective functions are represented with fuzzy numbers $\tilde{c}_{ijk}^{q}$, $q = 1, 2, \ldots, Q$.

8.5.2   Genetic Algorithm Approach

Gen, Ida, and Li proposed a genetic algorithm approach to the fuzzy multicriteria solid transportation problem [157, 451]. The basic implementation was the same as the one given in the previous section. The major effort was given to handling fuzziness. In multicriteria optimization, we are interested in finding Pareto solutions. When the coefficients of objectives are represented with fuzzy numbers, the objective values become fuzzy numbers. Since a fuzzy number represents many possible real numbers, it is not easy to compare among solutions to determine which is the Pareto solution. Fuzzy ranking techniques can help us to compare fuzzy numbers [58]. In Gen's approach, Pareto solutions are determined based on the ranked values of fuzzy objectives, and genetic algorithms are used to search for Pareto solutions.

Ranking Fuzzy Numbers with Integral Value. Liou and Wang proposed a method of ranking fuzzy numbers with integral values [276]. In general, a membership function can be divided into the left part and the right part [114, 432]. For the minimization problem the integral value of the inverse function of the left part reflects the optimistic viewpoint, and the integral value of the inverse function of the right part reflects the pessimistic viewpoint of the decision maker. A convex combination of right and left integral values through an index of optimism, called the *total integral value*, is used to rank fuzzy numbers.

Let $\tilde{A}$ be a triangular fuzzy number (TFN), denoted by (a_1, a_2, a_3). Its membership function $\mu_{\tilde{A}}$ is given by

$$
\mu_{\tilde{A}}(x) = \begin{cases} (x - a_1)/(a_2 - a_1), & a_1 \leq x \leq a_2 \\ (x - a_3)/(a_2 - a_3), & a_2 \leq x \leq a_3 \\ 0, & \text{otherwise} \end{cases} \tag{8.44}
$$

where a_1, a_2, and a_3 are real numbers (see Figure 8.6). Denote the left and right membership functions as $\mu_{\tilde{A}}(x)^L = (x - a_1)/(a_2 - a_1)$, $\mu_{\tilde{A}}(x)^R = (x - a_3)/(a_2 - a_3)$, respectively. The corresponding inverse functions of $\mu_{\tilde{A}}(x)^L$ and $\mu_{\tilde{A}}(x)^R$ can be expressed as follows:

$$
g_{\tilde{A}}(y)^L = a_1 + (a_2 - a_1)y, \tag{8.45}
$$

$$
g_{\tilde{A}}(y)^R = a_3 + (a_2 - a_3)y \tag{8.46}
$$

Thus the left and right integral values are

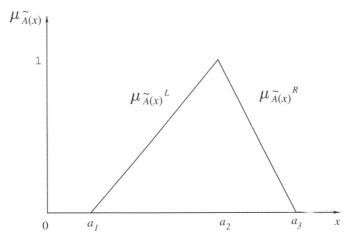

Figure 8.6. The membership function $\mu_{\tilde{A}}$.

$$I(\tilde{A})^L = \int_0^1 g_{\tilde{A}}(y)^L \, dy = \tfrac{1}{2}(a_1 + a_2) \qquad (8.47)$$

$$I(\tilde{A})^R = \int_0^1 g_{\tilde{A}}(y)^R \, dy = \tfrac{1}{2}(a_2 + a_3) \qquad (8.48)$$

The total integral value of the TFN $\tilde{A}$ is

$$
\begin{aligned}
I^\alpha(\tilde{A}) &= \alpha I(\tilde{A})^R + (1 - \alpha)I(\tilde{A})^L \\
&= \tfrac{1}{2}[\alpha a_3 + a_2 + (1 - \alpha)a_1]
\end{aligned}
\qquad (8.49)
$$

when a parameter $\alpha \in [0, 1]$ is used to adjust a decision makers' degree of optimism. When the decision degree of optimism α is 0.5, the above integral value is the same as ordinary representatives [244].

For example, total integral values of two triangular fuzzy numbers $\tilde{A}_1 = (3, 4, 7)$ and $\tilde{A}_2 = (4, 5, \frac{51}{8})$ are

$$I^\alpha(\tilde{A}_1) = 4 + 2\alpha \quad \text{and} \quad I^\alpha(\tilde{A}_2) = 4.5 + 1.2\alpha$$

In minimizing problems, for a pessimistic decision maker with $\alpha = 1$, the ranking order has $\tilde{A}_1 > \tilde{A}_2$; for an optimistic decision maker with $\alpha = 0$, the ranking order has $\tilde{A}_1 < \tilde{A}_2$; for a moderate decision maker with $\alpha = 0.5$, the ranking order has $\tilde{A}_1 > \tilde{A}_2$.

With this ranking method, the fuzzy objective values can be converted into integral values and Pareto solutions can be determined based on these integral values.

Evaluation. How to assess the merit for each chromosome in multicriteria context is a relatively difficult task in the evaluation phase of a genetic algorithm. In Gen, Ida, and Li's method, the weighted-sum method was used to calculate the fitness of chromosomes. It works as follows:

Procedure: Evaluation

Step 1. Calculate fuzzy objective values $\tilde{z}_q(X_s)$, $q = 1, 2, \ldots, Q$, for each chromosome X_s, $s = 1, 2, \ldots, pop\_size$.

Step 2. Convert fuzzy numbers $\tilde{z}_q(X_s)$ into integral values $I_q^\alpha(\tilde{z}_q(X_s))$ with ranking method.

Step 3. Determine the maximal and minimal integral values for each objective as follows:

$$I_q^{\min} = \min_s \{I_q^\alpha(\tilde{z}_q(X_s))|s = 1, 2, \ldots, pop\_size\}, \qquad q = 1, 2, \ldots, Q$$

$$I_q^{\max} = \max_s \{I_q^\alpha(\tilde{z}_q(X_s))|s = 1, 2, \ldots, pop\_size\}, \qquad q = 1, 2, \ldots, Q$$

Step 4. Calculate the weight coefficients as follows:

$$\delta_q = I_q^{\max} - I_q^{\min}, \qquad q = 1, 2, \ldots, Q$$

$$\beta_q = \frac{\delta_q}{\sum_{q=1}^{Q} \delta_q}, \qquad q = 1, 2, \ldots, Q$$

Step 5. Calculate the fitness value for each chromosome as follows:

$$eval(X_s) = \sum_{q=1}^{Q} \beta_q I_q^\alpha(\tilde{z}_q(X_s)), \qquad s = 1, 2, \ldots, pop\_size$$

The Best Compromise Solution. To determine the best compromise solution among Pareto solutions, the TOPSIS method given by Yoon and Hwang was used in the genetic algorithm. TOPSIS stands for technique for order preference by similarity to ideal solution, which is based upon the concept that the chosen alternative should have the shortest distance from the positive ideal solution and the farthest from the negative ideal solution [225]. Let X_k, $k = 1, 2, \ldots, m$, be the Pareto solutions in the last generation of genetic algorithm. The TOPSIS method works as follows:

Procedure: TOPSIS Method

Step 1. Determine the approximate positive and negative ideal solutions:

$$I_q^+ = \min \{I_q^\alpha(\tilde{z}_q(X_s))|s = 1, 2, \ldots, m\}, \qquad q = 1, 2, \ldots, Q$$

$$I_q^- = \max \{I_q^\alpha(\tilde{z}_q(X_s))|s = 1, 2, \ldots, m\}, \qquad q = 1, 2, \ldots, Q$$

Step 2. Construct the normalized values:

$$r_q^k = \frac{I_q^\alpha(\tilde{z}_q(X_k))}{\sqrt{\sum_{s=1}^{m} (I_q^\alpha(\tilde{z}_q(X_s)))^2 + (I_q^+)^2 + (I_q^-)^2}}$$

Step 3. Calculate the separation measures:

$$s_k^+ = \sqrt{\sum_{q=1}^{Q} w_q^2 (r_q^+ - r_q^k)^2}, \qquad k = 1, 2, \ldots, m$$

$$s_k^- = \sqrt{\sum_{q=1}^{Q} w_q^2 (r_q^k - r_q^-)^2}, \qquad k = 1, 2, \ldots, m$$

where r_q^+ and r_q^- are the normalized I_q^+ and I_q^-, and w_q is the relative importance (weight) of objectives, which satisfies the following condition:

$$\sum_{q=1}^{Q} w_q = 1, \qquad w_q \in [0, 1]$$

Step 4. Calculate the relative closeness to the ideal solution:

$$d_k = \frac{s_k^-}{s_k^+ + s_k^-}, \qquad k = 1, 2, \ldots, m$$

Step 5. Rank the preference order. The Pareto solutions can now be ranked according to the descending order of d_k:

Procedure: Fuzzy Multicriteria Solid Transportation Problem

begin
  $t \leftarrow 0$;
  initialize $P(t)$;
  evaluate $P(t)$;
  create Pareto solution $E(t)$;
  while (not termination condition) **do**
    recombine $P(t)$;
    evaluate $P(t)$;
    select $P(t + 1)$;
    update Pareto solution $E(t)$;
    $t \leftarrow t + 1$;
  end
  determine the best compromise solution;
end

8.5.3 Numerical Example

Consider the following example:

$$\min \ \tilde{z}_q = \sum_{i=1}^{3} \sum_{j=1}^{3} \sum_{k=1}^{3} \tilde{c}_{ijk}^{q} x_{ijk}, \qquad q = 1, 2, 3$$

$$\text{s.t.} \ \sum_{j=1}^{3} \sum_{k=1}^{3} x_{1jk} = 8, \qquad \sum_{j=1}^{3} \sum_{k=1}^{3} x_{2jk} = 9, \qquad \sum_{j=1}^{3} \sum_{k=1}^{3} x_{3jk} = 5$$

$$\sum_{i=1}^{3} \sum_{k=1}^{3} x_{i1k} = 7, \qquad \sum_{i=1}^{3} \sum_{k=1}^{3} x_{i2k} = 6, \qquad \sum_{i-1}^{3} \sum_{k=1}^{3} x_{i2k} = 5$$

$$\sum_{i=1}^{3} \sum_{j=1}^{3} x_{ij1} = 10, \qquad \sum_{i=1}^{3} \sum_{j=1}^{3} x_{ij1} = 5, \qquad \sum_{i=1}^{3} \sum_{j=1}^{3} x_{ij1} = 6$$

$$x_{ijk} \geq 0, \qquad \forall i, j, k$$

This is adapted from Bit, Biswal, and Alam [36]. Table 8.1 shows the fuzzy coefficients of $\tilde{c}_{ijk}^{q}$.

In the experiment, parameters were fixed as $max\_gen = 1000$, $pop\_size = 30$, $p_m = 0.2$, $p_c = 0.4$, $w_1 = 0.5$, $w_2 = 0.3$, and $w_3 = 0.2$. The Pareto solutions over 10 runs are summarized in Table 8.2 for both optimistic and moderate cases.

The best compromise solutions for each case are marked with $*$ notation. The solution for the optimistic case ($\alpha = 0$) is

$$x_{121} = 6, \qquad x_{331} = 4, \qquad x_{132} = 2$$
$$x_{232} = 2, \qquad x_{332} = 1, \qquad x_{213} = 7$$

for the solution for the moderate case ($\alpha = 0.5$) is

$$x_{121} = 5, \qquad x_{331} = 5, \qquad x_{122} = 1,$$
$$x_{132} = 2, \qquad x_{232} = 2, \qquad x_{213} = 7$$

Table 8.1. Fuzzy Coefficients in Objective Functions

q	1			2			3		
k	1	2	3	1	2	3	1	2	3
$\tilde{c}_{11k}^q$	(8, 9, 10)	(10, 12, 14)	(7, 9, 11)	(3, 6, 9)	(8, 9, 10)	(5, 7, 9)	(2, 3, 4)	(6, 7, 8)	(5, 7, 9)
$\tilde{c}_{12k}^q$	(4, 5, 6)	(5, 6, 7)	(3, 5, 7)	(7, 9, 11)	(8, 11, 14)	(1, 3, 5)	(5, 6, 7)	(6, 8, 10)	(5, 6, 7)
$\tilde{c}_{13k}^q$	(1, 2, 3)	(1, 2, 3)	(1, 1, 1)	(1, 2, 3)	(6, 7, 8)	(6, 7, 8)	(1, 1, 1)	(8, 9, 10)	(1, 3, 5)
$\tilde{c}_{21k}^q$	(1, 2, 3)	(8, 9, 10)	(6, 8, 10)	(1, 1, 1)	(2, 4, 6)	(1, 1, 1)	(7, 9, 11)	(7, 9, 11)	(4, 5, 6)
$\tilde{c}_{22k}^q$	(1, 2, 3)	(7, 8, 9)	(1, 1, 1)	(3, 4, 5)	(3, 5, 7)	(1, 2, 3)	(6, 8, 10)	(5, 6, 7)	(8, 9, 10)
$\tilde{c}_{23k}^q$	(4, 5, 6)	(1, 2, 3)	(5, 7, 9)	(6, 8, 10)	(8, 9, 10)	(6, 7, 8)	(3, 5, 7)	(1, 2, 3)	(3, 5, 7)
$\tilde{c}_{31k}^q$	(1, 2, 3)	(2, 4, 6)	(5, 6, 7)	(2, 3, 4)	(4, 6, 8)	(3, 4, 5)	(6, 8, 10)	(2, 4, 6)	(7, 9, 11)
$\tilde{c}_{32k}^q$	(1, 2, 3)	(3, 5, 7)	(1, 3, 5)	(3, 5, 7)	(4, 6, 8)	(4, 6, 8)	(7, 9, 11)	(4, 6, 8)	(1, 3, 5)
$\tilde{c}_{33k}^q$	(1, 1, 1)	(8, 9, 10)	(1, 1, 1)	(7, 8, 9)	(2, 3, 4)	(7, 9, 11)	(3, 5, 7)	(5, 7, 9)	(10, 11, 12)

Table 8.2. Obtained Pareto Optimal Solution

No.	$\tilde{z}_1$	$\tilde{z}_2$	$\tilde{z}_3$	$I = (I_1, I_2, I_3)$
\multicolumn{5}{c}{Optimistic Case ($\alpha = 0$)}				
1	(67, 100, 133)	(97, 123, 149)	(82, 120, 158)	(83.5, 110, 101)
2	(75, 98, 121)	(87, 112, 137)	(70, 107, 139)	(86.5, 99.5, 88.5)
3	(71, 103, 135)	(78, 106, 134)	(70, 107, 144)	(87, 92, 91)
4*	(75, 114, 153)	(50, 65, 80)	(48, 86, 124)	(94.5, 57.5, 67)
5	(89, 129, 169)	(95, 113, 131)	(52, 78, 104)	(109, 104, 65)
6	(89, 130, 171)	(48, 63, 78)	(90, 126, 162)	(109.5, 55.5, 108)
7	(103, 141, 179)	(43, 59, 75)	(83, 118, 153)	(122, 51, 100.5)
8	(109, 139, 169)	(115, 143, 171)	(47, 78, 109)	(124, 129, 62.5)
9	(106, 143, 180)	(42, 56, 70)	(61, 94, 127)	(124.5, 49, 77.5)
10	(117, 152, 187)	(38, 55, 72)	(76, 110, 144)	(134.5, 46.5, 93)
11	(134, 158, 182)	(116, 139, 162)	(40, 71, 102)	(146, 127.5, 55.5)
12	(136, 159, 182)	(114, 138, 162)	(47, 78, 109)	(147.5, 126, 62.5)

$$I^+ = (83.5, 46.5, 55.5) \qquad I^- = (147, 129, 108)$$

No.	$\tilde{z}_1$	$\tilde{z}_2$	$\tilde{z}_3$	$I = (I_1, I_2, I_3)$
\multicolumn{5}{c}{Moderate Case ($\alpha = 0.5$)}				
1	(73, 78, 123)	(132, 162, 192)	(105, 147, 189)	(98, 162, 147)
2	(74, 103, 132)	(100, 130, 160)	(86, 125, 164)	(103, 130, 125)
3	(71, 104, 137)	(94, 121, 148)	(80, 118, 156)	(104, 121, 118)
4*	(73, 109, 145)	(53, 71, 89)	(48, 87, 126)	(109, 71, 87)
5	(75, 114, 153)	(50, 65, 80)	(48, 86, 124)	(114, 65, 86)
6	(81, 122, 163)	(46, 59, 72)	(60, 98, 136)	(122, 59, 98)
7	(128, 153, 178)	(125, 149, 173)	(41, 74, 107)	(153, 149, 74)
8	(130, 156, 182)	(100, 125, 150)	(36, 69, 102)	(156, 125, 69)

$$I^+ = (98, 59, 69) \qquad I^- = (156, 162, 147)$$

9

FACILITY LAYOUT DESIGN PROBLEMS

9.1 INTRODUCTION

Facility layout design has been the subject of an interdisciplinary interest for over two decades. The process of developing facility layouts contains the elements of both art and science. Recently this subject has received considerable attention in operations research and management science. It is usually viewed as optimization problems, and the best facility layout is found out by optimizing some measure of performance subjecting to some constraints [138].

To date, significant progress has been made in manufacturing technology, and a large number of flexible manufacturing systems (FMS) have been implemented around the world. Optimal design of the physical layout of a manufacturing system is one of the most important issues that must be resolved in the early stages of the system design. Cost consequences of the layout decision can be observed during the implementations as well as the operation of the system. Good solutions to these problems provide a necessary foundation for effective utilization of the system. But coming up with such solution is a complicated task. The layout designer faces the difficult task of developing a system that is capable of handling a variety of products with variable demands at a reasonable cost.

In a manufacturing environment, layout design is concerned with the optimum arrangement of physical facilities, such as departments or machines, in a certain area. Usually the design criterion is considered as the minimizing material handling costs. Because of the combinatorial nature of facility layout problems, the heuristic technique is the most promising approach for solving practical size layout problems. Recently, the interest in application of genetic algorithms to facility layout design has been growing rapidly. Tate and Smith applied genetic algorithms to shape-constrained unequal area facility layout problems [399]. Cohoon et al. proposed a distributed genetic algorithm for the floorplan design problem [81]. Tam reported his experiences of applying genetic algorithms to facility layout problems [395]. In this chapter, we will show how genetic algorithms can be successfully applied to various machine layout and facility layout problems.

9.2   MACHINE LAYOUT PROBLEM

A flexible machining system is one of the forms of implementing computer-aided manufacturing (CAM). The design of a flexible machining system involves the layout of machines and work stations. Kusiak and Heragu published a survey paper on machine layout problems [259]. The layout of machines in a flexible machining system is typically determined by the type of material handling devices used. The most used material handling devices are as follows:

- Material handling robot
- Automated guided vehicle (AGV)
- Gantry robot

Figure 9.1 shows the four basic types of machine layouts commonly seen in practice. Machines served by an AGV tend to be arranged along a straight line [Figures 9.1(a) and 9.1(b)], whereas the reach constraints of a robot force machines to be arranged in a circle or in a cluster [Figures 9.1(c) and 9.1(d) respectively]. The four basic patterns of machine layout are named respectively as follows:

- Linear single-row
- Linear double-row
- Circular single-row
- Multiple-row

For the purpose of modeling the layout problem, only the single-row pattern and the multiple-row pattern need to be considered. This is because among the four patterns of layout shown in Figure 9.1 the circular single-row and linear single-row layouts can be considered as the single-row layout pattern. The linear double-row layout is a special case of the multiple-row layout. In a single-row pattern, machines are arranged in one row; and in a multiple-row pattern, machines are arranged linearly in two or more rows. Simple examples are shown in Figure 9.2.

A majority of conventional approaches formulate layout design as the quadratic assignment problem (QAP), in which the location of each site is specified in advance and the problem involves assigning facilities to each site so that the total material handling cost is minimized. Heragu and Kusiak [211] have pointed out that the machine problem for flexible manufacturing systems can be different from the traditional facility layout problem in one impor-

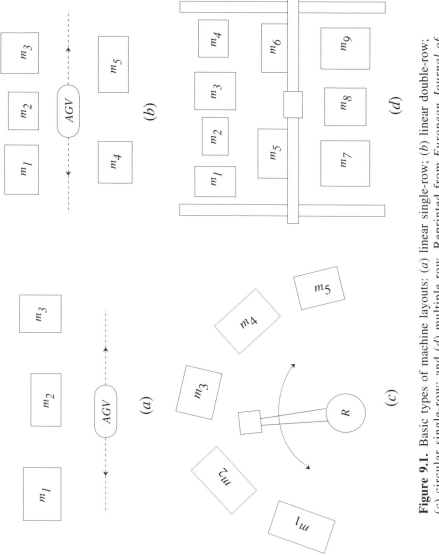

Figure 9.1. Basic types of machine layouts: (*a*) linear single-row; (*b*) linear double-row; (*c*) circular single-row; and (*d*) multiple row. Reprinted from *European Journal of Operational Research*, vol. 29, A. Kusiak and S. Heragu, "The facility layout problem," p. 230, 1992, with kind permission.

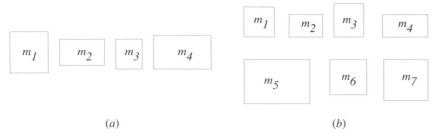

Figure 9.2. Machine layout patterns: (*a*) single-row layout and (*b*) multiple-row layout.

tant respect. Machine sizes are generally not equal, and if the distance between two machines is determined by their respective dimensions and the clearance required between them, then distances between locations are not equal and cannot be predetermined. They developed new models with absolute values in the objective function and constraints for machine layout problems.

9.3 SINGLE-ROW MACHINE LAYOUT PROBLEM

9.3.1 Mathematical Model

In order to model the single-row machine layout problem, the following assumptions are made [258]:

- Machines are rectangular in shape.
- Orientation of machines is known; for example, all machines are to be oriented lengthwise.

The single-row machine layout problem can be formulated as follows:

$$\min \quad \sum_{i=1}^{n-1} \sum_{j=i+1}^{n} c_{ij} f_{ij} |x_i - x_j| \tag{9.1}$$

$$\text{s.t.} \quad |x_i - x_j| \geq \tfrac{1}{2}(l_i + l_j) + d_{ij}, \qquad i = 1, \ldots, n-1, \quad j = i+1, \ldots, n \tag{9.2}$$

$$x_i \geq 0, \qquad i = 1, \ldots, n \tag{9.3}$$

where

n is the number of machines

f_{ij} is the frequency of trips between machines i and j

c_{ij} is the handling cost per unit distance traveled between machines i and j

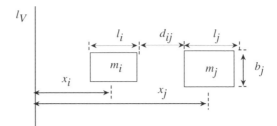

Figure 9.3. Illustration of parameters and decision variables.

l_i is the length of machine i

d_{ij} is the minimum clearance between machines i and j

x_i is the distance between the center of machine i and the vertical reference line l_V

The parameters l_i and d_{ij}, the decision variable x_i, and the vertical reference line l_V are illustrated in Figure 9.3. Constraint (9.2) ensures that no two machines in the layout overlap. Constraint (9.3) ensures non-negativity. The objective is to minimize the total cost involved in making the required number of trips between the machines.

9.3.2   Genetic Algorithm for Single-Row Machine Layout Problem

Single-row machine layout is a very special case. The essentials of this problem can be viewed as the sequencing problem of machines, so it can be solved in two separate steps:

- Sequencing machines
- Generating actual layout

In the first step, the sequence of machines is generated; and in the second step, the actual layout is created according to the sequence and geometric requirements of machines. As we know, genetic algorithms are very efficient for dealing with such a permutation problem. The basic idea of the algorithm given in this section is to use a genetic algorithm to find a better sequence of machines and then to create the actual layout according to the sequence and geometric requirements of machines.

Representation. A straightforward way to encode the machine layout into a chromosome for a single-row case is to use the permutation of machines. For example, machines are arranged in the following order:

$$[m_2 \quad m_1 \quad m_4 \quad m_6 \quad m_3 \quad m_5]$$

where m_i represents the ith machine. The chromosome $\boldsymbol{v}_k$ will be

$$\boldsymbol{v}_k = [2 \quad 1 \quad 4 \quad 6 \quad 3 \quad 5]$$

There are many genetic operators available for such encoding. We use PMX crossover and inverse mutation. For details, we refer to Chapter 3.

Evaluation. Generally, for an n-machine problem, a chromosome $\boldsymbol{v}_k$ is given as follows:

$$\boldsymbol{v}_k - [m_1^k, m_2^k, \dots, m_n^k]$$

where m_i^k represents a machine (or a machine symbol) which is in the ith position of the kth chromosome. According to the sequence and the geometric requirements of machines, we can calculate the x-axis coordinates of all machines (the positions of machines or actual layout of machines) as follows:

$$[x_1^k, x_2^k, \dots, x_n^k]$$

Then we can calculate the total cost for the kth chromosome as follows:

$$f_k = \sum_{i=1}^{n-1} \sum_{j=i+1}^{n} c_{ij} f_{ij} |x_i^k - x_j^k| \tag{9.4}$$

Because the layout design is a minimization problem, we must convert the objective function value for each chromosome to the fitness value, such that a fitter chromosome has a larger fitness value. The conversion is done by the following evaluation function:

$$eval(\boldsymbol{v}_k) = \frac{1}{f_k} \tag{9.5}$$

where $eval(\boldsymbol{v}_k)$ is the fitness function for the kth chromosome.

The roulette wheel method was used as the basic selection mechanism to reproduce the next generation. Elitist strategy was employed in the phase to preserve the best chromosome in the next generation and overcome the stochastic errors of sampling.

Example. The test problem, given by Kusiak [258], is a six-machine layout problem. The machine sizes, the frequency matrix, the cost matrix, and clearance between machines for the problem are as follows:

MACHINE SIZES

Machine i	Dimension $(l_i \times b_i)$
1	5.0×3.0
2	2.0×2.0
3	2.5×2.0
4	6.0×3.5
5	3.0×1.5
6	4.0×4.0

$$[f_{ij}] = \begin{pmatrix} 0 & 40 & 80 & 21 & 62 & 90 \\ 40 & 0 & 72 & 12 & 24 & 28 \\ 80 & 72 & 0 & 14 & 41 & 9 \\ 21 & 12 & 14 & 0 & 21 & 12 \\ 62 & 24 & 41 & 21 & 0 & 31 \\ 90 & 28 & 9 & 12 & 31 & 0 \end{pmatrix}$$

$$[c_{ij}] = \begin{pmatrix} 0 & 4 & 4 & 6 & 4 & 5 \\ 4 & 0 & 2 & 5 & 2 & 3 \\ 4 & 2 & 0 & 5 & 3 & 3 \\ 6 & 5 & 5 & 0 & 5 & 8 \\ 4 & 2 & 3 & 5 & 0 & 4 \\ 5 & 3 & 3 & 8 & 4 & 0 \end{pmatrix}, \qquad [d_{ij}] = \begin{pmatrix} 0 & 1 & 1 & 1 & 2 & 1 \\ 1 & 0 & 1 & 1 & 1 & 1 \\ 1 & 1 & 0 & 1 & 1 & 1 \\ 1 & 1 & 1 & 0 & 3 & 1 \\ 2 & 1 & 1 & 3 & 0 & 2 \\ 1 & 1 & 1 & 1 & 2 & 0 \end{pmatrix}$$

The evolutionary environment of our implementation is given as follows: *pop_size*, 20; *maxgen*, 50; p_c, 0.4; and p_m, 0.2. The best chromosome is listed as follows:

Generation the best solution occurred: 17

Cost: 19531.00

The sequence of machines: [6 1 3 5 2 4]

The layout based on the sequence is depicted in Figure 9.4, and the evolution process is shown in Figure 9.5.

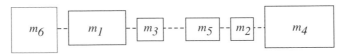

Figure 9.4. Single-row layout for the example problem.

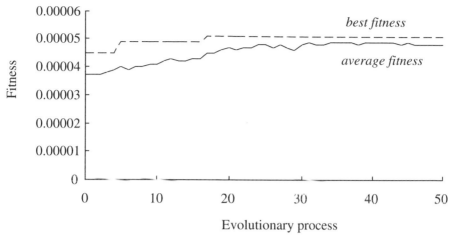

Figure 9.5. Evolution process for test problem.

9.4   MULTIPLE-ROW MACHINE LAYOUT PROBLEM

9.4.1   Mathematical Model

Most of the traditional methods for solving the facility layout problem assign facilities to the location of prespecified sites. This is a discrete optimization approach. In machine layout problems, the site location cannot be predetermined. Heragu and Kusiak have formulated a multiple-row machine layout problem as a two-dimensional continuous space allocation problem, which belongs to the continuous optimization approach [211]. However, in many practical problems, the machines are arranged along well-defined rows because in most cases the separation between rows can be predetermined according to the feature of material handling system; that is, this problem can be viewed as discrete in one dimension and continuous in another dimension. Let x_i and y_i be the distances from the center of machine i to vertical and horizontal reference lines, respectively. Let the decision variable z_{ik} be

$$z_{ik} = \begin{cases} 1, & \text{if machine } i \text{ is allocated to row } k \\ 0, & \text{otherwise} \end{cases} \tag{9.6}$$

The multiple-row machine layout problem with unequal area can be formulated as a mixed-integer programming problem.

$$\min \quad \sum_{i=1}^{n-1} \sum_{j=i+1}^{n} c_{ij} f_{ij}(|x_i - x_j| + |y_i - y_j|) \tag{9.7}$$

$$\text{s.t.} \quad |x_i - x_j| z_{ik} z_{jk} \geq \tfrac{1}{2}(l_i + l_j) + d_{ij}, \qquad i, j = 1, \ldots, n \tag{9.8}$$

$$y_i = \sum_{k=1}^{m} l_0(k-1) z_{ik}, \qquad i = 1, \ldots, n \tag{9.9}$$

$$\sum_{k=1}^{m} z_{ik} = 1, \qquad i = 1, \ldots, n, \tag{9.10}$$

$$\sum_{i=1}^{n} z_{ik} < n, \qquad k = 1, \ldots, m \tag{9.11}$$

$$x_i, y_i \geq 0, \qquad i = 1, \ldots, n \tag{9.12}$$

$$z_{ik} = 0, 1, \qquad i = 1, \ldots, n, k = 1, \ldots, m \tag{9.13}$$

where

n is the number of machines

m is the number of rows

f_{ij} is the frequency of trips between machines i and j

c_{ij} is the handling cost per unit distance traveled between machines i and j

l_i is the length of machine i

l_0 is the separation between two adjacent rows,

d_{ij} is the minimum clearance between machines i and j

x_i is the distance between center of machine i and the vertical reference line l_V

y_i is the distance between center of machine i and the horizontal reference line l_H

The objective is to minimize the total cost involved in making the required trips between the machines. Constraint (9.8) ensures that no two machines overlap. Constraints (9.10)–(9.11) ensure that one machine is assigned only to one row. From constraint (9.9) we know that variable y_i is not necessary in this model because it can be determined by z_{ik}, but with this notation the model can be easily understood. The parameters, decision variables, and reference lines are illustrated in Figure 9.6. From the model we know that the essentials of this problem comprise two different tasks:

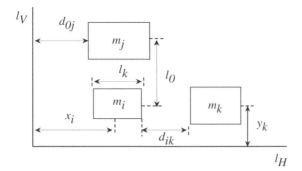

Figure 9.6. Illustration of parameters, decision variables, and reference lines.

- Allocate machines to rows (to determine y coordinates).
- Find the best positions of machines within each row (to determine x coordinates).

The first one is a combinatorial optimization problem. Although the second one is the single-row layout problem, it is easy to see that the best solution for each row may not be good for global solution of the problem due to the existence of traffic cost among rows. Thus we cannot simply handle this problem as several single-row layout problems.

9.4.2 Representation

For the multiple-row machine layout problem, a representation scheme is given in [59, 67], which can be viewed as the extended permutation representation, containing three lists of *separator/machine symbol/neat clearance*. The list of machine symbols represents all possible permutations of machines (or sequence of machines), the separator designates the allocation of machines to rows, and the list of neat clearance records the x-axis positions of all machines. For n machines and the two-row case, we have

$$[s, \{m_{i_1}, m_{i_2}, \ldots, m_{i_n}\}, \{\Delta_{i_1}, \Delta_{i_2}, \ldots, \Delta_{i_n}\}]$$

where m_{i_j} represents the machine in the jth position and Δ_{i_j} denotes the neat clearance between machines $m_{i_{j-1}}$ and m_{i_j}. The separator s, generally $1 < s < n$, denotes the cutting position in which the list is separated into two parts. A machine in the first part is assigned to the first row; otherwise, it is assigned to the second row.

This representation scheme belongs to the indirect encoding method. A natural question may arise: Why do we not directly code both x-axis and y-axis values into the chromosome? There are several advantages when we use this special approach.

1. First, this encoding technique can significantly reduce the illegal offspring generated during the evolutionary process. In this representation, the *x*-axis coordinate is represented by the neat clearance but not its absolute value. If we use the *x*-axis coordinate in encoding the representation, it is liable to yield a large amount of illegal offspring in the sense of machine overlapping along horizontal dimension. With the neat clearance encoding, the overlapping of machines will never occur. Similarly, to avoid the occurrence of machine overlapping along the vertical dimension, instead of encoding the *y*-axis value into the representation scheme, we use a separator list and a machine permutation list to represent the allocation of machines among rows.

2. Second, we know that the essentials of the multiple-row machine layout problem is the combination of discrete and continuous optimization problems in nature; that is,

- Continuous optimization problem along *x*-axis
- Combinatorial optimization problem along *y*-axis

When making a layout, the allocation of machines among rows and the permutation of machines within each row are the leading factors, and the positions of machines conduct the fine-tuning on the arrangement of machines determined by the allocation and sequencing. With this encoding scheme, we can treat such thorny problems with three separate parts:

- Allocate machines to rows.
- Sequence machines within each row.
- Perform fine-tuning of positions (*x* and *y* coordinates) for each machine.

3. Third, because we use a list to represent the machine permutation, some of the traditional genetic operators for ordering representation can be applied to manipulate it.

The proposed representation scheme meets the nature of the problem and can make our genetic algorithms very efficient.

Calculation of x-Axis Coordinates. The *x*-axis coordinate can be determined uniquely according to the neat clearance list and the machine permutation list. Suppose machines m_k and m_i are arranged within the same row as shown in Figure 9.7. The clearance and decision variables can be calculated as follows:

$$\Delta_i = \Delta l_{ki} - d_{ki}$$
$$x_i = x_k + d_{ki} + \Delta_i + \tfrac{1}{2}\,(l_i + l_k)$$
$$x_k = d_{k0} + \Delta_k + \tfrac{1}{2}\,l_k$$

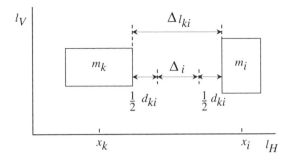

Figure 9.7. Illustration of neat clearance and decision variables.

where

Δ_i is the neat clearance associated with machine m_i

Δl_{ki} is the separation between two machines

d_{ki} is the required clearance between machines k and i

From Figure 9.7 we can see that the separation between two adjacent machines is composed of two parts: required clearance and neat clearance. Because neat clearance will lead to the increase of total travel cost among machines, one may say that zero neat clearance for all machines is best. This is true for the single-row case but is not true for the multiple-row case because of the existence of travel cost among rows.

Calculation of y-Axis Coordinates. The y-axis coordinate can be determined based on the separations of rows. The separation between rows can be predetermined according to the features of the material handling system. Let us consider the two-row case. Let l denote the value of separation between two rows as shown in Figure 9.8.

If we suppose that the position of the first row is 0, then the y-axis coordinates can be calculated as follows:

$$y_i = \begin{cases} 0 & \text{if } m_i \text{ is in the first row} \\ l & \text{if } m_i \text{ is in the second row} \end{cases}$$

Now we give an example to show how to construct an actual layout from this indirect encoding scheme. Suppose that we have a chromosome as follows:

$$[4, \{3, 6, 1, 5, 4, 2\}, \{1.2, 0.2, 0.4, 0.3, 2.2, 4.6\}]$$

The actual layout according to this encoding is depicted in Figure 9.9.

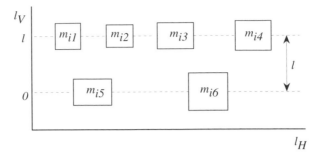

Figure 9.8. Illustration of *y*-axis coordinates.

9.4.3 Initialization

A random approach is used to generate the initial population—that is, the separator, the machine permutation list, and the neat clearance list. The overall procedure is shown below:

Procedure: Initialization

begin
  $i \leftarrow 0$;
  while $(i \leq pop\_size)$ **do**
    generate separator;
    generate machine permutation list;
    generate neat clearance list;
    $i \leftarrow i + 1$;
  end
end

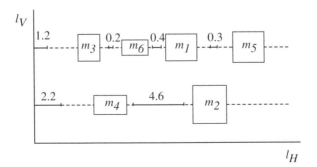

Figure 9.9. Actual machine layout.

Separator. For an *n*-machine and two-row case, the separator is simply a random integer within the open region $(1, n)$.

Machine permutation. Let Σ_0 be the set of available machines and let P be the list of machine permutation; then the machine permutation is randomly generated by the following procedure:

Procedure: Machine Permutation

begin
    $i \leftarrow 0$;
    $\Sigma_0 \leftarrow \{1, 2, \ldots, n\}$;
    $P \leftarrow \phi$;
    while $(i \leq n)$ **do**
      pick up a random machine m' from Σ_0;
      $P \leftarrow P \cup m'$;
      $\Sigma_0 \leftarrow \Sigma_0 \setminus m'$;
      $i \leftarrow i + 1$;
    end
    return permutation list P;
end

Neat Clearance. The procedure for generating a neat clearance list is essentially a random approach. Because of the existence of an allowable space constraint, the neat clearance is generated within an allowable region. Let L' be available space and let L be the upper bound of width of the allowable working area. Suppose the machine sequence is as follows:

$$\lfloor m_1, m_2, \ldots, m_n \rfloor$$

Then the initial available space can be calculated as follows:

$$L' = 2L - \left(\sum_{i=1}^{n} l_i + \sum_{i=1}^{n-1} d_{i,i+1} + d_{10} + d_{n0} \right) \tag{9.14}$$

Let Δ denote the list of neat clearance. The overall procedure is given below:

Procedure: Neat Clearance

begin
    $i \leftarrow 0$;
    $\Delta \leftarrow \phi$;
    calculate initial available space L';
    while $(i \leq n)$ **do**

pick up a random neat clearance Δ_i within $[0, L']$;
$\Delta \leftarrow \Delta \cup \Delta_i$;
$L' \leftarrow L' - \Delta_i$;
$i \leftarrow i + 2$;
end
return neat clearance list Δ;
end

9.4.4 Crossover

The basic elements of the proposed crossover consist of three parts: a random way to determine separator, an ordinary PMX to manipulate machine permutation list, and an arithmetical crossover to manipulate neat clearance list. The scheme of the crossover operator is depicted below:

Procedure: Crossover

begin
  $i \leftarrow 0$;
  while $(i \leq pop\_size \times p_c)$ **do**
    select two random individuals;
    generate new separator;
    generate new machine permutation with ordinary PMX;
    generate new neat clearance list with arithmetical crossover;
    $i \leftarrow i + 1$;
  end
end

For example, we have two parents shown as follows:

$$p_1 = [s^1, \{m_1^1, m_2^1, \ldots, m_n^1\}, \{\Delta_1^1, \Delta_2^1, \ldots, \Delta_n^1\}]$$
$$p_2 = [s^2, \{m_1^2, m_2^2, \ldots, m_n^2\}, \{\Delta_1^2, \Delta_2^2, \ldots, \Delta_n^2\}]$$

Now we explain how to produce an offspring.

New Separator. The procedure for generating a new separator is very simple, which contains two major steps:

1. Determination of the upper and lower bounds of the separator

2. Selection of a random integer from the interval formed with the upper and lower bounds

The upper and lower bounds can be directly calculated as follows:

$$s_U = \max \{s^1, s^2\}$$
$$s_L = \min \{s^1, s^2\}$$

Then we can make a closed interval with s_U and s_L as $[s_L, s_U]$. The new separator is a random integer within this interval.

New Neat Clearance List. We define a new arithmetical crossover to manipulate the neat clearance list. Suppose there are two neat clearance lists as follows:

$$\{\Delta_1^1, \Delta_2^1, \ldots, \Delta_n^1\}$$
$$\{\Delta_1^2, \Delta_2^2, \ldots, \Delta_n^2\}$$

The new neat clearance associated with each machine can be calculated as follows:

$$\Delta_i' = \alpha_1 \Delta_i^1 + \alpha_2 \Delta_i^2, \qquad i = 1, 2, \ldots, n \tag{9.15}$$
$$\alpha_1, \alpha_2 \in (0, 1) \tag{9.16}$$

The main difference between the proposed crossover and the conventional arithmetical crossover is that the conventional one requires

$$\alpha_1 + \alpha_2 - 1 \tag{9.17}$$

and the proposed one requires

$$\alpha_1 + \alpha_2 \leq 2 \tag{9.18}$$

The reason is that if we take the conventional approach, the generated neat clearances will be gradually decreasing along the evolutionary process. In this case, search space is highly dependent on the initial solution space. We force these parameters to follow inequality (9.18); then the search space can be greatly enlarged and also is independent of initial search space.

9.4.5 Mutation

Mutation is designed with the neighbor search technique to perform fine-tuning on the positions of machines.

Generating Neighbor. Suppose that the neat clearance list for a given chromosome is

$$\{\Delta_1, \Delta_2, \ldots, \Delta_i, \ldots, \Delta_n\}$$

and that a nonzero gene, say the ith gene Δ_i, is selected for mutation. If we let r be a given integer, then we can get $2r$ neat clearances as follows:

$$\Delta_{i'}^1 = \frac{\Delta_i}{r} \tag{9.19}$$

$$\Delta_{i'}^j = \Delta_{i'}^{j-1} + \frac{\Delta_i}{r}, \qquad j = 2, 3, \ldots, 2r \tag{9.20}$$

These neat clearances vary from Δ_i/r to $2\Delta_i$. Then the set of neat clearance lists is given below:

$$\{\Delta_1, \Delta_2, \ldots, \Delta_{i'}^1, \ldots, \Delta_n\}$$
$$\{\Delta_1, \Delta_2, \ldots, \Delta_{i'}^2, \ldots, \Delta_n\}$$
$$\cdots$$
$$\{\Delta_1, \Delta_2, \ldots, \Delta_{i'}^j, \ldots, \Delta_n\}$$
$$\cdots$$
$$\{\Delta_1, \Delta_2, \ldots, \Delta_{i'}^{2r}, \ldots, \Delta_n\}$$

The set of chromosomes formed with the above set of neat clearance lists, together with the separator and machine permutation list of the given chromosome, is regarded as the neighborhood of the given chromosome. A chromosome is said to be $2r$-*optimum* if it is better than any others in the neighborhood.

Mutation Procedure. The mutation procedure consists of three steps: First it selects a chromosome for mutation and then picks up a random neat clearance of the selected chromosome; second, it generates $2r$ neighbors for the selected chromosome; third, it evaluates all neighbors and the best neighbor is used as the offspring. The scheme of the proposed mutation is depicted below:

Procedure: Mutation

begin
  give an integer r;
  $i \leftarrow 0$;
  while $(i \le pop\_size \times p_m)$ **do**
    pick up a random chromosome v';
    pick up a random gene Δ_j from neat clearance list of v';
    generate $2r$ neighbors of v';
    evaluate all neighbors;
    select the best neighbor as the offspring;
    $i \leftarrow i + 1$;
  end
end

9.4.6   Evaluation Function

The evaluation function has two terms: total cost and penalty to illegality. The total cost is calculated using the objective function. Let v_k be the kth chromosome in the enlarged population,

$$[s^k, \{m_1^k, m_2^k, \ldots, m_n^k\}, \{\Delta_1^k, \Delta_2^k, \ldots, \Delta_n^k\}]$$

where m_j^k represents the machine in the jth position, and Δ_j^k denotes the neat clearance between machines m_{j-1} and m_j. The separator s^k, generally $1 < s^k < n$, denotes the cutting position to separate the list into two parts according to the two-row requirement. According to the procedures given above, we can calculate the x-axis and y-axis coordinates of all machines (the positions of machines or actual layout of machines) as follows:

$$\{(x_1^k, y_1^k), (x_2^k, y_2^k), \ldots, (x_n^k, y_n^k)\}$$

Then we obtain the total cost for kth chromosome as follows:

$$f_k = \sum_{i=1}^{n-1} \sum_{j=i+1}^{n} c_{ij} f_{ij}(|x_i^k - x_j^k| + |y_i^k - y_j^k|) \tag{9.21}$$

In general, two kinds of illegal solutions may occur in the machine layout problem:

(1) overlapping of machines and
(2) violation of working area.

Because the x-axis coordinates are represented as neat clearance, overlapping illegality will never occur in this encoding scheme. The violation of the working area can be measured in the following manner: Let L_k^1 and L_k^2 be the necessary working areas required by machines which are arranged in the first row and second row, respectively, for the given chromosome v_k, and let $L_k^u = \max\{L_k^1, L_k^2\}$; thus the penalty coefficient λ_k is calculated as follows:

$$\lambda_k = \begin{cases} 0, & \text{if } L_k^u - L \leq 0 \\ L_k^u - L, & \text{otherwise} \end{cases} \tag{9.22}$$

Then the evaluation function can be expressed as follows:

$$eval(v_k) = \frac{1}{f_k + \lambda_k P}, \qquad k = 1, 2, \ldots, pop\_size \qquad (9.23)$$

where P is the big positive penalty value and f_k is the total cost. After evaluating each chromosome, we can make a roulette wheel and use it as the basic selection mechanism to reproduce the next generation based on the current generation.

9.4.7   Example

Test problem consists of six machines. The frequency matrix and cost matrix are the same as those given for single-row case in Section 9.3.2. The clearance between machines is

$$[d_{ij}] = \begin{array}{c} d_{10} \quad 1 \quad 2 \quad 3 \quad 4 \quad 5 \quad 6 \quad d_{60} \\ \begin{pmatrix} 2 & 0 & 1 & 1 & 1 & 2 & 1 & 2 \\ 2 & 1 & 0 & 1 & 1 & 1 & 1 & 2 \\ 2 & 1 & 1 & 0 & 1 & 1 & 1 & 2 \\ 2 & 1 & 1 & 1 & 0 & 3 & 1 & 2 \\ 2 & 2 & 1 & 1 & 3 & 0 & 2 & 2 \\ 2 & 1 & 1 & 1 & 1 & 2 & 0 & 2 \end{pmatrix} \end{array} \qquad (9.24)$$

The separation between two rows is 8. The width restriction of the working area is 22. The evolutionary environment is given as follows: *popsize*, 20; *maxgen*, 500; p_c, 0.4; p_m, 0.4; and r, 10. The best chromosome is described as follows:

Generation the best solution occurred: 48

Fitness: 0.5217

Machines in row 1:   2 3 1 6

Machines in row 2:   4 5

Machine positions for row 1:   4.49, 7.74, 12.49, 17.99

Machine positions for row 2:   5.00, 12.5

The layout based on the sequence is depicted in Figure 9.10, and the evolution process is shown in Figure 9.11.

9.5   FUZZY FACILITY LAYOUT PROBLEM

Most traditional formulations for the facility layout problem have assumed that the volume of flow between facility pairs is a known fixed quantity, and the facility layout is treated as a static problem over the planning horizon [259]. However, such a deterministic assumption is not suitable for modern manufacturing systems. The flows among facilities may change from period to period

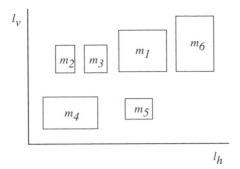

Figure 9.10. Multiple-row layout for the example problem.

due to the dynamic nature of businesses, growth, and demand fluctuation, and product mix. If some of these changes are planned and hence known *a priori*, the *dynamic layout model* of Rosenblatt can be used to handle such changes [357]. Unfortunately, changes in product mix, machine breakdowns, seasonal fluctuations, and demand are uncertain in nature. Under these circumstances, designers tend to obtain a satisfactory layout rather than an optimal layout.

There are two approaches proposed for modeling such uncertainties. One is the flexible or probabilistic approach [380] and the other is the robustness approach [252, 358]. If adopting the flexible approach, we must specify probability distribution for material flow; if adopting the robustness approach, we must provide several demand scenarios and the optimal solutions for each scenario. As we know, the optimal design of the physical layout must be resolved in the early stage of the manufacturing system design, and the product mix is highly uncertain during the design phase. Giving an exact probability distri-

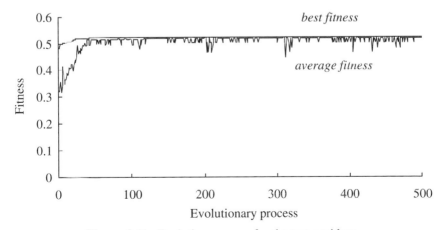

Figure 9.11. Evolution process for the test problem.

bution or giving some precise demand scenarios is as difficult as the optimal design of the facility layout itself. In this case, the estimation of material flows among facilities depends highly on the subjective judgments and/or preference of the manager or designer, which are imprecise in nature.

Cheng and Gen have proposed an alternative for handling such uncertainties; this is called the *fuzziness approach* [70]. Fuzzy set theory is a perfect means for modeling uncertainty or imprecision arising from mental phenomena. Its capability of handling uncertainty and imprecision is crucial to all the situations when the available information is not precise and the associated uncertainty cannot be ignored [336]. The subjective or preference information can usually (but not always) be represented as a convex fuzzy number [325]. Thus we can use the mathematics of fuzzy calculus to handle this uncertainty and imprecision. The practical merit of the fuzziness approach is that there is no need to force a manager or designer to give a precise product mix or a precise probability distribution.

9.5.1   Facility Layout Problem

The facility layout design problem is formulated as follows [420]: there is a set of m facilities, denoted by $\{M_i\}$, $i = 1, 2, \ldots, m$. Each facility is restricted to be rectangular and characterized by a triple (A_i, l_i, u_i), where A_i is the area of the facility, and l_i and u_i are the lower and upper bounds on the aspect ratio, respectively. The aspect ratio is defined as the ratio of the height h_i to the width w_i of the facility. This specification leads to the following relationship:

$$w_i h_i = A_i, \qquad i = 1, 2, \ldots, m \qquad (9.25)$$

$$l_i \le \frac{h_i}{w_i} \le u_i, \qquad i = 1, 2, \ldots, m \qquad (9.26)$$

The aspect ratio can also be defined as the inverse. Figure 9.12 shows the relationship between the height, width, and aspect ratio for a facility.

A facility layout for given m facilities consists of a bounding rectangle, R, partitioned by horizontal and vertical line segments into m nonoverlapping rectangular regions, denoted by $\{r_i\}$, $i = 1, 2, \ldots, m$. Each region r_i, with width x_i and height y_i, must be large enough to accommodate its facility M_i.

9.5.2   Fuzzy Interflow

Generally, the uncertainty of material flows among facilities can be represented as a convex fuzzy number, which is called as *fuzzy interflow*. Here, a trapezoidal fuzzy number (TrFN) is used to represent the fuzzy interflow [243]. A TrFN can be defined by a quadruple (a, b, c, d) and is shown in Figure 9.13.

Since a fuzzy number represents many possible real numbers that have different membership values, it is not easy to compare the final ratings to determine

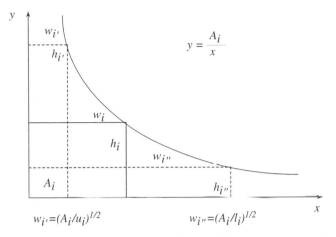

$$w_{i'} = (A_i/u_i)^{1/2} \qquad\qquad w_{i''} = (A_i/l_i)^{1/2}$$

Figure 9.12. The relationship between the height, width, and aspect ratio for a facility.

which alternatives are preferred. Many fuzzy ranking methods have been proposed to compare fuzzy numbers [56]. Lee and Li's approach suggests the use of generalized mean and standard deviation based on the probability measures of fuzzy events to rank fuzzy number [267]. When fuzzy number $\tilde{M}$ is a TrFN, the generalized mean value with a uniform density is calculated as follows:

$$m(\tilde{M}) = \frac{-a^2 b^2 + c^2 + d^2 - ab + cd}{3(-a - b + c + d)} \tag{9.27}$$

The standard deviation is defined as follows:

$$\sigma(\tilde{M}) = \left[\frac{1}{b-a} \left(\frac{b^4}{4} - \frac{ab^4}{3} + \frac{a^4}{12} \right) + \frac{1}{3}(c^3 - b^3) \right.$$
$$\left. + \frac{1}{d-c} \left(\frac{d^4}{12} - \frac{c^3 d}{3} + \frac{c^4}{4} \right) \right] \Big/ \left[\frac{1}{2}(-a - b + c + d) \right] - m(\tilde{M})^2 \tag{9.28}$$

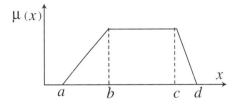

Figure 9.13. Trapezoidal fuzzy number.

9.5.3   Representation

A facility layout can be represented as a slicing structure constructed by recursively partitioning a rectangle R. Let the operations of a horizontal cut and a vertical cut be denoted by the operators + and *, respectively. Slicing structures comprising m given facilities (called *operands*) can be represented by slicing trees or Polish expressions over the alphabet set $\sum = \{1, 2, \ldots, m, *, +\}$. In a slicing tree, operators are internal nodes and operands are leaves. An example of a slicing structure is shown in Figure 9.14(a), and its slicing tree and Polish expression are shown in 9.14(b).

Polish expression is adopted as the coding scheme of chromosome in the genetic algorithms. A chromosome must have m different operands and $(m-1)$ operators where m is the number of facilities.

9.5.4   Initialization

A guided random approach is proposed to initialize genetic system. It constructs a chromosome from left to right by picking one element at a time from the set $\sum$ until a complete chromosome is formed. Because the random permutation of operands and operators may yield an illegal chromosome, we must check its legality at each random picking.

First let us define the concept of *prefix* for a chromosome. For a given chromosome with a total size of $(2m-1)$, a prefix is a partial expression (not a legal Polish expression) containing the first i elements of the chromosome with the same order as they are in the chromosome, where $i \le 2m - 1$. An example is shown in Figure 9.15.

Let Ψ denote the set of all possible prefixes for a given chromosome. For a prefix $\psi \in \Psi$, let $N(\psi)$ denote the total number of operands the prefix ψ contains and let $M(\psi)$ denote the total number of operators the prefix ψ contains.

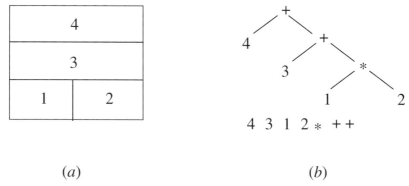

(a) (b)

Figure 9.14. A slicing structure and its representations: (a) example of a facility layout and (b) slicing tree and Polish expression.

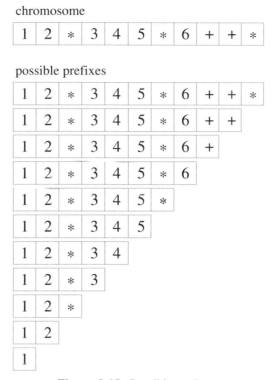

Figure 9.15. Possible prefixes.

The condition of legality for a chromosome (or a Polish expression) is given in the following proposition and corollary.

Proposition 9.1. For a given chromosome containing m operands and $(m-1)$ operators, if the equation

$$N(\psi) \geq M(\psi) + 1, \qquad \forall \psi \in \Psi \tag{9.29}$$

holds, then the chromosome is a legal Polish expression.

It is easy to verify that the chromosome given in Figure 9.15 is a legal one.

Corollary 9.1. For a given prefix ψ, if it does not meet the condition of equation (9.29), it is impossible to develop a legal Polish expression from the prefix ψ with a left-to-right generating procedures.

For example, prefix (1 2 *3 4) meets equation (9.29), so we can develop a legal Polish expression from it with the left-to-right generating procedure as (1 2 * 3 4 + +); prefix (1 2 * + +) does not meet equation (9.29), so we cannot develop a legal Polish expression from it with the left-to-right generating procedure.

Initial procedure constructs a chromosome from left to right by picking up a random element from the set $\sum = \{1, 2, \ldots, m, *, +\}$ once a time until a complete chromosome is formed. The legality check is performed at each random picking. Suppose we have a prefix (or a partially generated chromosome) ψ. According to Corollary 9.1, if $N(\psi) = M(\psi)+1$, the next element to be picked up must be an operand; otherwise, the next one can be an operand or an operator. The overall procedure is as follows:

Procedure: Guided Random Initialization

begin
   $\sum_1 \leftarrow \{*, +\}$;
   $i \leftarrow 1$;
   while $(i \leq pop\_size)$ **do**
     $\sum_0 \leftarrow \{1, 2, \ldots, m\}$;
     $\psi_i \leftarrow \phi$;
     select a random element σ_1 from $\sum_0$;
     $\sum_0 \leftarrow \sum_0 \backslash \{\sigma_1\}$;
     $\psi_i \leftarrow \psi_i \cup \{\sigma_1\}$;
     $j \leftarrow 1$;
     while $(j \leq 2m - 1)$ **do**
      if $N(\psi_i) = M(\psi_i) + 1$ **then**
       select σ_j from $\sum_0$;
      else
       select σ_j from $\sum_0 \cup \sum_1$;
      end
      $\sum_0 \leftarrow \sum_0 \backslash \{\sigma_j\}$;
      $\psi_i \leftarrow \psi_i \cup \{\sigma_j\}$;
      $j \leftarrow j + 1$;
     end
     $i \leftarrow i + 1$
   end
end

9.5.5   Crossover

The crossover operator used here was proposed by Cohoon et al. [81]. It performs the following two steps:

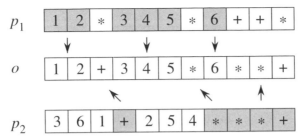

Figure 9.16. Crossover operator.

- It obtains operands from one parent.
- It obtains operators from the other parent.

Suppose we have two parents p_1 and p_2. The crossover operator copies the operands from parent p_1 into the corresponding positions in an offspring o; and then it copies operators from p_2, by making a left-to-right scan, to complete the offspring o. Crossover performed in this way can guarantee to generate legal offspring. Figure 9.16 gives an illustration of the crossover operation, and Figure 9.17 shows the layouts of parents and offspring.

9.5.6 Mutation

Random *swapping*, *inverting*, and *altering* are used as a mutation operation—that is, swapping two adjacent operands or two adjacent operators as shown in Figure 9.18(a), inverting a sequence of adjacent operators or a sequence of adjacent operands as shown in Figure 9.18(b), and altering an operator to the opposite one as shown in Figure 9.18(c). Mutation performed in this way can also guarantee to generate legal offspring. Figure 9.19 shows the layouts of parents and offspring. The overall procedure is shown below:

Procedure: Mutation

begin
    $i \leftarrow 0$;
    while ($i \leq pop\_size \times p_m$) **do**
        select a chromosome (not mutated) randomly;
        pick up a gene randomly;
        if (it is an operator) **then**
          select *swapping*, or *inverting* or *altering* mutation with equal chance;
        else
          select *swapping* or *inverting* mutation with equal chance;

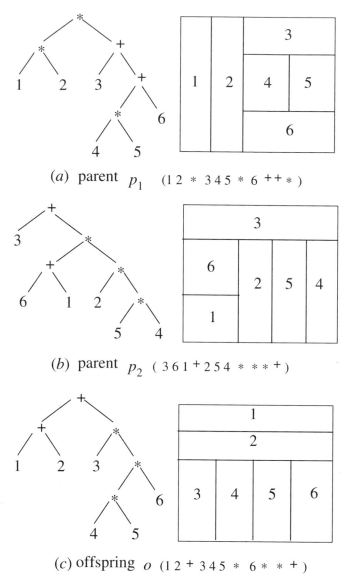

(a) parent p_1 (1 2 * 3 4 5 * 6 ++ *)

(b) parent p_2 (3 6 1 + 2 5 4 * ** +)

(c) offspring o (1 2 + 3 4 5 * 6 * * +)

Figure 9.17. Layouts after crossover.

```
    end
perform the selected mutation;
    i ← i + 1;
  end
end
```

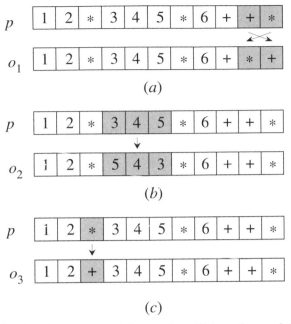

Figure 9.18. Mutation operator: (a) swapping; (b) inverting; and (c) altering.

9.5.7 Constructing a Layout from a Chromosome

Before evaluating each chromosome, we must convert it to a slicing structure. This is done in a top-down fashion starting from the root of the slicing tree and passing down to the children nodes recursively. For a given chromosome (or a slicing tree), first we separate it into two subtrees: a left subtree and a right subtree. The subtrees are also legal slicing trees. For each subtree we further separate it into two subtrees. The process is repeated until each subtree contains only one node.

For a chromosome, there is a position at which the chromosome will be separated into a left subtree and a right subtree. We call such position a *cut-point*. For a slicing tree ψ with total size $2m - 1$, let i $(1 \le i \le 2m - 2)$ be an arbitrary cut-point (or position) in the tree, let $n_i(\psi)$ denote the total number of operands contained in the right part from the cut-point to the rightmost part of the tree, and let $m_i(\psi)$ denote the total number of operators contained in the same right part of the tree. The condition for searching such a cut-point in a chromosome (or a slicing tree) is given by the following proposition:

Proposition 9.2. If a cut-point i is the first position, scanning from right to left ranging from $2m - 1$ to 1, which lets the equation

$$n_i(\psi) = m_i(\psi) \tag{9.30}$$

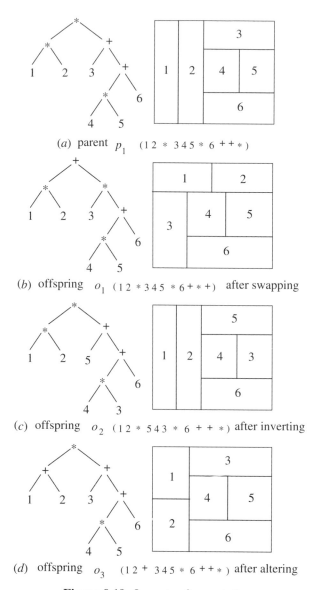

(a) parent p_1 (1 2 * 3 4 5 * 6 + + *)

(b) offspring o_1 (1 2 * 3 4 5 * 6 + * +)   after swapping

(c) offspring o_2 (1 2 * 5 4 3 * 6 + + *)   after inverting

(d) offspring o_3 (1 2 + 3 4 5 * 6 + + *)   after altering

Figure 9.19. Layouts after mutation.

holds then at cut-point i the slicing tree can be separated into (1) a left subtree containing the elements from 1 to $i - 1$ of the given slicing tree and (2) a right subtree containing elements from i to $2m - 2$. Each subtree is a legal slicing tree.

The set of possible dimensions (x_i, y_i) of region r_i accommodating M_i can be determined along with the recursively separating process using area requirement information. For a region r_i, we use x_i^l and x_i^r to denote the left and right boundary of the region, respectively, and y_i^u and y_i^b to denote the upper and bottom boundary of the region, respectively. The relation between dimensions and boundary lines for a region is calculated as follows:

$$x_i = x_i^r - x_i^l, \qquad i = 1, 2, \ldots, m \qquad (9.31)$$

$$y_i = y_i^u - y_i^b, \qquad i = 1, 2, \ldots, m \qquad (9.32)$$

Note that a simple tree is a special slicing tree with a length of 3. The overall procedure for constructing a layout from a chromosome is as follows:

Procedure: Construct (*tree*, *length*)

begin
  $i \leftarrow length$;
  while $(i > 0)$ **do**
    separate *tree* to two subtrees;
    calculate x_i^r, x_i^l, y_i^u, y_i^b;
    if the right subtree is a simple tree **then**
      calculate x_i^r, x_i^l, y_i^u, y_i^b;
      $i \leftarrow i - 2$;
      if neighbor left tree is a simple tree **then**
        calculate x_i^r, x_i^l, y_i^u, y_i^b;
        $i \leftarrow i - 2$;
      else
        construct (*lefttree*, *length*);
        $i \leftarrow i - length$;
    else
      construct (*righttree*, *length*);
      $i \leftarrow i - length$;
      if neighbor left is a simple tree **then**
        calculate x_i^r, x_i^l, y_i^u, y_i^b;
        $i \leftarrow i - 2$;
      else
        construct (*lefttree*, *length*);
        $i \leftarrow i - length$;
      end
    end
  end
  calculate x_i and y_i using x_i^r, x_i^l, y_i^u, y_i^b;
end

Figure 9.20 shows an example of converting chromosome (1 2 * 3 4 5 * 6 + * +) into a layout.

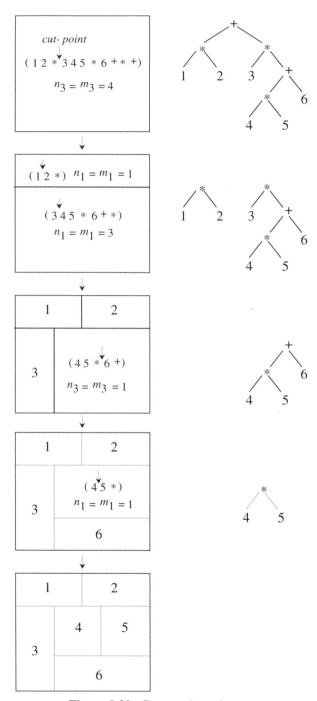

Figure 9.20. Constructing a layout.

9.5.8   Evaluation and Selection

Each layout in the enlarged population (parents and offspring) is then evaluated by an objective function of total traffic cost and assigned a fitness value. Let v_k denote the kth chromosome in the enlarged populations, let $\tilde{C}_k(v_k)$ be the total cost, let $d_{ij}(v_k)$ be a real number to denote the Manhattan distance between the centers of each pair of facilities i and j, and let $\tilde{c}_{ij}(v_k)$ be a trapezoidal fuzzy number to denote the fuzzy interflow between the facilities i and j. The overall evaluation of a particular layout is given as follows:

$$\tilde{C}_k(v_k) = \sum_{i=1}^{m} \sum_{j=1}^{m} \tilde{c}_{ij}(v_k)d_{ij}(v_k) \qquad (9.33)$$

Because the total cost is a trapezoidal fuzzy number, it is impossible to directly use it as fitness. We use Lee and Li's ranking approach to calculate the generalized mean values of the total costs for each layout. Because we treat with a minimization problem, we must convert this mean value into a fitness value, such that a fitter chromosome has a larger fitness value. The conversion can be simply done by the inverse of the mean value.

Because there are aspect ratio constraints on facilities, a layout generated in this manner may be illegal in the sense that there may exist one or more than one facility which violates the aspect ratio constraints. A penalty approach is adopted to handle such violation [382]. Let $m(\tilde{C}_k(v_k))$ denote the generalized mean value, let λ_k denote the total number of facilities which violate the aspect ratio constraints within the kth chromosome, and let P be a large penalty value. We add the penalty term into the fitness function and then we have

$$eval(v_k) = \frac{1}{m(\tilde{C}_k(v_k)) + \lambda_k P} \qquad (9.34)$$

where $eval(v_k)$ is the fitness function for the kth chromosome.

We then make a roulette wheel with the fitness values to reproduce the next generation. In our experiment, the *elitist way* way embedded within the roulette wheel selection in order to enforce preserving the best chromosome in the next generation and overcome the stochastic errors of sampling.

9.5.9   Numerical Example

Example 9.1. The test problem contains 15 facilities [59]. The area and aspect ratio of each facility is given in Table 9.1, and the fuzzy interflow represented as TrFNs is given in Table 9.2.

The evolutionary enviroment of our experiment is given as follows: The population size is 40, the ratio of crossover is 0.4, the ratio of mutation is 0.4, the penalty value is 5000, and the maximum generation is 200. A typical solution

Table 9.1. Geometric Constraints of Facilities

Facility Identification	Area	Aspect Ratio Lower Bound	Upper Bound
1	100	0.7	1
2	80	1	1
3	50	0.7	1.3
4	60	0.5	0.8
5	120	0.9	1
6	40	0.6	1
7	20	0.7	1.4
8	40	1	1
9	150	0.8	1.1
10	120	0.5	1.5
11	50	0.7	1.1
12	10	0.8	1.2
13	20	0.95	1.5
14	30	0.75	1.25
15	50	0.9	1.1

(the best chromosome obtained from one running of the genetic algorithm) is shown as follows:

(13 1 + 2 6 8 3 + +15 11 14 9 * + + * *4 7 * 5 10 + 12 * * + *)

The layout is depicted in Figure 9.21(a) and its tree representation is depicted in Figure 9.21(b). The evolutionary process is shown in Figure 9.22.

Table 9.2. Fuzzy Interflow Among Facilities

	1	2	3	4	5
1	(0, 0, 0, 0)	(0, 10, 15, 18)	(0, 0, 0, 0)	(2, 5, 8, 10)	(0, 1, 4, 12)
2	(8, 10, 15, 18)	(0, 0, 0, 0)	(0, 1, 2, 3)	(2, 3, 6, 8)	(1, 2, 5, 6)
3	(0, 0, 0, 0)	(0, 1, 2, 3)	(0, 0, 0, 0)	(8, 10, 14, 16)	(0, 2, 5, 6)
4	(2, 5, 8, 10)	(2, 3, 6, 8)	(8, 10, 14, 16)	(0, 0, 0, 0)	(0, 1, 4, 6)
5	(0, 1, 4, 12)	(1, 2, 5, 6)	(0, 2, 5, 6)	(0, 1, 4, 6)	(0, 0, 0, 0)
6	(0, 0, 0, 0)	(0, 2, 4, 5)	(0, 0, 0, 0)	(0, 1, 3, 5)	(2, 3, 5, 8)
7	(0, 1, 2, 3)	(0, 2, 3, 6)	(1, 2, 3, 5)	(2, 5, 6, 9)	(4, 5, 7, 9)
8	(1, 2, 5, 10)	(2, 3, 5, 7)	(3, 5, 6, 7)	(0, 0, 0, 0)	(3, 5, 7, 9)
9	(1, 2, 3, 4)	(1, 2, 3, 5)	(2, 4, 6, 8)	(0, 0, 0, 0)	(0, 5, 8, 10)
10	(1, 2, 4, 5)	(0, 0, 0, 0)	(1, 5, 8, 10)	(0, 2, 4, 6)	(0, 1, 4, 6)
11	(0, 2, 3, 6)	(0, 2, 7, 8)	(0, 2, 4, 5)	(0, 1, 2, 3)	(0, 0, 0, 0)
12	(0, 0, 0, 0)	(0, 0, 0, 0)	(1, 2, 3, 6)	(0, 0, 0, 0)	(2, 3, 6, 7)
13	(2, 4, 5, 9)	(6, 10, 12, 15)	(3, 5, 7, 9)	(1, 2, 5, 8)	(0, 0, 0, 0)
14	(0, 0, 0, 0)	(4, 5, 8, 9)	(0, 5, 8, 9)	(2, 5, 8, 9)	(3, 5, 8, 10)
15	(0, 0, 0, 0)	(0, 0, 0, 0)	(3, 5, 10, 12)	(0, 0, 0, 0)	(2, 5, 9, 10)

Table 9.2. *Continued*

	6	7	8	9	10
1	(0, 0, 0, 0)	(0, 1, 2, 3)	(1, 2, 5, 10)	(1, 2, 3, 4)	(1, 2, 4, 5)
2	(0, 2, 4, 5)	(0, 2, 3, 6)	(2, 3, 5, 7)	(1, 2, 3, 5)	(0, 0, 0, 0)
3	(0, 0, 0, 0)	(1, 2, 3, 5)	(3, 5, 6, 7)	(2, 4, 6, 8)	(1, 5, 8, 10)
4	(0, 1, 3, 5)	(2, 5, 6, 9)	(0, 0, 0, 0)	(0, 0, 0, 0)	(0, 2, 4, 6)
5	(0, 1, 3, 5)	(4, 5, 7, 9)	(3, 5, 7, 9)	(0, 5, 8, 10)	(0, 1, 4, 6)
6	(0, 0, 0, 0)	(1, 2, 3, 4)	(0, 2, 3, 4)	(0, 1, 2, 4)	(4, 5, 6, 8)
7	(1, 2, 3, 4)	(0, 0, 0, 0)	(4, 6, 8, 10)	(0, 0, 0, 0)	(0, 1, 2, 3)
8	(0, 2, 3, 4)	(4, 6, 8, 10)	(0, 0, 0, 0)	(3, 5, 7, 8)	(0, 2, 4, 6)
9	(0, 1, 2, 44)	(0, 0, 0, 0)	(3, 5, 7, 8)	(0, 0, 0, 0)	(0, 0, 0, 0)
10	(4, 5, 6, 8)	(0, 1, 2, 3)	(0, 2, 4, 6)	(0, 0, 0, 0)	(0, 0, 0, 0)
11	(0, 0, 0, 0)	(4, 5, 8, 10)	(7, 10, 15, 18)	(2, 10, 12, 15)	(0, 0, 0, 0)
12	(0, 0, 0, 0)	(3, 5, 7, 9)	(0, 0, 0, 0)	(1, 5, 8, 9)	(2, 4, 6, 9)
13	(1, 2, 4, 5)	(2, 5, 6, 7)	(2, 5, 9, 10)	(5, 10, 15, 18)	(0, 0, 0, 0)
14	(3, 5, 8, 9)	(0, 1, 3, 4)	(0, 0, 0, 0)	(0, 0, 0, 0)	(0, 0, 0, 0)
15	(8, 10, 12, 15)	(0, 0, 0, 0)	(0, 0, 0, 0)	(0, 2, 7, 8)	(4, 5, 8, 10)

	11	12	13	14	15
1	(0, 2, 3, 6)	(0, 0, 0, 0)	(2, 4, 5, 9)	(0, 0, 0, 0)	(0, 0, 0, 0)
2	(0, 2, 7, 8)	(0, 0, 0, 0)	(6, 10, 12, 15)	(4, 5, 8, 9)	(0, 0, 0, 0)
3	(0, 2, 4, 5)	(1, 2, 3, 6)	(3, 5, 7, 9)	(0, 5, 8, 9)	(3, 5, 10, 12)
4	(0, 1, 2, 3)	(0, 0, 0, 0)	(1, 2, 5, 8)	(2, 5, 8, 9)	(0, 0, 0, 0)
5	(0, 0, 0, 0)	(2, 3, 6, 7)	(0, 0, 0, 0)	(3, 5, 8, 10)	(2, 5, 9, 10)
6	(0, 0, 0, 0)	(0, 0, 0, 0)	(1, 2, 4, 5)	(3, 2, 8, 9)	(8, 10, 12, 15)
7	(4, 5, 8, 9)	(3, 5, 7, 9)	(2, 5, 6, 7)	(0, 1, 3, 4)	(0, 0, 0, 0)
8	(7, 10, 15, 18)	(0, 0, 0, 0)	(2, 5, 9, 10)	(0, 0, 0, 0)	(0, 0, 0, 0)
9	(2, 10, 12, 15)	(1, 5, 8, 9)	(5, 10, 15, 18)	(0, 0, 0, 0)	(0, 2, 7, 8)
10	(0, 0, 0, 0)	(2, 4, 6, 9)	(0, 0, 0, 0)	(0, 0, 0, 0)	(4, 5, 8, 10)
11	(0, 0, 0, 0)	(2, 5, 8, 10)	(0, 0, 0, 0)	(3, 5, 8, 12)	(0, 0, 0, 0)
12	(2, 5, 8, 10)	(0, 0, 0, 0)	(2, 3, 5, 6)	(1, 3, 5, 7)	(0, 0, 0, 0)
13	(0, 0, 0, 0)	(2, 3, 5, 6)	(0, 0, 0, 0)	(6, 10, 14, 17)	(0, 2, 5, 8)
14	(3, 5, 8, 12)	(1, 3, 5, 7)	(6, 10, 14, 17)	(0, 0, 0, 0)	(3, 4, 6, 7)
15	(0, 0, 0, 0)	(0, 0, 0, 0)	(0, 2, 5, 8)	(3, 4, 6, 7)	(0, 0, 0, 0)

From the evolutionary process, we can see that before the 10th generation, all chromosomes are illegal in the sense of violating the aspect ratios. The penalty function guides the genetic search to a feasible region. After entering the feasible region, the penalty loses its function and natural selection forces chromosomes to evolve gradually. In the 63rd generation the best solution is found.

The fuzzy cost measure of the best layout is $\tilde{C} = (2946.91, 5841.40, 9561.81, 12613.53)$. We can use possibility theory and fuzzy integrals to interpret the fuzzy performance. The possibility of a member of the support set of a fuzzy number taking on a certain exact value is defined to be the degree of ease which that member can take on that certain value [429]. In most cases, the possibility is exactly the membership function value of that member. For example, the possibility that the layout will have total cost of 4394 units is 0.5, as illustrated in Figure 9.23. The concept of possibility may be difficult to understand in an

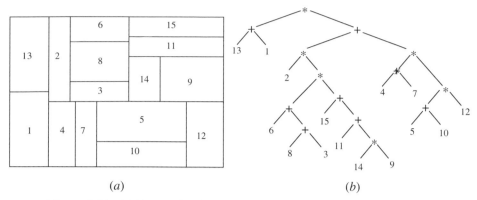

(a)   (b)

Figure 9.21. (a) Layout for the best chromosome. (b) Tree representation.

absolute sense, but it can be very useful in a relative context. For example, when comparing one support set value to another, the manager can identify which value is more possible based on their possibilities. Genetic algorithms can provide a set of superior solutions to managers for their judgment.

Perhaps of more use to the managers is a proportional measure of the distribution to the density of the membership function. It is calculated by taking a fuzzy integral over the interest area and dividing it by the integral of fuzzy number over a support set shown as follows:

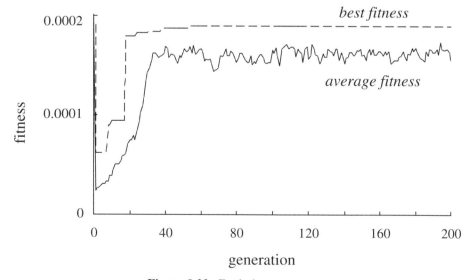

Figure 9.22. Evolutionary process.

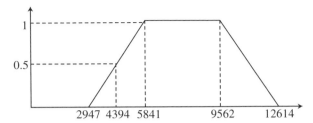

Figure 9.23. Possibility of $\tilde{C}$ = 4394.

$$\tilde{I}(x \le \beta) = \frac{\int_a^\beta \mu(x)\,dx}{\int_S \mu(x)\,dx} \tag{9.35}$$

For example, the fuzzy integral $\tilde{I}$ of 4394 is 0.13 using equation (9.35). The manager can interpret this result as the fuzzy expectation of the layout yielding total cost less than or equal to 4394 is 0.13. The fuzzy integral $\tilde{I}$ of 9562 is 0.91, so the fuzzy expectation of the layout yielding total cost less than or equal to 9562 is 0.91.

Fuzzy interflow is characterized by a quadruple (a, b, c, d), where a represents the optimistic estimation on material flow among facilities (the best case), d the pessimistic estimation (the worst case), b the average estimation (the near-best case), and c another average estimation (the near-worst case). According to the four cases, we can make four equivalent nonfuzzy problems using the same fuzzy data and solved by the proposed algorithms (with slight modification to make it suitable for the nonfuzzy case). We use the same parameter setting for each running of the four problems, and the best solutions for each case are given below:

Best case: (5 14 3 * 9 * *15 10 + 12 13 * 1 * *6 7 8 11 + * * 4 * 2 * + +)
Near best: (6 7 + 8 * 9 1 2 + * * 5 15 11 13 * 10 3 * 14 * 4 12 * * + * * +)
Near worst: (9 15 * 14 * 4 13 1 * * * 10 12 6 * + 11 3 8 7 + + 5 * 2 * * * +)
Worst case: (7 3 * 14 * 4 * 11 6 * + 2 * 12 * 9 13 8 * * + 5 * 10 15 1 + + *)

Figure 9.24 depicts the layout and its tree representation for the best case, Figure 9.25 is for the near-best case, Figure 9.26 is for the near-worst case, and Figure 9.27 is for the worst case.

For each layout, we calculate the total costs under four kinds of material flow (best, near-best, near-worst, and worst) and the comparison with fuzzy solution is summarized in Table 9.3.

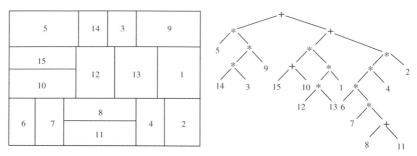

Figure 9.24. Layout and tree representation for the best case.

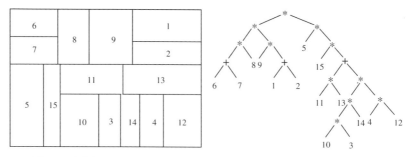

Figure 9.25. Layout and tree representation for the near-best case.

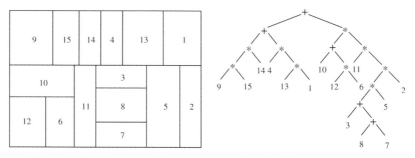

Figure 9.26. Layout and tree representation for the near-worst case.

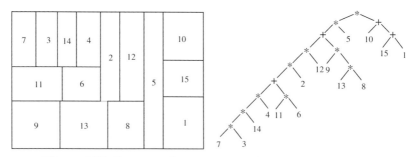

Figure 9.27. Layout and tree representation for the worst case.

Table 9.3. Comparative Results

Solutions	a	b	c	d
Fuzzy	2947	5841	9562	12614
Best-case	2795	6068	9953	13215
Near-best case	2895	5083	9625	12922
Near-worst case	2971	5869	9581	12685
Worst case	3012	6257	10024	13134

From the comparative results we know that for an equivalent nonfuzzy case we can obtain a layout with a minimum cost under the considered material flows, but for the other cases this layout will yield large costs than the solution obtained by the fuzzy approach. That is, the nonfuzzy approach can obtain a layout suitable for its considered case while the fuzzy approach can get a reasonable solution suitable for all cases ranging from the best case to the worst case. For example, the total cost of an equivalent nonfuzzy problem for the best case of material flows is 2794, which is smaller than the one in the fuzzy case (2947). Using the same layout, we calculate the costs for three other cases: the near-best case, the near-worst case, and the worst case. The costs (6068, 9953, 13215) are larger than those in the fuzzy case. The relative error with respect to the solution obtained by the fuzzy approach is given in Figure 9.28.

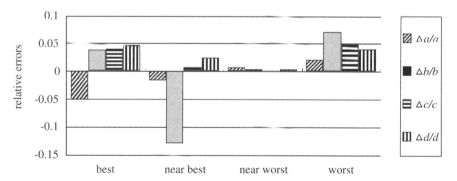

Figure 9.28. Relative error with respect to the fuzzy solution.

10

SELECTED TOPICS IN ENGINEERING DESIGN

10.1 RESOURCE-CONSTRAINED PROJECT SCHEDULING PROBLEMS

Since the pioneering work of Kelley, the resource-constrained project scheduling problem has been addressed by a great number of researchers [245]. Generally, scheduling decisions are subject to both precedence constraints and resource constraints. Many research works have dealt with a variety of situations in which one or both of these types of constraints are relaxed, or at least simplified. Because the resource-constrained project scheduling problem must deal with both precedence and resource constraints, it is much harder than other scheduling problems [21, 386].

The general resource-constrained project scheduling problem can be stated as follows: The problem has a set of interrelated activities (precedence relations) where each activity can be performed in one of several modes (ways) and each mode is characterized by a known duration and given resource requirements. A solution is to make a schedule (when each activity should begin and which resource/duration mode should be adopted) with respect to precedence and resource constraints so as to optimize some managerial objectives [42].

The earliest attempts were made to find an exact optimal solution to the problem by using standard solution techniques of mathematical programming such as integer programming [75, 341, 388], bounded enumeration [97], branch-and-bound [333], and implicit enumeration [394]. Using the results of Blazewicz [40], it is easy to show that the resource-constrained project scheduling problem is NP-hard. For large projects, the size of the problem may render optimal methods computationally impracticable. In such cases, the problem is most amenable to heuristic problem solving, using fairly simple scheduling rules capable of producing reasonably good suboptimal schedules. Over the past 30 years, a larger number of heuristic algorithms have been developed and tested; we refer to Alvarez-Valdés for a survey and computational comparison [7].

Most of the heuristic methods known so far can be viewed as priority dispatching rules which assign activity priorities in making sequencing decisions for resolution of resource conflicts according to either temporally related heuristic rules or resource-related heuristic rules. Patterson et al. presented a backtracking algorithm which is more flexible, considering various objective function forms and activities which can be accomplished in various modes [334]. Bell and Han developed a two-phase heuristic that performs better than all other heuristics [31]. Sampson and Weiss used the local search techniques to solve the generalized resource-constrained project scheduling problem and show that the local search heuristic is superior to existing heuristic techniques [365]. Jeffcoat and Bulfin reported their works on resource-constrained scheduling with simulated annealing in which the ready times and deadlines are considered [235]. Drexl and Gruenewald considered the nonpreemptive multiple mode resource-constrained project scheduling problem and presented a stochastic scheduling method [113].

10.1.1   Problem Statement

We consider the usual version of the resource-constrained project scheduling problem as follows:

- A single project consists of a number of activities with known duration.
- Start time of each activity is dependent upon the completion of some other activities (precedence constraints).
- Resource consumptions are constant over the scheduling horizon.
- Resources are available in limited quantities but renewable from period to period.
- There is no substitution between resources.
- Activities cannot be interrupted.
- There is only one execution mode for each activity.
- The managerial objective is to minimize the project duration.

The problem can be stated mathematically as follows:

$$\min \quad t_n \tag{10.1}$$

$$\text{s.t.} \quad t_j - t_i \geq d_i, \qquad \forall j \in S_i \tag{10.2}$$

$$\sum_{t_i \in A_{t_i}} r_{ik} \leq b_k, \qquad k = 1, 2, \ldots, m \tag{10.3}$$

$$t_i \geq 0, \qquad i = 1, 2, \ldots, n \tag{10.4}$$

where

t_i is the starting time of activity i

d_i is the duration (processing time) of activity i

S_i is the set of successors of activity i

r_{ik} is the amount of resource k required by activity i

b_k is the total availability of resource k

A_{t_i} is the set of activities in process at time t_i

m is the number of different resource types

Activities 1 and n are dummy activities which mark the beginning and the ending of the project. The objective is minimizing total project duration. Constraint (10.2) ensures that none of the precedence constraints is violated. Constraint (10.3) ensures that in any period, the amount of resource k used by all activities does not exceed the limited quantity of resource k. Associated with the problem we can define a *precedence graph* $G = (V, H)$. H is the set of precedence constraints and V is the set of activities of the project. G is directed and acyclic. A simple network representation is given in Figure 10.1.

10.1.2 Hybrid Genetic Algorithms

Cheng and Gen have proposed a hybride genetic algorithm to the resource-constrained project scheduling problem [62, 64]. A genetic algorithm is used here as a meta-strategy while local search techniques and problem-specific heuristics are incorporated within it in order to improve the performance. There are two essential issues to be dealt with for the problem:

- Determine the order of activities without violating the precedence constraints.
- Determine the earliest start time for each activity according to available resources.

The basic idea of the proposed approach is to use genetic algorithms to find an appropriate order of activities and problem-specific heuristics to determine the earliest start times.

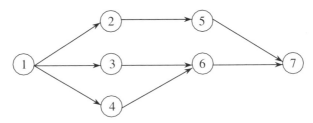

Figure 10.1. Network representation of project.

Representation. Many representational schemes have been proposed for scheduling problems. To some extent, all of these representations can be viewed as the extension of *permutation* representation [71].

Generally, the representation schemes proposed for a scheduling problem can be classified into *direct representation* and *indirect representation* [15]. Almost all previous genetic algorithm studies on scheduling problems have used an indirect representation; that is, genetic algorithms work on a population of encoded solutions, and a transition from chromosome representation to a legal production schedule has to be performed by a schedule builder prior to evaluation. The examples are Nakano's *binary representation* [313], Syswerda's *list of orders representation* [392], Bagchi's *list of order/process plan/resource representation* [15], and Davis' *preference list representation* [100].

In a direct problem representation, the schedule itself is used as a chromosome, and no decoding procedure is therefore necessary, but it takes a great deal of effort to develop complicated genetic operators. The examples are Kanet's *list of order/machine/start time representation* [241] and Bruns' *complete schedule representation* [50].

The resource-constrained project scheduling problem can essentially be viewed as a kind of ordering problem subject to some constraints because once the order of activities is given, the possible earliest start time for each activity can be easily determined according to available resources. What we are concerned about is how to find out an appropriate order of activities with genetic algorithms. In the indirect representation, we can use a list of ordered activities to encode the problem into a chromosome. In this way, a chromosome represents one of the possible permutations of activities. Let v_k represent the kth chromosome in the current population and let i_j be the activity in the j position of the chromosome. The chromosome can be represented as follows:

$$v_k = [i_1 \, i_2 \cdots i_j \cdots i_n]$$

The search space of genetic algorithms is the space of all possible permutations of activities, but the feasible space is just a part of the whole space because of the binding of precedence constraints.

Initialization. As noted by Davis and Steenstrup, the initialization process can be executed with either a randomly created population or a well adapted population [102]. For the resource-constrained project scheduling problem, because of the existence of a precedence constraint, the random way to initialize the genetic system will yield a considerable number of infeasible schedules.

A *single-pass* heuristic procedure was proposed to initialize population. It considers as yet unscheduled activities one at a time in a fixed order. A schedule is filled *from left to right*. At each stage, it maintains a set of schedulable activities and assigns priority among competing activities on a purely random basis. The process is repeated until all activities have been scheduled. It can guarantee to yield a precedence feasible schedule. Let v_k^t denote a partial chromosome of v_k containing the first t activities, Q_t denote the set of schedulable activities at stage t, corresponding to a given v_k^t, and P_i denote the set of all direct predecessors of activity i. The set of schedulable activities Q_t corresponding to a given v_k^t is defined by

$$Q_t = \{ j | P_j \subset v_k^t \}$$

which contains the activities with their tail nodes in the partial schedule v_k^t and is the collection of all competitive activities. The procedure to generate initial schedules is as follows:

Procedure: Initial Schedule Generation

begin
  $k \leftarrow 1$;
  while $(k \leq pop\_size)$ **do**
    $t \leftarrow 1$;
    $v_k \leftarrow \{1\}$;
    while $(t \leq n)$ **do**
      update the set of schedulable activities Q_t;
      select an activity j from Q_t randomly;
      $v_k \leftarrow v_k \cup \{j\}$;
      $t \leftarrow t + 1$;
    end
    $k \leftarrow k + 1$;
  end
end

Evaluation. Usually two main steps are included in evaluation:

- Evaluate the original objective function.
- Convert the objective function's value to fitness.

As we know, the objective of the problem is to minimize project duration. First, we must calculate the possible earliest start times of activities for all chromosome. Because each chromosome specifies one of the possible orders of activities, the start times for a given chromosome can be calculated from left to right according to the given order of activities and the available resources. Let σ_i denote the earliest time at which activity i could be started, ϕ_i denote the earliest time at which activity i could be completed, d_i denote the duration of activity i, and b_{kt} denote the current available resource k at stage t. For a given chromosome v_k and an activity j in an arbitrary position, the possible start time

σ_j is determined by

$$\sigma_j = \max \{t \,|\, t > \sigma_{j\,\min} \text{ and } b_{kt}, b_{k(t+1)}, b_{k(t+2)}, \dots, b_{k(t+d_j)} \geq r_{jk}, k = 1, 2, \dots, m\}$$

where

$$\sigma_{j\,\min} = \max \{\phi_i \,|\, i \in P_j\}$$

The finishing time ϕ_j is simply $\sigma_j + d_j$.

Since we deal with a minimization problem, we have to convert the original objective function to a fitness function in order to ensure that the fitter individual has a larger fitness value. This can be done by the following transformation:

$$g(\boldsymbol{v}_k) = \frac{f_{\max} - f(\boldsymbol{v}_k)}{f_{\max} - f_{\min}}$$

where $g(\boldsymbol{v}_k)$ is the fitness function, $f(\boldsymbol{v}_k)$ is the original objective function (i.e., the project duration), $\boldsymbol{v}_k$ is the kth chromosome in the current generation, and $f_{\max}$ and $f_{\min}$ are the maximum and minimum values of the original objective function in current generation.

A γ-augmented transformation was applied to the fitness function as follows:

$$g(\boldsymbol{v}_k) = \frac{f_{\max} - f(\boldsymbol{v}_k) + \gamma}{f_{\max} - f_{\min} + \gamma} \tag{10.5}$$

where γ is a positive real number which is usually restricted within the open interval $(0,1)$. The purpose of using such transformation is twofold: (1) to prevent equation (10.5) from zero division and (2) to make it possible to adjust the selection behavior from fitness-proportional selection to pure random selection.

Genetic Operators. Crossover and mutation are two main genetic operators for the genetic algorithm. In conventional implementation, crossover acts as the main genetic operator while the mutation just acts as a supplementary means to introduce the smallest possible change in the search space. As we know, a desired genetic search is one that can strike a balance between exploitation and exploration of the search space [105]. Cheng and Gen adopted a new approach to guide the design of genetic operators. The ideal implementation of two operators is that one is designed to perform a widespread search to try to explore the area beyond local optima whereas the other is designed to perform an intensive search to try to exploit the best solution for possible improvement. If we try to design an intensive genetic operator, we have to augment it with domain-specific knowledge. For many combinatorial optimization problems, either-or genetic operators (crossover or mutation) can be designed with the intensive strategy according to the nature of the problem at hand. With this approach, mutation plays the same important role as crossover does. In their implementation, crossover was to perform a widespread search whereas mutation was to perform an intensive search.

Crossover. A number of recombination operators have been investigated for ordering or permutation representation. These crossover operators cannot be directly applied to the problem because of the existence of precedence constraint among activities. The proposed crossover operator consists of two main steps: First it performs an ordinary PMX (partially mapped crossover) operation [174] to rearrange the order of activities, and then it performs a repairing procedure to solve precedence conflict. Because a PMX operator may scramble the precedence relationship of activities, it is necessary to solve the precedence conflict in order to ensure that the offspring are the precedence feasible schedules. The scheme of the crossover operator is given as follows:

Procedure: Crossover

begin
  $i \leftarrow 0$;
  while ($i \leq pop\_size \times p_c$) **do**
    select two random individuals;
    generate an offspring with ordinary PMX;
    solve precedence conflicts in offspring;
    $i \leftarrow i + 1$;
  end
end

Let S_j denote the set of all direct successors of activity j. For a given offspring $\boldsymbol{v}_k = [i_1, i_2, \ldots, i_j, \ldots, i_n]$, let $\boldsymbol{v}_k^j$ denote the partial schedule containing the first j activities; that is, $\boldsymbol{v}_k^j = [i_2, i_2, \ldots, i_j]$ ($j \geq n$). The repairing procedure is as follows:

Procedure: Repairing

begin
  $j \leftarrow 2$;
  while ($j \leq n - 1$) **do**
  $j \leftarrow j + 1$;
  check each element in $\boldsymbol{v}_k^{j-1}$ whether or not it belongs to S_{i_j};
  if (there are m elements which belong to S_{i_j}) **then**
    put them in a random positions between $[j + 1, n - 1]$;
    $j \leftarrow j - m$;
  end
  end
end

Mutation. A mutation operator was designed with an intensive search strategy. It is incorporated with a neighborhood technique to try to find an improved offspring. A set of schedules transformable from a schedule x by exchanging no more than λ genes are regarded as the neighborhood of schedule x. A schedule x is said to be λ-*optimum* if x is better than any other solutions in the neighborhood in this sense. The scheme of the mutation operator is depicted below.

Procedure: Mutation

begin
  $i \leftarrow 0$;
  while $(i \leq pop\_size \times p_m)$ **do**
    select an unmutated chromosome randomly;
    pick up a random subschedule with λ genes in the chromosome;
    generate neighbor schedules by pairwise interchange;
    solve precedence conflicts for illegal neighbors;
    evaluate all neighbor schedules;
    select the best neighbor as offspring;
    $i \leftarrow i + 1$;
  end
end

Let us see an example. Assume that the subschedule with length of $\lambda = 3$ randomly selected is (3 4 5). The possible permutation of the genes in the subschedule together with remaining genes of the chromosome form the neighbor chromosomes (suppose all permutations are legal) as shown in Figure 10.2.

Selection. The *roulette wheel* approach was adopted as the selection procedure which is one of the fitness-proportional selections. The *elitist way* was combined

parent chromosome

1	2	3	4	5	6	7	8	9

neighbor chromosome

1	2	3	5	4	6	7	8	9
1	2	5	3	4	6	7	8	9
1	2	5	4	3	6	7	8	9
1	2	4	5	3	6	7	8	9
1	2	4	3	5	6	7	8	9

Figure 10.2. Mutation: neighbor chromosomes.

with this approach in order to preserve the best chromosome in the next generation and overcome the stochastic errors of sampling. With the elitist selection, if the best individual in the previous generation is not reproduced to the new generation, remove one individual randomly from the new population and add the best one of the previous population to the new population.

10.1.3   Examples

Experiments were conducted with two standard test problems in references [96 and 98]. The proposed algorithm has been coded with C and run on NEC EWS 4800/260 Workstation.

Example 10.1. The problem given in reference [98] has 27 activities, including two dummy activities; each activity has fixed multiple unit requirements of three different resources; and resource availabilities are constant over the project duration. The problem is shown in Figure 10.3.

The evolutionary environment of our experiment is as follows: The population size is 40, the ratio of crossover is 0.3, and the ratio of mutation is 0.3. The best chromosome in the 12th generation is as follows:

generation for best chromosome: 12

* objective function *

f^*: 64

* best schedule *

$v^* =$ [1 2 4 3 7 8 6 10 11 5 16 20 12 9 14 15 17 13 24 22 18 19
        23 21 25 26 27]

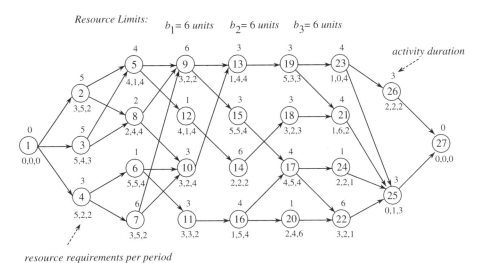

Figure 10.3. First test problem. (From Davis and Patterson [98].)

* starting time *

(0 0 5 8 13 15 21 25 26 27 27 33 36 36 39 43 47 48 48 49 52 55
57 57 61 61 64)

* finishing time *

(0 5 8 13 15 21 25 26 27 33 33 36 39 39 43 47 48 49 54 52 55
57 61 61 64 64 64)

Figure 10.4 gives the comparison with other known heuristic results. The heuristics are *minimum late finish time* (LFT), *greatest resource utilization* (GRU), *shortest imminent operation* (SIO), *minimum job slack* (MINSLK), *resource scheduling method* (RSM), *select jobs randomly* (RAN), *most jobs possible* (MJP), and *Greatest Resource Demand* (GRD). From these results we can see that the proposed algorithm is superior to all existing heuristic techniques.

Figure 10.5 shows the evolutionary process for the first test problem. From these results we can see that the proposed algorithm can find the known optimum very rapidly.

Example 10.2. The problem given in reference [96] has 27 activities, including two dummy activities; each activity has fixed multiple unit requirements of three different resources; and resource availabilities are constant over the project duration. The problem is shown in Figure 10.6.

The evolutionary environment for the second test problem is the same as that of the first one. The best chromosome in the 29th generation is

generation for best chromosome: 29

* objective function *

f^*: 35

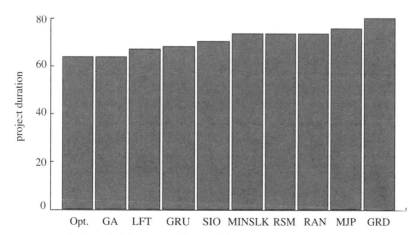

Figure 10.4. Comparison with other heuristic results.

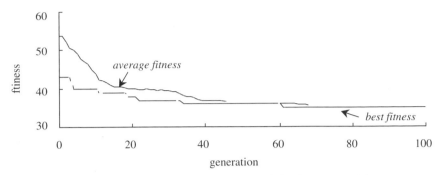

Figure 10.5. Evolutionary process of the first test problem.

* best schedule *

$v^* = [1 \ \ 3 \ \ 2 \ \ 4 \ \ 7 \ \ 8 \ \ 5 \ \ 12 \ \ 6 \ \ 10 \ \ 11 \ \ 14 \ \ 9 \ \ 15 \ \ 16 \ \ 13 \ \ 18 \ \ 19 \ \ 20 \ \ 24 \ \ 23 \ \ 17$
$22 \ \ 21 \ \ 25 \ \ 26 \ \ 27]$

* starting time *

(0 0 0 3 4 6 6 10 11 14 14 18 18 18 20 21 22 22 24 24 27 27
27 30 30 32 35)

* finishing time *

(0 4 3 6 10 11 10 18 14 16 18 20 20 21 22 24 24 27 26 27 30
30 32 32 33 35 35)

Figure 10.7 shows the evolution process for the second test problem. The
solution distribution over 100 runs are given Figure 10.8.

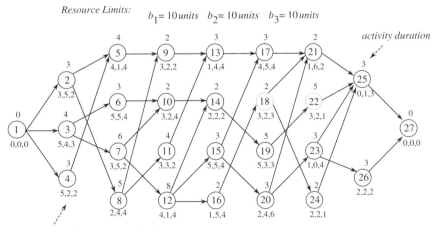

Figure 10.6. Second test problem. (From Davis [96].)

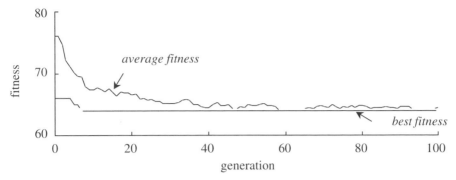

Figure 10.7. Evolutionary process of the second test problem.

10.2 FUZZY VEHICLE ROUTING AND SCHEDULING PROBLEM

The vehicle routing problem (VRP) involves the design of a set of minimum cost routes, originating and terminating at a central deport, for a fleet of vehicles which services a set of customers with known demands. Each customer is serviced exactly once and, furthermore, all customers must be assigned to vehicles without exceeding vehicle capacities [43].

The vehicle routing problem with time window constraints (VRPTW) is an important generalization of the VRP. In the VRPTW, each customer has a time window (or time interval formed with deadline and earliest time constraints) in which he must be served. When time windows are present, the problem involves a combination of routing and scheduling components. Routes must be designed in order to minimize total transportation cost, but, at the same time, scheduling must be performed in order to ensure time feasibility.

In many realistic applications, the concept of time window does not model the customers' preference very well. Even though customers are asked to provide a fixed time window for service, they really hope to be served at a desired

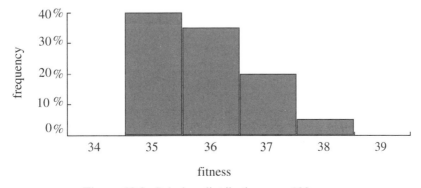

Figure 10.8. Solution distribution over 100 runs.

time if possible. We call such a desired time the due time. In such a case, the preference information of customers can be naturally represented as a convex fuzzy number with respect to the satisfaction for service time. Cheng and Gen have suggested that we use the concept of fuzzy due time to replace the concept of time window because it can describe customers preference better than fixed time window. When seeking a scheduling under the consideration of fuzzy due time, we are interested in not only the feasibility of service time for all customers as the conventional one does, but also the reasonability of service time in the sense that we enforce the service time for each customer to approach his due time as close as possible. Cheng and Gen formulated the fuzzy vehicle routing problem (FVRP) and proposed a genetic algorithm to solve it [65, 69].

10.2.1 Problem Formulation

Fuzzy vehicle routing problem is formulated based on the concept of fuzzy due time, where the membership function of fuzzy due time corresponds to the grade of satisfaction of a service time. The objectives considered here are to minimize the fleet size of vehicles, maximize the average grade of satisfaction over customers, and minimize the total travel distance and total waiting time for vehicles [63].

Fuzzy Due Time. Customers' preferences for service can be classified into two types:

- The tolerable interval of service time
- The desirable time for service

The tolerable interval of service time for customer i is described by $[e_i, l_i]$, where e_i is the earliest time and l_i is the latest time. The desirable service time is described by the due time u_i and generally, $e_i \leq u_i \leq l_i$.

Conventional approaches just consider the tolerances of customers and represent it as a time window, which is shown in Figure 10.9. The grade of satisfaction of service is 1 (full satisfaction) if the service time falls within the

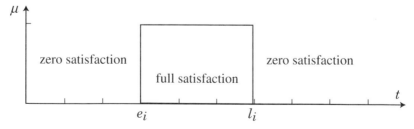

Figure 10.9. Time window.

range of a time window; otherwise, the grade of satisfaction of service is 0 (no satisfaction). That is, the grade of satisfaction of service $\mu_i(t)$ can be defined for any service time t $(t > 0)$ as

$$\mu_i(t) = \begin{cases} 1, & \text{if } e_i \leq t \leq l_i \\ 0, & \text{otherwise} \end{cases} \tag{10.6}$$

As we can see, if we try to manipulate both kinds of customers' preferences simultaneously, it is natural to represent the customers' preferences as a triangular fuzzy number (TFN) with respect to the grade of satisfaction for service time, which is defined by the triple (e_i, u_i, l_i). For example, if a customer is served at his desired time, the grade of satisfaction for him is 1 (full satisfaction); otherwise, the grade of satisfaction gradually decreases along with the increase of difference between the service time and his desired time. The grade of satisfaction will be 0 (no satisfaction) if the service time falls outside the time interval. We call such a TFN as a fuzzy due-time associated with customer i, which is shown in Figure 10.10.

Let us denote the membership function of the fuzzy due time of customer i by $\mu_i(t_i)$, which represents the grade of satisfaction when service time is t_i. The membership function is defined as follows:

$$\mu_i(t_i) = \begin{cases} 0, & t_i < e_i \\ \dfrac{t_i - e_i}{u_i - e_i}, & e_i \leq t_i \leq u_i \\ \dfrac{l_i - t_i}{l_i - u_i}, & u_i \leq t_i \leq l_i \\ 0, & t_i > l_i \end{cases} \tag{10.7}$$

Figure 10.11 shows four typical customers' preferences with the same desired service time but different tolerances for earliness and lateness—that is,

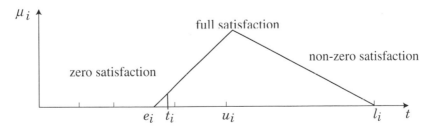

Figure 10.10. Fuzzy due time.

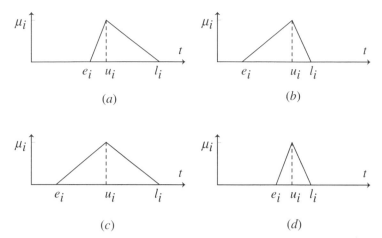

Figure 10.11. Typical customers' preferences: (*a*) tolerable for lateness but less tolerable for earliness; (*b*) tolerable for earliness but less tolerable for lateness; (*c*) tolerable for both earliness and lateness; and (*d*) less tolerable for both earliness and lateness.

- Tolerable for lateness but less tolerable for earliness
- Tolerable for earliness but less tolerable for lateness
- Tolerable for both earliness and lateness
- Less tolerable for both earliness and lateness

If vehicle k travels directly from customer i to j and arrives too early at j, it will wait. The waiting time for the vehicle is determined by the following equations:

$$w_j = t_j - (t_i + r_{ij}) \tag{10.8}$$

where r_{ij} is the travel time from customer i to customer j and w_j is the waiting time at customer j for the vehicle. The relationship is shown in Figure 10.12.

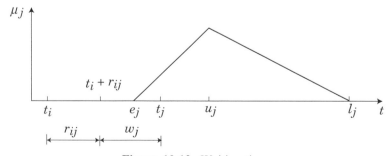

Figure 10.12. Waiting time.

Model. To provide a precise statement of this problem, we first introduce some notations and then specify a multiple objective optimization formulation.

Constants

n is the number of customers

m is the number of vehicles

c_k is the capacity of vehicle k

a_i is the size of the delivery to customer i

d_{ij} is the distance of direct travel from customer i to j

$w_i(t_i)$ is the waiting time for a vehicle at customer i when his service time is t_i

$\mu_i(t_i)$ is the degree of satisfaction for customer i

Variables

t_i is the service time for customer i

$$y_{ik} = \begin{cases} 1, & \text{if customer } i \text{ is serviced by vehicle } k \\ 0, & \text{otherwise} \end{cases}$$

$$x_{ijk} = \begin{cases} 1, & \text{if vehicle } k \text{ travels directly from customer } i \text{ to } j \\ 0, & \text{otherwise} \end{cases}$$

Then the F_VRP can be formulated as follows:

Formulation of Fuzzy Vehicle Routing Problem

$$\min \quad \sum_{j=1}^{n} \sum_{k=1}^{m} x_{0jk} \tag{10.9}$$

$$\max \quad \frac{1}{n} \sum_{i=1}^{n} \mu_i(t_i) \tag{10.10}$$

$$\min \quad \sum_{k=1}^{m} \sum_{i=0}^{n} \sum_{j=0}^{n} d_{ij} x_{ijk} \tag{10.11}$$

$$\min \quad \sum_{i=1}^{n} w_i(t_i) \tag{10.12}$$

$$\text{s.t.} \quad \mu_i(t_i) > 0, \qquad \forall i \tag{10.13}$$

$$\sum_{i=1}^{n} a_i y_{ik} \le c_k, \qquad \forall k \tag{10.14}$$

$$\sum_{k=1}^{m} y_{ik} = 1, \qquad \forall i \tag{10.15}$$

$$\sum_{i=0}^{n} x_{ijk} = y_{jk}, \qquad \forall j, k \tag{10.16}$$

$$\sum_{j=0}^{n} x_{ijk} = y_{ik}, \qquad \forall i, k \tag{10.17}$$

$$\sum_{ij \in S \times S} x_{ijk} \le |S| - 1, \qquad \forall k, \quad S \subseteq \{1, 2, \ldots, n\} \tag{10.18}$$

$$x_{ijk} = 0 \text{ or } 1, \qquad \forall i, j, k \tag{10.19}$$

$$y_{ik} = 0 \text{ or } 1, \qquad \forall i, k \tag{10.20}$$

$$t_i \ge 0, \qquad \forall i \tag{10.21}$$

where the subscript 0 stands for the center depot of the fleet of vehicles. Objective (10.9) minimizes fleet size, objective (10.10) maximizes the average grade of satisfaction over customers, objective (10.11) minimizes total travel cost, and objective (10.12) minimizes total waiting time for vehicles. Constraint (10.13) ensures that the service time for each customer is within a tolerable interval of time. Constraint (10.14) ensures that every vehicle is assigned to customers without exceeding its capacity. Constraint (10.15) ensures that each customer is serviced by one and only one vehicle. Constraints (10.16) and (10.17) ensure that for each customer, there are only two customers directly connected with him, one directly reaches him, and another he directly travels to by a vehicle. Constraint (10.18) shows the relationship between direct travels and the number of customers for vehicle k. It is easy to see that two well-known combinatorial optimization problems are embedded within this formulation. Constraints (10.14) and (10.15) are the constraints of a generalized assignment problem. If variables y_{ik} are fixed to satisfy (10.13)–(10.15), then for given k, constraints (10.16)–(10.18) define a traveling salesman problem over the customer assigned vehicle k.

The commonly considered objectives by conventional approaches are as follows:

- Minimize fleet size.
- Minimize total travel distance.

Usually the emphasis is placed on the objective of minimum fleet size, but the tradeoff between fleet size and travel cost must be considered.

For the fuzzy vehicle routing problem, besides these objectives, another two objectives must be considered:

- Maximize the average grade of customer satisfaction.
- Minimize total waiting time over vehicles.

The emphasis is placed on the objective of average grade of customer satisfaction. This may lead to the increase of total waiting time for vehicles, so we also need to make a tradeoff between average grade by customers and total waiting time for vehicles.

When making a decision for the crisp case, conventional approaches just can guarantee that the service time for each customer is within his allowable time window, but do not care for customers' desirable service time. Under the consideration of fuzziness of customers' preferences, we are interested in not only the feasibility of service time for all customers as the conventional one does, but also the reasonability of service time in the sense that we enforce the service time for each customer to approach his due time as close as possible so as to maximize the average grade of satisfaction.

10.2.2 Related Genetic Algorithm Studies

During the past few years, several studies have reported their experiences about applying genetic algorithms to the vehicle routing problem with time and capacity constraints. Thangiah, Vinayagamoorthy, and Gubbi [400] have investigated vehicle routing problems with time deadlines. The objective is to minimize the number of vehicles and the distance traveled for servicing a set of customers without exceeding the capacity of the vehicles or being tardy. They proposed a kind of cluster-first route-second heuristic procedure. Genetic algorithms are used to cluster customers (or assign vehicles to customers). The customers within each cluster (or vehicle) are then routed using a cheapest insertion heuristic. The solutions obtained from this heuristic are further improved using a standard 2-opting postprocessing of the customers to determine if some of the customers should be serviced by vehicles from neighboring clusters.

Blanton and Wainwright [39] investigated vehicle routing problem with time window and capacity constraints. The objective is to determine the minimum number of vehicles required to service all of the customers within their allowed time windows subject to vehicle capacity constraints. They intended to give a single global solution procedure rather than several separate phases as proposed by Thangiah et al. The list of customer permutation was adopted as a coding scheme, and a nearest-neighbor heuristic-based crossover operator was proposed. In the evaluation step, Davis' greedy adding heuristic was used to decode a customer permutation list with regard to vehicle schedules. The greedy heuristic takes the first k customers in the permutation list to initialize (seed) the k vehicles. The remaining $n - k$ customers are examined individually. As each new customer is selected, possible subtours are evaluated for each vehicle and the best subtour is selected. There are two problems in their implementation of genetic algorithms. One problem is that this decoding procedure will destroy subtours in chromosome, so evolution herein loses its meaning. Another problem is that the decoding procedure is highly dependent on the first k customers. Because the essential of crossover operator is a kind of nearest-neighbor heuristics, the first k customers are not suitable to initialize k vehicles.

Potvin and Dube [339] have applied genetic algorithms to the search of good parameter settings for a vehicle routing heuristic and have shown that it is possible to greatly improve the results of a parallel insertion heuristic via genetic search in the parameter space.

10.2.3 Hybrid Genetic Algorithm

Cheng and Gen have proposed a hybrid genetic algorithm to solve the fuzzy vehicle routing and scheduling problem [65, 69]. An insertion-heuristic-based crossover operator was designed to find out improved offspring. In order to handle the fuzzy feature of the problem, a *push–bump–throw* procedure was proposed. With the procedure, fuzzy vehicle routing problem can be treated with two main steps within crossover operation:

- We first just consider the problem as an ordinary problem with a time window and obtain a feasible solution with sequential insertion heuristic.
- We then turn our attention on its fuzzy characteristic and determine the best service time for each customer with the push–bump–throw procedure.

Representation. A list of *customer/service time/vehicle* was used as chromosome representation, which represents the permutation of customers associated with service time and assigned vehicle. A gene is an ordered triple (*customer*, *service time*, *vehicle*). This representation belongs to the direct way, which is sketched in Figure 10.13.

customer i_1	customer i_2	$\cdots\cdots$	customer i_n
service time t_{i_1}	service time t_{i_2}	$\cdots\cdots$	service time t_{i_n}
vehicle v_{i_1}	vehicle v_{i_2}	$\cdots\cdots$	vehicle v_{i_n}

Figure 10.13. Representation scheme of chromosome.

Initialization. For the fuzzy vehicle routing problem, because of the existence of allowable service time constraints, it is impossible to use a simple random procedure to generate initial population. The proposed initial procedure contains three main steps:

1. Generate a permutation of customers randomly.
2. Cluster customers to vehicles with *left-to-right scan* procedure.
3. Determine the best service time for each customer with *push–bump–throw* procedure.

Procedure: Initialization

begin
  $i \leftarrow 0$;
  while $(i \leq pop\_size)$ **do**
    generate customer permutation;
    cluster customers to vehicles;
    determine customers' best service times;
    $i \leftarrow i + 1$;
  end
end

Clustering Procedure. Clustering procedure is a sequential adding procedure which assigns customers to vehicles. When adding a new customer to a current vehicle, it is necessary to check the capacity feasibility for the current vehicle and service time feasibility for the customer. If both are feasible, the customer is assigned to current vehicle; otherwise, the customer is assigned to a new vehicle. And then the possible earliest service time for each customer is calculated. Repeat above procedure until all customers are assigned to a vehicle. The overall procedure is shown below:

Procedure: Clustering

begin (for a chromosome)
  assign the leftmost customer to a vehicle;
  $i \leftarrow 1$;
  while $(i \leq n)$ **do**
    check capacity/service time feasibilities;
    if both are feasible **then**
      add the customer to current vehicle;
    else
      open a new vehicle;
    end

determine his earliest service time;
$i \leftarrow i + 1$;
end
end

Suppose the kth vehicle has $i - 1$ customers denoted with $U_k = \{1, \ldots, i - 1\}$. Now we consider whether we can assign the next customer i into the vehicle. We need to check the capacity feasibility of the current vehicle with the following inequality:

$$\sum_{j \in U_k} a_j + a_i \leq c_k$$

and check the service time feasibility for the customer with the following inequality:

$$t_{i-1} + r_{i-1,i} \leq l_i$$

If both are feasible, customer i is assigned to the current vehicle, and the possible earliest service time for customer i is calculated as follows:

$$t_i = \begin{cases} e_i, & \text{if } t_{i-1} + r_{i-1,i} \leq e_i \\ t_{i-1} + r_{i-1,i}, & \text{otherwise} \end{cases}$$

If neither capacity nor service time is feasible, customer i is assigned to a new vehicle.

Push–Bump–Throw Procedure. After the above step, we get a population of chromosomes; each of them represents a capacity and service time feasible schedule. The service time for each customer is given as his possible earliest service time. We need further to determine the best service time for each customer to maximize the total grade of satisfaction. A push–bump–throw procedure is proposed here for determining the best service time for each customer. The basic idea of the procedure is rather simple: We try to make the service time of a customer approach his desired time (due time) as close as possible.

Usually a feasible chromosome contains several vehicle routing plans, and a routing plan for a vehicle is composed of some *tight paths* and/or *loose paths*.

Tight path: a list of ordered customers with zero waiting time between any adjacent customers.

Loose path: a list of ordered customers with no-zero waiting time between any adjacent customers.

Figure 10.14 illustrates the tight path and the loose path.

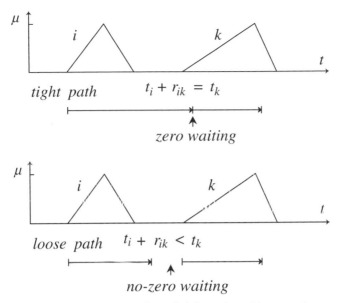

Figure 10.14. Illustration of tight path and loose path.

The push–bump–throw procedure first scans the feasible schedule from left to right and tries to find out a possible forward push within the tight path. Such a push will increase the total degree of satisfaction along the path without violating time constraints. Let P be a list of ordered customers to denote a tight path, and let w denote the first no-zero waiting time following the tight path. The possible forward push for customer i can be determined as follows:

$$\Delta_i = \begin{cases} u_i - t_i, & \text{if } e_i < t_i < u_i \\ l_i - t_i, & \text{if } u_i < t_i < l_i \end{cases} \tag{10.22}$$

The possible forward push for a tight path can be determined as follows:

$$\Delta_{\min} = \min\{\min_{i \in P}\{\Delta_i\}, w\} \tag{10.23}$$

If there is a such possible forward push, we push the related part with it. After each push, we should perform two kinds of checking:

- Unpushable checking
- Connection checking

1. In connection checking, we examine whether w becomes zero. If the waiting time becomes zero, it means that the next tight path is connected to the cur-

rent tight path. In the other words, the current light path *bumps into* the next tight path.

2. In unpushable checking, we examine the unpushable path among the pushed parts and then separate them from the consideration. Because any forward push on the unpushable path will cause either (a) violation of time constraints or (b) decrease of the total grade of satisfaction. In other words, we *throw off* the unpushable path from the current tight path.

Let $\mu_i'(t_i)$ denote the slope of the satisfaction function at time t_i for customer i. We sum all derivative terms of customers in the tight path as follows:

$$\mu' = \sum_{i \in P} \mu_i'(t_i) \tag{10.24}$$

If μ' is larger than zero, a possible forward push will cause the increase of total grade of satisfaction; otherwise, such a push will cause the decrease of the total grade of satisfaction. According to μ', we can decide whether we perform a push and examine the unpushable path and separate it from consideration. During the push–bump–throw processes, tight and loose paths vary frequently due to the forward push. Figure 10.15 illustrates the push–bump–throw procedure with a simple example.

From the figure we can see that the first tight path contains customers i, k, and j. The possible forward push for the tight path is $\Delta_{\min} = \Delta_k$. Perform a push with Δ_k to let the grade of satisfaction become 1 for customer k. Now customers i and k form the unpushable path. After *throwing off* customers i and k, the possible forward push for the remaining path is $\Delta_{\min} = w$. After *pushing* the remaining path with w, customer j *bumps into* customer l.

Let P be the current tight path under treatment, let P_k be formed with the first k customers of P, and let P' be a tight path following P. The overall procedure is as follows:

Procedure: Push–Bump–Throw

begin
    assign the leftmost customer to a vehicle;
    $P \leftarrow$ first tight path;
    while (all customers are treated) **do**
      unpushable checking:
        if $\sum_{i \in P_k} \mu_i' \leq 0$ **then**
          $P \leftarrow P \backslash P_k$ (throw off P_k);
      end
      empty checking:
      if P is empty **then**

before push

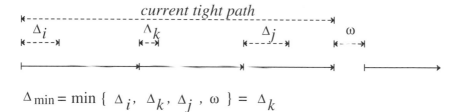

$$\Delta_{min} = \min \{ \Delta_i, \Delta_k, \Delta_j, \omega \} = \Delta_k$$

first push with Δ_k

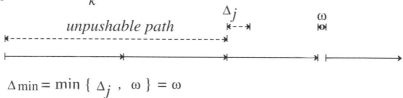

$$\Delta_{min} = \min \{ \Delta_j, \omega \} = \omega$$

second push with ω

Figure 10.15. Illustration of push–bump–throw procedure.

```
    P ← P';
  end
 find out Δmin for P;
 push with Δmin;
 connection checking:
  if w' = 0 then
    P ← P ∩ P' (expand P);
  end
 end
end
```

Genetic Operator. The heuristic algorithms for the vehicle routing and scheduling problem with time windows can be divided into *tour-building heuristics* and *tour-improvement heuristics*. The sequential insertion heuristic is one kind of tour-building heuristic. It first initializes a route with a "seed" customer, and the remaining unrouted customers are added to this route one by one until it is full with respect to the scheduling horizon and/or capacity constraint. If unrouted customers remain, the initialization and insertion procedures are then repeated until all customers are serviced. The sequential insertion algorithm has outperformed all other approaches on a set of routing test problems [384].

Feasible Insertion. Now we examine the necessary and sufficient condition for feasible insertion. Assume that we have a partially constructed feasible route $(i_1, i_2, \ldots, i_m)$. We want to insert a customer, say k, between customers i_{p-1} and i_p, $1 \le p \le m$. Let $t_{i_p}^{\text{NEW}}$ be the new service time for customer i_p after the insertion k. If we assume that triangle inequality holds for travel distances, this insertion defines a *forward push* in the schedule:

$$\Delta t_{i_p} = t_{i_p}^{\text{NEW}} - t_{i_p} \tag{10.25}$$

$$\Delta t_{i_r} = \max \{0, \Delta t_{i_{r-1}} - w_r\}, \qquad p < r \le m \tag{10.26}$$

If $\Delta t_{i_p} > 0$, some of the customers i_r, $p \le r \le m$, may become infeasible. It is easy to see that we should examine these customers sequentially for time feasibility until we find some customer, say i_r with $r < m$, for which $\Delta t_{i_r} = 0$ or where i_r is time infeasible. In the worst case all the customers i_r, $p \le r \le m$, are examined. We have just proved the following:

Lemma. The necessary and sufficient conditions for time feasibility when inserting a customer, say k, between i_{p-1} and i_p, $1 \le p \le m$, on a partially constructed feasible route $(i_1, i_2, \ldots, i_m)$ are

$$t_k \le l_k \text{ and } t_{i_r} + \Delta t_{i_r} \le l_{i_r}, \qquad p \le r \le m$$

Crossover Operator. The proposed crossover operator is primarily based on the sequential insertion heuristic, which contains two main steps:

- Generate a capacity and service time feasible schedule with the sequential insertion heuristic.
- Determine the best servicing time for each customer by using the push–bump–throw procedure.

The proposed procedure is as follows:

Step 1. First select a customer from the rightmost customers of two parents randomly as the initial customer of offspring.

Step 2. Exchange the related customers properly so as to let the selected customer be in the last position for both parents.

Step 3. Remove the selected one from the parents of the offspring, assign him to vehicle 1, and determine his possible earliest service time as described in the last section.

Step 4. Now two customers adjacent to the selected one become the candidate customers for insertion. Calculate the best inserting position on the partial route of offspring for each of them and determine the best insertion customer.

Step 5. Insert the best one into the current vehicle and determine his possible earliest service time.

Step 6. If neither of the two candidate customers can be inserted into the current vehicle, one of them is randomly put at the last position of the offspring to open a new vehicle.

Step 7. Repeat the above steps: calculating, comparing, exchanging, removing, and inserting until a complete offspring is generated.

Step 8. After an offspring is generated, use the push–bump–throw procedure to determine the best servicing time for each customer.

Let use see an example. For the ease of illustration, we just use the list of ordered customers as chromosome representation. Consider two parent chromosomes as follows:

$$p_1: \quad 5 \quad 1 \quad 3 \quad 2 \quad 4 \quad 6 \quad 7 \quad 9 \quad 8$$
$$p_2: \quad 8 \quad 1 \quad 2 \quad 9 \quad 3 \quad 5 \quad 4 \quad 6 \quad 7$$

Suppose customer 7 is selected as the initial customer for offspring. Assign him to vehicle 1 and determine his possible earliest service time. Now we have a partial offspring as follows:

$$
\begin{array}{llllllllll}
p_1: & 5 & 1 & 3 & 2 & 4 & 6 & 8 & 9 & \times \\
p_2: & 8 & 1 & 2 & 9 & 3 & 5 & 4 & 6 & \times \\
o: & 7 & \times & \times & \times & \times & \times & \times & \times & \times \\
\end{array}
$$

After several steps, we obtain the following partial offspring:

$$
\begin{array}{llllllllll}
p_1: & 5 & 1 & 3 & 2 & 8 & 6 & \times & \times & \times \\
p_2: & 8 & 1 & 2 & 6 & 3 & 5 & \times & \times & \times \\
o: & 4 & 9 & 7 & \times & \times & \times & \times & \times & \times \\
\end{array}
$$

The two candidate customers for insertion are 5 and 6. Calculate their insert costs and select the one with a low cost for insertion, say 5. Then we need to exchange the position of customers 5 and 6 for the first parent as follows:

$$
\begin{array}{llllllllll}
p_1: & 6 & 1 & 3 & 2 & 8 & 5 & \times & \times & \times \\
p_2: & 8 & 1 & 2 & 6 & 3 & 5 & \times & \times & \times \\
o: & 4 & 9 & 7 & \times & \times & \times & \times & \times & \times
\end{array}
$$

Suppose the best insert position is between customers 4 and 9 in the partial offspring; remove customer 5 from both parent chromosomes and insert it into partial offspring as follows:

$$
\begin{array}{llllllllll}
p_1: & 6 & 1 & 3 & 2 & 8 & \times & \times & \times & \times \\
p_2: & 8 & 1 & 2 & 6 & 3 & \times & \times & \times & \times \\
o: & 4 & 5 & 9 & 7 & \times & \times & \times & \times & \times
\end{array}
$$

We assign customer 5 to the same vehicle as customers 4 and 9, and we determine his possible earliest service time. Repeat these steps until a complete offspring is generated.

The best inserting customer is determined by the following equation:

$$
c(i,k,j) = \alpha_1 \left(1 - \frac{d_{ij}}{d_{ik} + d_{kj}} \right) + \alpha_2 \frac{w_k^{\text{NEW}} + w_j^{\text{NEW}}}{w_j^{\text{OLD}}} \tag{10.27}
$$

$$
\alpha_1 + \alpha_2 = 1, \qquad \alpha_1 \geq 0, \qquad \alpha_2 \geq 0 \tag{10.28}
$$

where w_k^{NEW} and w_j^{NEW} are the waiting times after inserting customer i. w_j^{OLD} is the waiting time before inserting customer i. The insertion measure function $c(i,k,j)$ is the combination of a weighted sum of the quasi-normalized detour (first part) and the quasi-normalized waiting time saved by the detour (second part). The best inserting customer is the one that minimizes the function.

Evaluation and Selection. Fitness is calculated based on the original objective functions. The weighted sum of objectives for a chromosome is given by

$$
g(\boldsymbol{v}_k) = \rho_1 \frac{u_k}{u_{\text{max}}^0} + \rho_2 \left(1 - \frac{1}{n} \sum_i \mu_{ki}(t_{ki}) \right) + \rho_3 \frac{1}{n} \sum_i \frac{w_{ki}}{w_{\text{max}}^0} + \rho_4 \frac{d_k}{d_{\text{max}}^0}
$$

$$
\sum_{i=1}^{4} \rho_i = 1, \qquad \rho_i \geq 0, \qquad i = 1, 2, 3, 4 \tag{10.29}
$$

where $\boldsymbol{v}_k$ is the kth chromosome, w_{ki} is the waiting time for customer i in chromosome $\boldsymbol{v}_k$, w_{max}^0 is the maximum waiting time among initial chromosomes, u_k is the total number of vehicles used in chromosome $\boldsymbol{v}_k$, u_{max}^0 is the maximum number of vehicles among initial chromosomes, d_k is the total distance

Table 10.1. Basic Setting of Function Parameters

ρ_1	ρ_2	ρ_3	ρ_4	α_1	α_2
0.5	0.3	0.1	0.1	0.5	0.5

traveled for chromosome v_k, and d_{max}^0 is the maximum distance among initial chromosomes. Then the fitness for a chromosome is calculated as follows:

$$eval(v_k) = \frac{g(v_{max}) - g(v_k)}{g(v_{max}) - g(v_{min})} \tag{10.30}$$

The *roulette wheel* approach is adopted as the selection mechanism, and the *elitist method* is combined with it to preserve the best chromosome in the next generation and overcome the stochastic errors of sampling.

10.2.4 Experimental Results

Primary computational experiments were conducted to test the effectiveness of the proposed procedure. The test problem consists of 30 customers. The fuzzy due time for each customer was randomly generated. The basic setting of weights for the evaluation function and the insert cost function is given in Table 10.1. The basic setting of parameters for genetic algorithms is given in Table 10.2.

Based on above basic settings, the crossover probability was varied from 0.1 to 0.5. Table 10.3 gives the average result over 10 runs for each setting. From the results we can see that crossover with a higher probability can yield a better solution.

Based on the above basic settings, the population size was varied from 10 to 50. The average results over 10 runs for each case are reported in Table 10.4. The results show that when the population size is larger than 30, its value has no significant influence on the performance of the genetic algorithm.

Based on the above basic settings, the weights of objective functions (10.29) were varied as shown in Table 10.5 to investigate how they impact on the final decision. The results over 10 runs are reported in Table 10.6.

From the results we can see that any emphasis on the objective of the grade of satisfaction (setting 2, $\rho_2 = 0.85$) or the total travel distance (setting 4, $\rho_4 = 0.85$) will lead to the increase of the fleet size and the total waiting time.

Table 10.2. Basic Setting of Genetic Algorithm Parameters

p_c	pop_size	max_gen
0.3	30	400

Table 10.3. Results with Varied p_c

p_c	μ	w	d	u
0.1	0.825	14.71	1659	3
0.2	0.841	15.41	1541	3
0.3	0.897	15.31	1603	3
0.4	0.883	16.39	1437	3
0.5	0.907	16.89	1457	3

μ, the average grade of satisfaction; w, the total waiting time; d, the total distance; u, total number of vehicles.

Table 10.4. Results with Varied *pop_size*

p_c	μ	w	d	u
10	0.909	40.04	977	10
20	0.908	17.42	811	5
30	0.893	14.50	1619	3
40	0.876	14.57	1537	3
50	0.889	14.48	1586	3

Table 10.5. Variation of Weights

Settings	ρ_1	ρ_2	ρ_3	ρ_4
1	0.85	0.05	0.05	0.05
2	0.05	0.85	0.05	0.05
3	0.05	0.05	0.85	0.05
4	0.05	0.05	0.05	0.85
5	0.50	0.30	0.10	0.10

Table 10.6. Results with Varied Weights

Settings	μ	w	d	u
1	0.854	16.20	1442	3
2	1.000	53.51	921	17
3	0.843	14.89	1557	3
4	0.976	54.41	822	19
5	0.881	15.75	1467	3

Table 10.7. Comparison Genetic Algorithm with Heuristic

Problem	Method[a]	μ	w	d	u
1	GA	0.779	11.83	1436	4
	H	0.694	5.7	1830	5
2	GA	0.866	12.05	1367	4
	H	0.660	13.20	1756	5
3	GA	0.799	15.26	1405	3
	H	0.610	12.00	1685	4
4	GA	0.765	5.78	1631	3
	H	0.460	6.30	1632	5
5	GA	0.728	11.32	1322	3
	H	0.480	8.10	1988	4
6	GA	0.871	15.80	1428	3
	H	0.700	10.50	1699	3

[a]GA, genetic algorithm; H, modified insertion heuristic.

The proposed genetic algorithm was further compared with modified insertion heuristic. Because Solomon's insertion heuristic cannot handle fuzzy data, it was augmented with the push–bump–throw procedure. The results are given in Table 10.7, which shows that the genetic algorithm is clearly superior to the modified insertion heuristic.

10.3   LOCATION-ALLOCATION PROBLEM

The *location-allocation problem*, also known as the *multi-Weber problem* or the *p-median problem*, arises in many practical settings [111]. The classical single Weber problem, in Euclidean space, is to find the one single location which minimizes the summed distance from a number of fixed points representing locations of customers. Cooper was the first author to formally recognize and state the multi-Weber problem. He proved that the objective function is neither concave nor convex and has many local optima [85]. The research done by Eilon et al. shows that for a single problem with $m = 5$ and $n = 50$, using 200 trials, 61 local minima were found, and the worst solution deviated from the best by 40.9% [117]. Recognizing the combinatorial size of the problem (a Sterling number of Type II), Cooper proposed a heuristic called *alternative location-allocation*, which is the best heuristic available. With the development of nonlinear programming techniques, relaxing integer allocation constraints and treating the location variables and allocation variables simultaneously, some new methods have been developed in the past two decades. Murtagh and Niwattisyawong presented an approach which relaxes the zero–one constraints on the allocations and which instead allows them to take on values anywhere on the interval [0, 1].

MINOS, a large-scale nonlinear programming package, is then used to solve for both the locations and the allocations simultaneously [304]. Bongartz presented an approach which also relaxes the zero–one allocation constraint, adopts the active set method, and exploits the special structure inherent in the location-allocation problem [45]. These methods can be applied to relatively large-scale practical problems involving several hundreds of customers and tens of facilities. But none of them can ensure a global solution or a near-global optimum.

Harris et al. recognized that while the number of possible solutions is a Sterling number of type II, the number of feasible solutions is much less [209]. They found that in any feasible solution the subsets of customers must be contained in nonoverlapping convex hulls, and they proposed an algorithm which exploits this convex hull property of the problem to generate all feasible solutions and then find the optimum by complete enumeration of these feasible solutions. Ostresh developed a different algorithm based on the passing line to solve the two-facility Weber problem [323]. Recently, Rosing developed an optimal method for the generalized multi-Weber problem based on Harris's idea [359]. However, these optimal methods are still only applicable to small size problems.

Gong, Gen, Xu, and Yamazaki have proposed a kind of hybrid evolutionary method for the location-allocation problem with capacity constraints [177, 465, 466], which utilizes some efficient traditional optimization techniques to some allocation subproblems and the capacity constraints handled by these techniques. In this section, we will introduce their studies.

10.3.1 Location-Allocation Model

There are m facilities to be located, and n customers with known locations to be allocated to the variable facilities. Each customer has the demand q_j, $j = 1, 2, \ldots, n$, and each facility has the capacity b_i, $i = 1, 2, \ldots, m$. We need to find the locations of facilities and allocations of customers to facilities so that total summed distance among customers and their facilities are minimized (see Figure 10.16). This problem is formulated mathematically as follows:

$$\min \quad \sum_{i=1}^{m} \sum_{j=1}^{n} z_{ij} \cdot \sqrt{(x_i - \bar{x}_j)^2 + (y_i - \bar{y}_j)^2} \tag{10.31}$$

$$\text{s.t.} \quad \sum_{i=1}^{m} z_{ij} = 1, \qquad j = 1, 2, \ldots, n \tag{10.32}$$

$$\sum_{j=1}^{n} q_j \cdot z_{ij} \leq b_i, \qquad i = 1, 2, \ldots, m \tag{10.33}$$

$$z_{ij} = 0 \text{ or } 1, \qquad i = 1, 2, \ldots, m, \quad j = 1, 2, \ldots, n \tag{10.34}$$

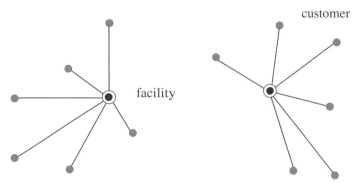

Figure 10.16. Location-allocation problem.

where

$(\bar{x}_j, \bar{y}_j)$ = location of customer $j, j = 1, 2, \ldots, n$

(x_i, y_i) = location of facility i, decision variables, $i = 1, 2, \ldots, m$

z_{ij} = 0–1 decision variable

$$z_{ij} = \begin{cases} 1, & \text{customer } j \text{ is served by facility } i \\ 0, & \text{otherwise} \end{cases}$$

Constraint (10.32) ensures that every customer should be served by only one facility; constraint (10.33) ensures that service capacity of each facility should not be exceeded.

The problem defined as above is an extension of the common location-allocation problem where the service capacity is imposed upon facilities. The allocation subproblem is a general assignment problem which is known as the *NP-hard problem*, and cannot be solved as easily by the nearest service rule like the uncapacitated problem. An extension of the ordinary alternative location-allocation method to this problem can be found in reference [177]. The alternative location-allocation (ALA) method is sensitive to the initial locations of facilities and sometimes unable to find the global or near-global optima because it improves locations by fixing the allocation of customers.

10.3.2   Hybrid Evolutionary Method

With the hybrid evolutionary method, the problem is divided into two levels: location and allocation. At the allocation level, the Lagrange relaxation method is adopted to solve this general assignment problem in order to obtain the solution quickly since this subproblem needs to be solved frequently. At the location level, the evolutionary method is adopted to search the whole locatable area.

By this method, the whole locatable area can be efficiently searched so that the global or near-global optima may be found.

Chromosome Representation. Since location variables are continuous, the float-value chromosome representation is used. A chromosome is given as follows:

$$c^k = [x_1^k, y_1^k, x_2^k, y_2^k, \ldots, x_m^k, y_m^k]$$

where (x_i^k, y_i^k) is the location of the ith facility in the kth chromosome, $i = 1, 2, \ldots, m$.

Initialization. Since the optimal locations should be within the rectangular region which contains all customers, the initial locations of facilities are randomly picked up from this rectangle and this procedure is repeated until all population members are generated.

Procedure: Initialization

Step 0. Find the rectangle $[x_{\min}, x_{\max}] \times [y_{\min}, y_{\max}]$.

Step 1. Set $k := 0$.

Step 2. If $k \geq pop\_size$, then stop; otherwise continue.

Step 3. Set $k := k + 1$ and $i := 1$.

Step 4. If $i > m$, then go to step 2; otherwise continue.

Step 5. Pick one value randomly in interval $[x_{\min}, x_{\max}]$ as x_i^k.

Step 6. Pick one value randomly in interval $[y_{\min}, y_{\max}]$ as y_i^k.

Step 7. Set $i := i + 1$, then go to step 4.

Evaluation. The sum of the optimal allocation distances among customers and their serving facilities $D(c^k)$ is used as the fitness value of chromosome c^k. The calculation of $D(c^k)$ needs solving the allocation subproblem. The Lagrange relaxation method was used to solve the allocation subproblem, which can be referred to [177].

Crossover. Crossover depends on how to choose parents and how parents produce their children. Two mating strategies are used: one is *free mating*, which selects two parents at random; the other is *dominating mating*, which uses the fittest individual as a fixed parent and randomly selects another parent from the population pool. These two strategies are used alternatively in the evolutionary process.

Suppose two parents are selected to produce a child:

$$c^{k1} = [x_1^{k_1}, y_1^{k_1}, x_2^{k_1}, y_2^{k_1}, \ldots, x_m^{k_1}, y_m^{k_1}]$$
$$c^{k2} = [x_1^{k_2}, y_1^{k_2}, x_2^{k_2}, y_2^{k_2}, \ldots, x_m^{k_2}, y_m^{k_2}]$$

Only one child is to be produced:

$$c = [x_1, y_1, x_2, y_2, \ldots, x_m, y_m]$$

The genes in the child are decided by the following equations:

$$x_i = r_i \cdot x_i^{k_1} + (1 - r_i) \cdot x_i^{k_2}$$
$$y_i = r_i \cdot y_i^{k_1} + (1 - r_i) \cdot y_i^{k_2}$$

where $r_1, r_2, \ldots, r_m$ are independent random numbers in $(0, 1)$.

Procedure: Crossover

Step 0. Set $l := 0$.

Step 1. If $l \geq child\_size \cdot (1 - p_m)$, then stop; otherwise $l := l + 1$ and continue.

Step 2. If generation t is odd, go to step 3; otherwise go to step 4.

Step 3. Select two parents randomly from the population pool to produce a child and then go to step 1.

Step 4. Select one parent randomly while choosing the fittest individual as the other parent to produce a child and then go to step 1.

Mutation. Two kinds of mutation operators are suggested. One is to give a small random perturbation to the chromosome of parent to form a new child chromosome, called *subtle mutation*. Another is to produce a child in the same way as in the initialization procedure, called *violent mutation*. These two kinds of mutation operators act alternatively in the evolutionary process.

Suppose the chromosome to be mutated is as follows:

$$c^k = [x_1^k, y_1^k, x_2^k, y_2^k, \ldots, x_m^k, y_m^k]$$

Then the child produced by the *subtle mutation* $c = [x_1, y_1, x_2, y_2, \ldots, x_m, y_m]$ is as follows:

$$x_i = x_i^k + \text{random value in } [-\varepsilon, \varepsilon]$$
$$y_i = y_i^k + \text{random value in } [-\varepsilon, \varepsilon]$$

where ε is a small positive real number. The child produced by violent mutation

$c = [x_1, y_1, x_2, y_2, \ldots, x_m, y_m]$ are as follows:

$$x_i = \text{random value in } [x_{min}, x_{max}]$$
$$y_i = \text{random value in } [y_{min}, y_{max}]$$

Procedure: Mutation

Step 0. Set $l := 0$.

Step 1. If $l \geq child\_size \cdot p_m$, then stop; otherwise $l := l + 1$.

Step 2. Select one parent randomly.

Step 3. If generation t is odd, go to step 4; otherwise go to step 5.

Step 4. Produce a child with subtle mutation and then go to step 1.

Step 5. Produce a child with violent mutation and then go to step 1.

Selection. ES-$(\mu + \lambda)$ selection is adopted to select the better individuals among parents and their children to form the next generation [10, 14]. However, the strategy usually leads to degeneration of the evolutionary process. In order to avoid this degeneration, a new selection strategy called *relative prohibition* is suggested.

Given two positive parameters α and γ, the neighborhood for a chromosome s_k is defined as follows:

$$\Omega(s_k, \alpha, \gamma) \triangleq \{s | \, \|s - s_k\| \leq \gamma, D(s_k) - D(s) < \alpha, s \in R^{2m}\} \tag{10.35}$$

In selection process, once s_k is selected into the next generation, any chromosome falling within its neighborhood is prohibited from selection. The value of γ defines the neighborhood of s_k in terms of location, which is used to avoid selecting individuals with very small difference in location. The value of α defines the neighborhood of s_k in terms of fitness, which is used to avoid selecting individuals with very small difference in fitness. The neighborhood defined by equation (10.35) is called a *prohibited neighborhood*.

Procedure: selection

Step 0. Let all parents and children be active. Select the best individual among parents and children as the first member of next generation, inactivate the selected one, and set $l := 1$.

Step 1. If $l > pop\_size$ then stop, else continue.

Step 2. If there is no active parent or child, produce $(pop\_size - l)$ new member in next generation randomly and stop, otherwise, continue.

Step 3. Take out the best individual among the remaining active parents and children

Step 4. Check if this individual falls into the *prohibited neighborhood* of the selected member, if not, save it as the new member of next generation and set $l := l + 1$.

Step 5. Inactivate the selected one and go to step 2.

10.3.3   Numerical Example

Cooper and Rosing's examples are used to test the effectiveness of the hybrid evolutionary method [359]. Cooper carefully constructed the first-half data which contains three natural groups, and Rosing increased the number of customers with random points. These examples provide a good standard to test the effectiveness of the proposed method because their global optimal solutions have already been found.

These examples include 30 customers whose coordinates of locations are shown in Table 10.8. The demands of customers are treated as equal and the capacities of facilities are assumed unlimited.

Both the ALA method and the hybride evolutionary method (HEM) were applied to solve these examples. When using the ALA method, it was run to solve each example 40 times from randomly generated initial locations. The environment parameters of HEM were set as follows: *max_gen*, 600; *pop_size*, 40;

Table 10.8. Coordinates of Cooper and Rosing's Example

Order Number	X	Y	Order Number	X	Y
1	5	9	16	53	8
2	5	24	17	1	34
3	5	18	18	33	8
4	13	4	19	3	26
5	12	19	20	17	9
6	13	39	21	53	20
7	28	37	22	24	17
8	21	45	23	40	22
9	25	50	24	22	41
10	31	9	25	7	13
11	39	2	26	5	17
12	39	16	27	39	3
13	45	22	28	50	50
14	41	30	29	16	40
15	49	31	30	22	45

Reprinted from *European Journal of Operational Research*, vol. 58, K. E. Rosing, "An optimal method for solving the (generalized) multi-Weber problem," pp. 420–423, 1992 with kind permission of Elsevier Science—NL, Sara Burgerhartstraat 25, 1055 KV Amsterdam, The Netherlands.

Table 10.9. Comparison Results of Copper and Rosing's Example

Problem n/m	Rosing's Method Optimal Objective	ALA		HEM	
		Best	Percent Error	Best	Percent Error
15/2	214.281	219.2595	2.32	214.2843	0.0015
15/3	143.197	144.8724	1.17	143.2058	0.0061
15/4	113.568	115.4588	1.69	113.5887	0.0182
15/5	97.289	99.4237	2.19	97.5656	0.2843
15/6	81.264	84.0772	3.46	83.0065	2.14
30/2	447.728	450.3931	0.5952	447.73	0.0004
30/3	307.372	310.3160	0.9578	307.3743	0.0007
30/4	254.148	258.4713	1.7010	254.2246	0.0301
30/5	220.057	226.8971	3.1083	220.4335	0.1711
30/6	—	208.4301	3.4940	201.4031	0.0

child_size, 80; p_m, 10%; α, 1.0; and γ, 1.0. The computing results are given in Table 10.9. In the table, the percent error was calculated as follows: (actual value − optimal value)/optimal value × 100%. Rosing did not give the result of the case $n = 30$ and $m = 6$ and estimated that the computing time for it might be more than 10 CPU hours on Convex 210 under a Convex UNIX operating system. The computing time of HEM for this case was about 4 minutes. As shown in Table 10.9, the HEM can always find the global solution and outperform ALA in all these examples.

10.4   OBSTACLE LOCATION-ALLOCATION PROBLEM

The location-allocation model given in the previous section considers the ideal case without obstacles for facility location design. In practice, obstacle or forbidden region constraint usually needs to be considered. For example, lakes, parks, buildings, cemeteries, rivers, and so on, provide obstacles to facility location design. There are two kinds of obstacles:

- The prohibition of locations
- The prohibition of connecting pathes

In the former case, a new facility should not be located within the region of obstacles. This problem is also called on *obnoxious location*. In the latter case, the connecting path between customer and facility should avoid passing through the region of obstacles. This problem is also called *location with barriers to travel*. When an obstacle constraint is considered, the location-allocation problem becomes much more complex and very difficult to solve. Because of its wide applicability, many researchers have concentrated on the obstacle location-allocation problem. A primary concern is to reduce the size of the set of feasible locations one must consider in seeking the optimal solution. When the entire problem is restricted to a given network, with users located only at nodes, Hakimi showed that the location-allocation problem always has an optimal solution with facilities located on the nodes [199]. This result reduced the problem of continuous search to a combinatorial one. This method depends on the selection of the possible candidates of locations which are not easily given.

The Manhattan metric l_1 (or rectilinear metric) obstacle location-allocation problem has been discussed thoroughly for the following two reasons: (1) The optimal solution can be restricted to a particular finite set of candidate points which can be determined by inspection, and (2) there is a wide application background, such as urban facility construction, communication network design, facility layout, circuit board design, and even location and routing of robots [138, 260, 278]. Larson and Sadig discussed facility location problem with l_1 metric in the presence of barriers to travel [261]. Hamacher and Nickel gave some theoretical results as well as algorithms for facility location problem where the location of facility is restricted only [204].

One of the typically used distance metrics in practice is the Euclidean distance l_2. However, with this distance metric, the obstacle location-allocation problem is very difficult to solve. The decision space for location variables is continuous, and the objective functions become very complex. It is not easy to calculate the gradient information. Katz and Cooper discussed l_2 distance facility location problem with only one forbidden circle [242]. The location-allocation problem with polygonal obstacles was discussed in reference [178]. Gong, Gen, Xu, and Yamazaki have extended their work to the obstacle location-allocation problem [175, 176]. Since the evolutionary method has many procedures that are similar to those discussed in the previous section, only the unique portions will be discussed in the following section.

10.4.1   Obstacle Location-Allocation Model

There are n customers with known locations, and m facilities must be built to supply some kind of service to all customers—for example, supplying materials or energy. There are also p obstacles representing some forbidden areas. The following assumptions are made to formulate the mathematical model:

- Customer j has service demand $q_j, j = 1, 2, \ldots, n$.
- Facility i has service capacity $b_i, i = 1, 2, \ldots, m$.
- Each customer should be served by only one facility.
- New facilities should not be built within any obstacle.

- Connecting paths between facilities and customers should not be allowed to pass through any of the obstacles.

The problem is to choose the best locations for facilities so that the sum of distances among customers and their serving facilities is minimal. This is illustrated in Figure 10.17. It is also assumed that all obstacles can be represented as a convex polygon. The obstacle location-allocation problem can be formulated as the following tri-level model:

Level 1

$$\min \quad D(S_1, S_2, \ldots, S_m) \tag{10.36}$$
$$\text{s.t.} \quad S_i \,\bar{\epsilon}\, inner(P_l), \qquad i = 1, 2, \ldots, m, \quad l = 1, 2, \ldots, p \tag{10.37}$$

Level 2

$$D(S_1, S_2, \ldots, S_m) \triangleq$$

$$\min \quad \sum_{i=1}^{m} \sum_{j=1}^{n} d(S_i, W_j) \cdot z_{ij} \tag{10.38}$$

$$\text{s.t.} \quad \sum_{i=1}^{m} z_{ij} = 1, \qquad j = 1, 2, \ldots, n \tag{10.39}$$

$$\sum_{j=1}^{n} q_j \cdot z_{ij} \leq b_i, \qquad i = 1, 2, \ldots, m \tag{10.40}$$

$$z_{ij} = 0 \text{ or } 1, \qquad i = 1, 2, \ldots, m, \quad j = 1, 2, \ldots, n \tag{10.41}$$

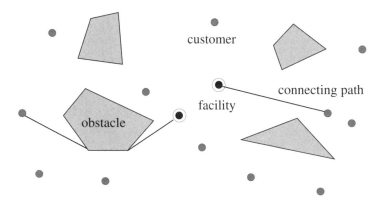

Figure 10.17. Obstacle location-allocation problem.

Level 3

$$d(S_i, W_j) \triangleq \min |l| \qquad (10.42)$$
$$\text{s.t.} \quad l \in L(S_i, W_j) \qquad (10.43)$$

where

$W_j = (\bar{x}_j, \bar{y}_j)$ is the location of jth customer

$(\bar{x}_j, \bar{y}_j)$ are x and y coordinates

P_l is the lth obstacle (polygon)

$inner(P_l)$ is the set of points which are in the inside of the lth obstacle, $l = 1, 2, \ldots, p$

$S_i = (x_i, y_i)$ is the location of ith facility

(x_i, y_i) are x and y coordinates, decision variables

$D(S_1, S_2, \ldots, S_m)$ is the minimal distance sum among all allocations of customers to facilities according to the certain fixed locations of facilities

$d(S_i, W_j)$ is the shortest connecting path between the facility S_i and customer W_j which can avoid obstacles

z_{ij} is the 0-1 decision variable; $z_{ij} = 1$ indicates that the jth customer is served by ith service center, $z_{ij} = 0$ otherwise

$L(S_i, W_j)$ is the set of possible paths which connect ith facility and jth customer and can avoid any obstacle P_l

Constraints (10.37) indicate that locations of facilities should not fall into the inside of any obstacle (polygon); constraints (10.39) indicate that every customer should be served by only one facility; constraints (10.40) indicate that the serving amount of each facility should not exceed its capacity; constraints (10.43) indicate that connecting paths between facilities and customers should avoid obstacles.

If obstacles are ignored, $d(S_i, W_j)$ becomes the direct Euclidean distance between facility S_i and customer W_j. It can be expressed as

$$d(S_i, W_j) = \sqrt{(x_i - \bar{x}_j)^2 + (y_i - \bar{y}_j)^2} \qquad (10.44)$$

When obstacles are considered, $d(S_i, W_j)$ is the function of S_i, W_j, and P_l, $l = 1, 2, \ldots, p$. This function cannot be expressed by a mathematical expression directly.

This model is a tri-level mixed integer programming problem. The objective functions and the constraints are very complex. Some of them cannot be expressed in a direct mathematical manner. The gradient information of the objective functions is also difficult to obtain, and this problem has many local optimal solutions. Traditional methods cannot solve this problem well, so the evolutionary method is much more suitable to find the global or near-global optima because of its potential capability in global search and its least stringent requirements with regard to conditions for computing.

10.4.2   Feasibility of Location

The constraints in Level 1 require that all locations of facilities should be out of obstacle regions. Testing these constraints is equivalent to testing whether a point in the plane falls into a polygon. This can be simply done by calculating the square of all triangles which consist of one edge of the obstacle polygon and the point and comparing it with the square of the obstacle polygon. The algorithm was given in reference [177].

10.4.3   Shortest Path of Avoiding Obstacles

In Level 3 it is necessary to find the shortest path connecting two specified points which can avoid all obstacles. Many researchers have concentrated on the *path of avoiding obstacle* problem, especially in robot navigation. Here the visible graph idea is adopted; however, some differences are considered in order to develop the special algorithm to this problem. According to experience, only a small number of obstacles have relations to the connecting path linking the specified two points on the plane. Thus it is not necessary to construct the whole visible graph including all obstacles, which would lead to a large-scale graph. An algorithm that only considers the relevant obstacles to construct a small-size visible graph is given in reference [177], and then the Dijikstra shortest path algorithm is applied to the visible graph to obtain the shortest path avoiding obstacles.

10.4.4   Hybrid Evolutionary Method

The evolutionary method to solve the obstacle location-allocation problem is almost the same as that used to solve the capacitated location-allocation problem, except for two differences. One is the feasibility adjustment of chromosomes, and the other is the calculation of the shortest path connecting customers and their service facilities, which can avoid obstacles.

Feasibility Adjustment of Chromosome. Since there are obstacles, the locations of the chromosome produced by initialization, crossover, and mutation procedures may become infeasible. Generally, there are three kinds of methods to treat an infeasible chromosome. The first is to discard it, but according to the experience of some researchers, this method may lead to very low efficiency. The second is to add penalty to an infeasible chromosome. The third is to repair the infeasible chromosome according to the characteristics of the specified problem. Here, based on the specificity of the obstacle location-allocation problem, a simple method is used to repair infeasible chromosomes. First the feasibility of a location is tested; once it is infeasible, it is repaired to be feasible by replacing it with its nearest vertex of the obstacle it falls into. This is shown in Figure 10.18.

Algorithm. The algorithm for the hybrid evolutionary method is described as follows:

Procedure: Hybrid Evolutionary Method

begin
  initialization;
  feasibility adjustment of chromosomes;
  evaluation of chromosomes;
  for $i = 1$ to *max_gen* **do**
  begin
    crossover;
    mutation;
    feasibility adjustment of chromosomes;
    evaluation;
    selection;
  end
end

10.4.5   Case Study

This problem deals with locating electric stations in an oil field. There are 14 oil wells whose coordinates of locations and electric demands are shown in Table 10.10. There exist four obstacles: two small towns, one large lake, and one large factory (see Table 10.11). The locations for three electric stations must be found. Two of them can supply electricity for four oil wells and one can serve six oil wells, as shown in Table 10.12. The electric stations should

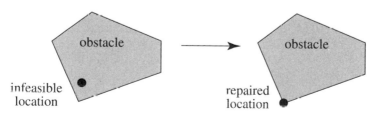

Figure 10.18. Feasibility adjustment of infeasible location.

Table 10.10. Coordinates and Demands of Oil Field

j	x_j	y_j	q_j	j	x_j	y_j	q_j
1	−2.0	0.0	1.0	8	22.0	20.0	1.0
2	2.0	4.0	1.0	9	26.0	30.0	1.0
3	8.0	2.0	1.0	10	20.0	38.0	1.0
4	5.0	12.0	1.0	11	10.0	36.0	1.0
5	12.0	14.0	1.0	12	6.0	40.0	1.0
6	24.0	16.0	1.0	13	2.0	36.0	1.0
7	30.0	8.0	1.0	14	−4.0	12.0	1.0

Table 10.11. Vertex Coordinates of Obstacles

	Obstacle 1	Obstacle 2	Obstacle 3	Obstacle 4
Number of obstacle vertices:	5	3	4	5
Vertex coordinates of obstacles:	(2.0, 1.0)	(8.0, 6.0)	(0.0, 8.0)	(7.0, 15.0)
	(6.0, 1.0)	(10.0, 3.0)	(3.0, 7.0)	(16.0, 18.0)
	(8.0, 4.0)	(9.0, 12.0)	(5.0, 14.0)	(14.0, 32.0)
	(4.0, 6.0)		(0.5, 16.0)	(12.0, 34.0)
	(1.0, 3.0)			(5.0, 23.0)

Table 10.12. Electric Capacity of Stations

Station i	Capacity b_i
1	4.0
2	6.0
3	4.0

Table 10.13. Comparison of HEM and Heuristic Method

Methods	Distance Sum	Facility Locations	Customer Allocations
HEM	97.435	1. (8.9509, 36,9761)	10, 11, 12, 13
		2. (3.1611, 63528)	1, 2, 3, 4, 5, 14
		3. (23.9536, 17.3541)	6, 7, 8, 9
Heuristic	107.0858	1. (9.5, 37.5)	10, 11, 12, 13
		2. (1.0, 4.59)	1, 2, 3, 14
		3. (19.8333, 16.6667)	4, 5, 6, 7, 8, 9

Evolutionary Process

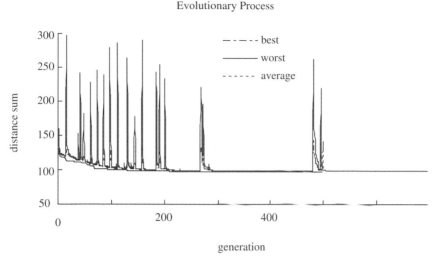

Figure 10.19. Evolutionary process of HEM.

not be built in any obstacle, and the electric lines should avoid passing through obstacles according to the safety and engineering possibility.

This problem has six continuous decision variables representing the location coordinates of three electric stations and 42 zero–one decision variables representing the allocation of oil wells to electric stations. With the hybrid evolutionary method proposed above, a more satisfactory solution is found than was suggested by the heuristic method in reference [177].

The environmental parameters for the hybrid method are set as follows: *max_gen*, 1200; *pop_size*, 10; *child_size*, 40; p_m, 10; α, 1.0; γ, 4.0. The comparative results are given in Table 10.13. The evolutionary processes are shown in Figure 10.19. It can be seen that the proposed hybride evolutionary method is more effective in finding a better solution than the heuristic method in this case study problem.

10.5 PRODUCTION PLAN PROBLEM

The production plan problem is to determine the rate of production under varying types of demand and cost. This problem has been considered by several authors such as Federgruen and Schechner [125], Liu [277], Luss [280], Moinzadeh and Nahmias [297], and Wang, Wilson and Odrey [411]. Previous studies are mainly done for solving the discrete time production plan problems, and the continuous-time case has received less attention from solution approaches. Recently, Gen and Liu have examined the continuous-time production plan problem and presented an evolution program for solving it [161, 444].

10.5.1   Formulation of Production Plan Problem

To provide a precise statement of this problem, we first introduce some notations as follows:

> $z(t)$ is the rate of production at time t
> $r(t)$ is the rate of demand at time t
> $y(t)$ is the inventory level at time t
> $c(z)$ is the cost of production per unit time when the rate of production is z
> $g(dz/dt)$ is the cost per unit time of changing the rate of production
> $l(y)$ is the inventory cost per unit time when the inventory level is y

If $y < 0$, $l(y)$ is the cost of backlogged sales of quantity y; and if $y > 0$, $l(y)$ is the storage cost of quantity y per unit time.

Then the total quantity of production at time t is

$$Z(t) = \int_0^t z(\tau)\,d\tau \tag{10.45}$$

the total demand at time t is

$$R(t) = \int_0^t r(\tau)\,d\tau \tag{10.46}$$

and, certainly,

$$y(t) = Z(t) - R(t) \tag{10.47}$$

Then the continuous-time production plan problem is to minimize

$$J(z) = \int_0^T \left\{ c(z(t)) + g\left(\frac{dz(t)}{dt}\right) + l(y(t)) \right\} dt \tag{10.48}$$

That is, we shall find a production plan $z(t)$ which is a function of t such that the return $J(z)$ is minimized.

In this model, we have assumed that the rate of production is differentiable in any interval of continuity. In the interpretation of the integral $\int_0^T g(dz/dt)\,dt$ of the cost function $J(z)$, it should be noted that we do not wish to confine ourselves to production policies which are differentiable or even continuous. If $z(t)$ is discontinuous at some point, then dz/dt is not defined and the difference $g(z(t+0)) - g(z(t-0))$ will be regarded as the contribution of that point to the integral. Thus the integral is the sum of the integral in which $z(t)$ is continuous and the contributions at the points of discontinuity.

If there is a positive setup cost in each time of changing the rate of production, then the optimal production rate must be a step function; that is, the rate of production is changed in finite times and constant in each interval.

10.5.2 Evolution Program for Production Plan Problem

Representation. As explained above, an optimal production plan takes the form of a step function. Thus it can be rewritten as follows:

$$z(t) = \begin{cases} z_i, & t_{i-1} \le t < t_i, \quad i = 1, 2, \ldots, n \\ 0, & t > T \end{cases} \tag{10.49}$$

Let us see an example given in Figure 10.20. From the figure we can see that a production plan is characterized by the following three factors:

1. The total number of changes of production rate n over planning horizon T, which separates the planning horizon into several periods
2. The value of production rate z_i at each period within T
3. The duration $t_i - t_{i-1}$ for each period at which production rate keeps unchanged

A chromosome must capture the three factors of production plan to represent a solution to the problem. Before the problem is solved, the total change number n is unknown. To accommodate this case, a variable-length string $(x_1, x_2, \ldots, x_{2n})$ is used as a chromosome to represent a solution of the problem, where the elements with even index x_{2i}, $i = 1, 2, \ldots, n$, represent production rate and those with odd index x_{2i-1}, $i = 1, 2, \ldots, n$, correspond to the proportion of duration to the planning horizon T for period i.

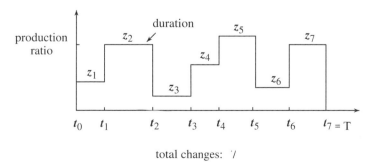

Figure 10.20. Example of production rate over planning horizon.

The deduction of solution from a chromosome is as follows:

$$t_0 = 0$$

$$t_i = \frac{x_1 + x_3 + \ldots + x_{2i-1}}{x_1 + x_3 + \ldots + x_{2n-1}} \cdot T, \qquad i = 1, 2, \ldots, n \qquad (10.50)$$

$$z_i = x_{2i}, \qquad i = 1, 2, \ldots, n$$

It is easy to see from equations (10.50) that a chromosome always generates feasible solution.

Initialization. Initial population is randomly generated by the following steps:

Step 1. Generate a random integer n the number of changes of production rate. Usually, the possible range for n can be given by problem-specific knowledge.

Step 2. Generate n random numbers $x_1, x_3, \ldots, x_{2n-1}$ on the interval $[\epsilon, 1]$.

Step 3. Generate n random numbers $x_2, x_4, \ldots, x_{2n}$ on the interval $[\epsilon, up\_bound]$

Step 4. Repeat above steps to produce *pop_size* chromosomes

where ϵ is a small positive number and *up_bound* is a possible upper bound of the rates of production.

Evaluation and Selection. A kind of rank-based evaluation function is used to assess the merit of each chromosome. The evaluation procedure consists mainly of three steps:

Step 1. Calculate the objective values according to objective (10.48).

Step 2. Sort chromosomes on the value of the objective in ascending order.

Step 3. Assign each chromosome a rank-based fitness value. Let r_k be the rank of chromosome $\boldsymbol{v}_k$. For a parameter $a \in (0, 1)$ specified by the user, the rank-based fitness function is defined as follows:

$$eval(\boldsymbol{v}_k) = a(1 - a)^{r_k - 1}$$

where $r_k = 1$ means the best chromosome and $r_k = pop\_size$ means the worst chromosome.

We have that

$$\sum_{k=1}^{pop\_size} eval(\boldsymbol{v}_k) \approx 1$$

The selection process is based on spinning a roulette wheel *pop_size* times; each time, a single chromosome is selected into a new population.

Crossover. Arithmetical crossover is used, which is defined as a linear combination of two vectors [283]. For each pair of parents v_1 and v_2, if they have the same dimension (same n), the crossover operator will produce two children v' and v'' as follows:

$$v' = c_1 \cdot v_1 + c_2 \cdot v_2$$
$$v'' = c_2 \cdot v_1 + c_1 \cdot v_2$$

where $c_1, c_2 \geq 0$ and $c_1 + c_2 = 1$. Since the constraint set is convex, the arithmetical crossover operation ensures that both children are feasible if both parents are.

Mutation. Two types of mutation operators are defined. The first, *hetero-dimensional mutation*, generates a new dimension n and then produces a new feasible solution $(x_1, x_2, \ldots, x_{2n})$ to replace the old one. The second, *homo-dimensional mutation*, mutates the chromosome in a negative gradient (subgradient) direction.

It is well known that the Taylor's expansion of a continuous differentiable function f is

$$f(x + \Delta x) = f(x) + (\nabla f(x + \theta \Delta x))' \Delta x, \qquad x \in R^n$$

where $0 \leq \theta \leq 1$, $\nabla f(x)$ denotes the gradient of the function f at point x, and Δx is a small perturbation in R^n. Because of the complexity of this problem, the gradient may not exist. In such a case, we can calculate the ith component of the gradient (regardless of the existence) approximately by using the following equation:

$$\frac{f(x_1, \ldots, x_i + \Delta x_i, \ldots, x_n) - f(x_1, \ldots, x_i, \ldots, x_n)}{\Delta x_i} \tag{10.51}$$

where Δx_i is a small real number.

Denote p_{m1} and p_{m2} as the probabilities of hetero-dimensional and homo-dimensional mutation, respectively. Mutation performs as follows: Generating a random real number r in $[0, 1]$, if $r < p_{m1}$, perform a hetero-dimensional mutation; if $p_{m1} < r < p_{m1} + p_{m2}$, perform a homo-dimensional mutation. For homo-dimensional mutation, a direction d of approximate negative gradient is generated with the equation (10.51). Let a parent be $v = (x_1, x_2, \ldots, x_{2n})$; an offspring is generated as follows:

$$v' = v + M \cdot d$$

If it is not feasible, then set M by a random real number in $(0, M)$ until $v + M \cdot d$ is feasible.

10.5.3 Example

Let us consider the following example: The rate of demand is

$$r(t) = 1 + \sin \left(\frac{t}{T} 4\pi \right), \qquad 0 \le t \le T \tag{10.52}$$

The cost function of production per unit time is

$$c(z) = c \cdot z \tag{10.53}$$

The cost function per unit time of changing the rate of production has the following form:

$$g\left(\frac{dz}{dt} \right) = k + w \cdot \left| \frac{dz}{dt} \right| \tag{10.54}$$

The inventory cost per unit time is defined by

$$l(y) = \begin{cases} h \cdot y, & y \ge 0 \\ s \cdot (-y), & y < 0 \end{cases}$$

Parameters were set as follows: $T = 100$; $c = 5$; $w = 20$; $k = 100$; $h = 3$; $s = 10$; initial inventory level, $y(0)$, equals 0; population size, 50; probability of crossover, p_c, equals 0.1; probability of hetero-dimensional mutation, p_{m1}, equals 0.1; probability of homo-dimensional mutation, p_{m2}, equals 0.4; parameter a in the rank-based evaluation function equals 0.1.

Figure 10.21 shows that in 3000 iterations the best objective is 2370.78. The best production plan with five changes in the rate of production is 1.93 on the interval [0, 1.78], 1.56 on the interval [1.78, 27.57), 0.50 on the interval [27.57, 46.03], 1.47 on the interval [46.03, 69.49), and 0.54 on the interval [69.49, 100] as shown in Figure 10.22. The accumulated quantities of production and demand are shown in Figure 10.23.

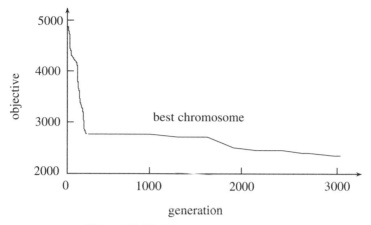

Figure 10.21. Progress of Evolution

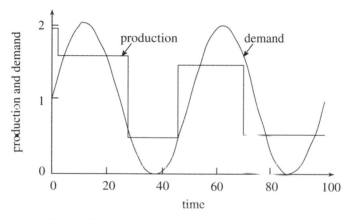

Figure 10.22. Rates of production and demand.

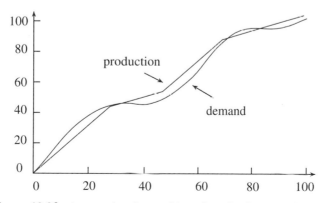

Figure 10.23. Accumulated quantities of production and demand.

BIBLIOGRAPHY

[1] Abuali, F., R. Wainwright, and D. Schoenefeld, A new encoding scheme for spanning trees applied to the probabilistic minimum spanning tree problem, in Eshelman [120], pp. 470–475.

[2] Ackley, D., *A Connectionist Machine for Genetic Hillclimbing*, Kluwer Academic Publishers, Boston, 1987.

[3] Adams, J., E. Balas, and D. Zawack, the shifting bottleneck procedure for job shop scheduling, *International Journal of Flexible Manufacturing Systems*, vol. 34, no. 3, pp. 391–401, 1987.

[4] Alander, J., Interval arithmetic and genetic algorithms in global optimization, In Pearson et al. [335], pp. 387–391.

[5] Albrecht, R., C. Reeves, and N. Steele, editors, *Artificial Neural Nets and Genetic Algorithms*, Springer-Verlag, New York, 1993.

[6] Alefeld, G. and J. Herzberger, *Introduction to Interval Computations*, Academic Press, New York, 1983 (translated by J. Rokne.)

[7] Alvarez-Valdés, R. and J. Tamarit, Heurisitic algorithms for resource constrained project scheduling: a review and an empirical analysis, in Slowinski, R. and J. Weglarz, editors, *Advances in Project Scheduling*, pp. 113–134, Elsevier Science Publishers, Amsterdam, 1989.

[8] Aneja, Y. and K. Nair, Bicriteria transportation problem, *Management Science*, vol. 25, pp. 73–78, 1978.

[9] Applegate, D. and W. Cook, A computational study of the job shop scheduling problem, *ORSA Journal of Computing*, vol. 3, no. 2, pp. 149–156, 1991.

[10] Bäck, T., Selective pressure in evolutionary algorithms: a characterization of selection mechanisms, in Fogel [132], pp. 57–62.

[11] Bäck, T., *Evolutionary Algorithms in Theory and Practice*, Oxford University Press, New York, 1996.

[12] Bäck, T., D. Fogel, and Z. Michalawecz, editors, *Handbook of Evolutionary Computation*, Oxford University Press, Oxford, 1997.

[13] Bäck, T. and F. Hoffmeister, Extended selection mechanisms in genetic algorithms, in Belew and Booker [30], pp. 92–99.

[14] Bäck, T. and S. Khuri, An evolutionary heuristic for the maximum independent set problem, in Fogel [132], pp. 531–535.

[15] Bagchi, S., S. Uckun, Y. Miyabe, and K. Kawamura, Exploring problem-specific recombination operators for job shop scheduling, in Belew and Booker [30], pp. 10–17.

[16] Bagchi, U., Simultaneous minimization of mean and variation of flow time and waiting time in single machine systems, *Operations Research*, vol. 37, pp. 118–125, 1989.

[17] Bagchi, U., L. Chang, and R. Sullivan, Minimizing absolute and squared deviations of completion times with different earliness and tardiness penalties and a common due date, *Naval Research Logistics Quarterly*, vol. 34, pp. 739–751, 1987.

[18] Bagchi, U., R. Sullivan, and L. Chang, Minimizing mean absolute deviation of completion times about a common due date, *Naval Research Logistics Quarterly*, vol. 33, pp. 227–240, 1986.

[19] Baker, J., Reducing bias and inefficiency in the selection algorithm, in Grefenstette [186], pp. 14–21.

[20] Baker, J., Adaptive selection methods for genetic algorithms, in Grefenstette [189], pp. 100–111.

[21] Baker, K., *Introduction to Sequencing and Scheduling*, John Wiley & Sons, New York, 1974.

[22] Baker, K. and G. Scudder, Sequencing with earliness and tardiness penalties: A review, *Operations Research*, vol. 38, pp. 22–36, 1990.

[23] Balas, E., Machine sequencing via disjunctive graphs: an implicit enumeration algorithm, *Operations Research*, vol. 17, pp. 941–957, 1969.

[24] Barlow, R., L. Hunter, and F. Proschan, Optimum redundancy when components are subject to two kinds of failure, *Journal of SIAM*, vol. 11, pp. 64–73, 1963

[25] Bauer, R., *Genetic Algorithms and Investment Strategies*, John Wiley & Sons, New York, 1994.

[26] Bazaraa, M., J. Jarvis, and H. Sherali, *Linear Programming and Network Flows*, 2nd ed., John Wiley & Sons, New York, 1990.

[27] Bazaraa, M., and O. Kirca, A branch and bound based heuristic for solving the QAP, *Naval Research Logistics Quarterly*, vol. 30, pp. 287–304, 1983.

[28] Bazaraa, M., H. Sherali, and C. Shetty, *Nonlinear Programming: Theory and Algorithm*, 2nd ed., John Wiley & Sons, New York, 1993

[29] Bean, J., Genetic algorithms and random keys for sequencing and optimization, *ORSA Journal on Computing*, vol. 6, no. 2, pp. 154–160, 1994.

[30] Belew, R. and L. Booker, editors, *Proceedings of the Fourth International Conference on Genetic Algorithms*, Morgan Kaufmann Publishers, San Mateo, CA, 1991.

[31] Bell, C. and J. Han, A new heuristic solution method in resource constrained project scheduling, *Naval Research Logistics*, vol. 38, pp. 315–331, 1991.

[32] Bertsimas, D., The probabilistic minimum spanning tree problem, *Networks*, vol. 20, pp. 245–275, 1990.

[33] Bhanu, B. and S. Lee, *Genetic Learning for Adaptive Image Segmentation*, Kluwer Academic Publishers, Norwell, MA, 1994.

[34] Biethahn, J., *Evolutionary Algorithms in Management Applications*, Springer-Verlag, Berlin, 1995.

[35] Bit, A., M. Biswal, and S. Allam, Fuzzy programming approach to multicriteria decision making transportation problem, *Fuzzy Sets and Systems*, vol. 50, pp. 135–141, 1992.

[36] Bits, A., M. Biswal, and S. Alam, Fuzzy programming approach to multiobjective solid transportation problem, *Fuzzy Sets and Systems*, vol. 57, pp. 183–194, 1993.

[37] Bjorndal, M., A. Caprara, P. Cowling, P. Croce, H. Lourence, F. Malucelli, A. Orman, D. Pisinger C. Rego, and J. Salazar, Some thoughts on combinatorial optimization, *European Journal of Operational Research*, vol. 83, pp. 253–270, 1995.

[38] Blackstone, J., D. Phillips, and G. Hogg, A state-of-the-art survey of dispatching rules for manufacturing job shop operations, *International Journal of Production Research*, vol. 20, pp. 26–45, 1982.

[39] Blanton, J. and R. Wainwright, Multiple vehicle routing with time and capacity constaints using genetic algorithm, in Forrest [137], pp. 452–459.

[40] Blazewicz, J., Complexity of computer scheduling algorithms under resource constraints, in *Proceedings, First Meeting of the AFCET-SMF on Applied Mathematics*, pp. 169–178, Palaiseau, Poland, 1978.

[41] Blazewicz, J., K. Ecker, G, Schmidt, and J. Weglarz, *Scheduling in Computer and Manufacturing Systems*, 2nd ed., Springer-Verlag, New York, 1994.

[42] Boctor, F., Some efficient multi-heuristic procedures for resource constrained project scheduling, *European Journal of Operational Research*, vol. 49, pp. 3–13, 1990.

[43] Bodin, L., B. Golden, A. Assad, and M. Ball, Routing and scheduling of vehicles and crews: the state of the art, *Computers and Operations Research*, vol. 10, pp. 62–212, 1983.

[44] Bolc, L. and J. Cytowski, *Search Methods for Artificial Intelligence*, Academic Press, London, 1992.

[45] Bongartz, I., A projection method for norm location-allocation problems, *Mathematical Programming*, vol. 66, pp. 283-312, 1994.

[46] Booker, L., Improving search in genetic algorithms, in Davis [102].

[47] Bracken, J., and G. McCormick, *Selected Applications of Programming*, John Wiley & Sons, New York, 1968

[48] Brindle, A., Genetic Algorithms for Function Optimization, Ph.D. thesis, Universiry of Alberta, Edmonton, 1981.

[49] Bruno L., J. Coffman, and R. Sethi, Scheduling independent tasks to reduce mean finishing time, *Communications on ACM*, vol. 17, pp. 382–387, 1974.

[50] Bruns, R., Direct chromosome representation and advanced genetic operators for production scheduling, in Forrest [137], pp. 352–359.

[51] Budnick, F., D. McLeavey and R. Mojena, *Principles of Research for Management*, 2nd ed., Irwin Press, Homewood, IL, 1988.

[52] Burkard, R. and T. Bonninger, A heuristic for quadratic boolean program with applications to quadratic assignment problems, *European Journal of Operational Research*, vol. 13, pp. 374–386, 1983.

[53] Campbell, H., R. Dudek, and M. Smith, A heuristic algorithm for the n-job m-machine sequencing problem, *Management Science*, vol. 16B, pp. 630–637, 1970.

[54] Chambers, L., *Practical Handbook of Genetic Algorithms*, vols. 1 and 2, CRC Press, New York, 1995.

[55] Charnes, A. and W. W. Cooper, *Management Models and Industrial Applications of Linear Programming*, John Wiley & Sons, New York, 1961

[56] Chartrand, G., and L. Lesniak, *Graphs and Digraphs*, Wadsworth & Brooks, Monterey, CA, 1979.

[57] Chen, C., V. Vempati, and N. Aljaber, An application of genetic algorithms for flow shop problems, *European Journal of Operational Research*, vol. 80, pp. 389–396, 1995.

[58] Chen, S and C. Hwang, *Fuzzy Multiple Attribute Decision Making*, Springer-Verlag, Berlin, 1992.

[59] Cheng, C., Study on Optimal Design of Fuzzy Facility Layout, Master's thesis, Ashikaga Institute of Technology, Ashikaga, Japan, March 1995.

[60] Cheng, R. and M. Gen, On film copy deliverer problem, In Zheng, W., editor, *Proceedings of the Second International Conference on Systems Science and Systems Engineering*, pp. 542–547, Bejing, 1993.

[61] Cheng R. and M. Gen, Crossover on intensive search and traveling salesman problem, in Gen and Kobayashi [160], pp. 568–571.

[62] Cheng, R. and M. Gen, Evolution program for resource constrained project scheduling problem, in Fogel [132], pp. 736–741.

[63] Cheng, R. and M. Gen, Vehicle routing problem with fuzzy due-times using genetic algorithm, in *The Third Conference of Asian-Pacific Operational Research Society*, Fukuoka, Japan, 1994.

[64] Cheng, R. and M. Gen, Resource constrainted project scheduling problem using genetic algorithms, *International Journal of Intelligent Automation and Soft Computing*, 1996 (to appear).

[65] Cheng, R. and M. Gen, Fuzzy vehicle routing and scheduling problem using genetic algorithms, in Herrera, F. and J. Verdegay, editors, *Genetic Algorithms and Soft Computing*, pp. 683–709, Springer-Verlag, 1996.

[66] Cheng, R., M. Gen, and M. Sasaki, Film-copy deliverer problem using genetic algorithms, *International Journal of Computers and Industrial Engineering*, vol. 29, no. 1–4, pp. 549–553, 1995.

[67] Cheng, R. and M. Gen, Genetic algorithms for multi-row machine layout problem, *Engineering Design and Automation*, 1996 (to appear).

[68] Cheng, R, and M. Gen, Minmax earliness/tardiness scheduling in identical parallel machine syetem using genetic algorithm, *International Journal of Computers and Industrial Engineering*, vol. 29, no. 1–4, pp. 513–517, 1995.

[69] Cheng, R. and M. Gen, Vehicle routing problem with fuzzy due-time using genetic algorithms, *Japanese Journal of Fuzzy Theory and Systems*, vol. 7, no. 5, pp. 1050–1061, 1995.

[70] Cheng, R. and M. Gen, Genetic search for facility layout design under interflows uncertainty, *Japanese Journal of Fuzzy Theory and Systems*, vol. 8, no. 2, pp. 335–346, 1996.

[71] Cheng, R., M. Gen, and Y. Tsujimura, A tutorial survey of job-shop scheduling problems using genetic algorithms: part I. representation, *International Journal of Computers and Industrial Engineering*, vol. 30, no. 4, pp. 983–997, 1996.

[72] Cheng, T. and X. Cai, On the complexity of completion time variance minimization problem, Working paper, University of Western Australia, Perth, 1990.

[73] Cheng, T. and M. Gupta, Survey of scheduling research involving due date determination decision, *European Journal of Operational Research*, vol. 38, pp. 156–166, 1989.

[74] Cheng, T. and C. Sin, A state-of-the-art review of parallel-machine scheduling research, *European Journal of Operational Research*, vol. 47, pp. 271–292, 1990.

[75] Christofides, N., R. Alvarez-Valdés and J. Tamarit, Project scheduling with resource constrained: A branch and bound approach, *European Journal of Operational Research*, vol. 29, pp. 262–273, 1987.

[76] Cleveland, G. and S. Smith, Using genetic algorithms to schedule flow shop releases, in Schaffer [370], pp. 160–169.

[77] Climaco, J., C. Antunes, and M. Alves, Interactive decision support for multiobjective transportation problem, *European Journal of Operational Research*, vol. 65, pp. 58–67, 1993.

[78] Coffman, E., *Computer and Job-shop Scheduling Theory*, John Wiley & Sons, New York, 1976.

[79] Cohen, J., R. Church and D. Sheer, Generating multiobjective trade-off: an algorithm for bicriteria problems, *Water Resource Research*, vol. 15, pp. 1001–1010, 1979.

[80] Cohon, J., *Multiobjective Programming and Planning*, Academic Press, New York, 1978.

[81] Cohoon, J., S. Hegde, and N. Martin, Distributed genetic algorithms for the floorplan design problem, *IEEE Transactions on Computer-Aided Design*, vol. 10, pp. 483–491, 1991.

[82] Coit, D. and A. Smith, Penalty guided genetic search for reliability design optimization, *International Journal of Computers and Industrial Engineering*, vol. 30, no. 4, pp. 895–904, 1996.

[83] Coit, D. and A. Smith, Reliability optimization of series-parallel systems using a genetic algorithm, *IEEE Transactions on Reliability*, 1996 (to appear).

[84] Conway, R., W. Maxwell, and L. Miller, *Theory of Scheduling*, Addison-Wesley, Reading, MA, 1967.

[85] Cooper, L., Location-allocation problems, *Operations Research*, vol. 11, no. 3, pp. 331–344, 1963.

[86] Croce, F., R. Tadei, and G. Volta, A genetic algorithm for the job shop problem, *Computers and Operations Research*, vol. 22, pp. 15–24. 1995.

[87] Crowder, H. and M. Padberg, Solving large-scale symmetic traveling salesman problems to optimality, *Management Science*, vol. 26, pp. 495–509, 1980.

[88] Current, J. and M. Marsh, Multobjective design of transportation networks and routing problems: taxonomy and annotation, *European Journal of Operational Research*, vol. 65, pp. 4 19, 1993.

[89] Current, J., and H. Min, Multiobjective design of transportation networks: taxonomy and annotation, *European Journal of Operational Research*, vol. 26, pp. 187–201, 1986.

[90] Dagtzig, G., D. Fulkerson, and S. Johnson, Solution of a large scale traveling salesman problems, *Operations Research*, vol. 2, pp. 393–410, 1954.

[91] Dannenbring, D., An evaluation of flow shop sequencing heuristics, *Management Science*, vol. 23, pp. 1174–1182, 1977.

[92] Dauzere-Pers, S. and J. Lasserre, A modified shifting bottleneck procedure for job-shop scheduling, *International Journal of Production Researches*, vol. 31, pp. 923–932, 1993.

[93] Davidor, Y., A genetic algorithm Applied to robot trajectory generation, in Davis [101], pp. 923–932, 1991

[94] Davidor, Y., *Genetic Algorithms and Robotics*, World Scientific Publishing, Singapore, 1991.

[95] Davidor, Y., H. Schwefel, and R. Männer, editors, *Parallel Problem Solving from Nature: PPSN III*, Springer-Verlag, Berlin, 1994.

[96] Davis, E, Computational experience with a new multi-resource algorithm, in Lombaers, H., editor, *Project Planning by Network Analysis*, pp. 256–260, North-Holland, New York, 1969.

[97] Davis, E. and G. Heidorn, An algorithm for optimal project scheduling under multiple resources constraints, *Management Science*, vol. 17, pp. B803–B816, 1971

[98] Davis, E. and J. Patterson, A comparison of heuristic and optimum solutions in resource-constained project scheduling, *Management Science*, vol. 21, pp. 944–955, 1975.

[99] Davis, L., Applying adaptive algorithms to domains, In *Proceedings of the International Joint Conference on Artificial Intelligence*, pp. 162–164, 1985.

[100] Davis, L., Job shop scheduling with genetic algorithms, in Grefenstette [186], pp. 136–140.

[101] Davis, L., editor, *Handbook of Genetic Algorithms*, Van Nostrand Reinhold, New York, 1991.

[102] Davis, L., editor, *Genetic Algorithms and Simulated Annealing*, Morgan Kaufmann Publishers, Los Altos, CA, 1987.

[103] Dawkins, R., *The Selfish Gene*, Oxford University Press, Oxford, 1976.

[104] De Jong, K., An Analsis of the Behavoir of a Class of Genetic Adaptive Systems, Ph.D. thesis, University of Michigan, Ann Arbor, 1975.

[105] De Jong, K., Genetic algorithms: a 25 year perspective, in Zurada et al. [433], pp. 125–134.

[106] De Jong, K. and W. Spears, On the state of evolutionary computation, in Forrest [137], pp. 618–623, 1993.

[107] Dell'Amico, M. and M. Trubian, Applying tabu search to the job shop scheduling problem, *Annals of Operations Research*, vol. 40, pp. 231–252, 1993.

[108] deSilva, C., editor, *Proceedings of the Second IEEE Conference on Evolutionary Computation*, IEEE Press, Perth, 1995.

[109] Dijkstra, E., A note on two problems in connection with graphs, *Numerische Mathematik*, vol. 1, pp. 269–271, 1959.

[110] Dileepan, P. and T. Sen, Bicriterion static scheduling research for a single machine, *Omega*, vol. 16, pp. 53–59, 1988.

[111] Domschke, W. and A. Drex, *An International Bibliography on Location and Layout Planning*, Springer, Heidelberg, 1984.

[112] Dorndorf, U. and E. Pesch, Evolution based learning in a job shop scheduling environment, *Computers and Operations Research*, vol. 22, pp. 25–40, 1995.

[113] Drexl, A. and J. Gruenewald, Nonpreemptive multi-mode resource constrained project scheduling, *IIE Transaction*, vol. 25, pp. 74–81, 1993.

[114] Dubois, D. and H. Prade, Fuzzy real algebra: some results, *Fuzzy Sets and Systems*, vol. 2, 327–348, 1979.

[115] Dudek, R., S. Panwalkar, and M. Smith, The lessons of flow shop scheduling research, *Operations Research*, vol. 40, pp. 7–13, 1992.

[116] Eilon, S. and I. Chowdhury, Minimizing waiting time variance in the single machine problem, *Management Science*, vol. 23, pp. 567–575, 1977.

[117] Eilon, S. et al., *Distribution Management: Mathematical Modeling and Practice Analysis*, Hafner, New York, 1971.

[118] El-Sayed, M. E. M., B. J. Ridgely, and Sandgren, Nonlinear structural optimization using goal programming, *Computers and Structures*, vol. 32, no. 1, pp. 69–73, 1989.

[119] Emmons, H., Scheduling to a common due date on parallel common processors, *Naval Research Logistics Quarterly*, vol. 34, pp. 803–810, 1987.

[120] Eshelman, L. J., editor, *Proceedings of the Sixth International Conference on Genetic Algorithms*, Morgan Kaufmann Publishers, San Francisco, 1995.

[121] Eshelman, L. and J. Schaffer, Real-coded genetic algorithms and interval-schemata, in Whitley [416], pp. 187–202.

[122] Falkenauer, E. and S. Bouffoix, A genetic algorithm for job shop, in *Proceedings*

of the IEEE International Conference on Robotics and Automation, pp. 824–829, 1991.

[123] Falkenauer, E. and Delchambre, A genetic algorithm for bin packing and line balancing, in *Proceeding of the IEEE International Conference on Robotics and Automation*, pp. 1186–1193, 1992.

[124] Fang, H., P. Ross, and D. Corne, A promising genetic algorithm approach to job-shop scheduling, rescheduling, and open-shop scheduling problems, in Forrest [137], pp. 375–382.

[125] Federgruen, A. and Z. Schechner, Cost formulas for continuous review inventory models with fixed delivery lags, *Operations Research*, vol. 31, no. 5, pp. 957–965, 1983.

[126] Fisher, H. and G. Thompson, Probabilistic learning combinations of job-shop scheduling rules, in Muth and Thompson [307], Chapter 15, pp. 1225–1251, 1963.

[127] Floudas, C. and P. Paralos, *Recent Advances in Global Optimization*, Princeton University Press, Princeton, 1992.

[128] Fogarty, T., editor, *Evolutionary Computing*, Springer-Verlag, Berlin, 1994.

[129] Fogel, D., editor, *Proceedings of the Third IEEE Conference on Evolutionary Computation*, IEEE Press, Nagoya, Japan, 1996

[130] Fogel, D., An Introduction to simulated evolutionary optimization, *IEEE Transactions on Neural Networks*, vol. 5, pp. 3–14, 1994.

[131] Fogel, D., *Evolutionary computation: toward a new philosophy of machine intelligence*, IEEE Press, Piscataway, NJ, 1995.

[132] Fogel, D., editor, *Proceedings of the First IEEE Conference on Evolutionary Computation*, IEEE Press, Orlando, FL, 1994.

[133] Fogel, D. and W. Atmar, editors, *Proceeding of the First Annual Conference on Evolutionary Programming*, Evolutionary Programming Society, San Diego, 1992.

[134] Fogel, D. and W. Atmar, editors, *Proceeding of the Second Annual Conference on Evolutionary Programming*, Evolutionary Programming Society, La Jolla, 1993.

[135] Fonseca, C. and P. Fleming, Genetic algorithms for multiobjective optimization: formulation, discussion and generalization, in Forrest [137], pp. 416–423.

[136] Forrest, S., Documentation for prisoners dilemma and norms programs that use the genetic algorithm, working paper, University of Michigan, Ann Arbor, 1985.

[137] Forrest, S., editor, *Proceedings of the Fifth International Conference on Genetic Algorithms*, Morgan Kaufmann Publishers, San Mateo, CA, 1993.

[138] Francis, R., L. McGinnis, and J. White, *Facility Layout and Location: An Analytical Approach*, 2nd ed., Prentice Hall, Englewood Cliffs, NJ, 2nd ed., 1992.

[139] Fraser, A., Simulation of genetic systems by automatic digital computers: I. introduction, *Australian Journal of Biological Science*, vol. 10, pp. 484–491, 1957.

[140] Fraser, A., Simulation of genetic systems by automatic digital computers: II. effects of linkage on rates of advance under selection, *Australian Journal of Biological Science*, vol. 10, pp. 492–499, 1957.

[141] Fraser, A., Simulation of genetic systems by automatic digital computers:VI. epistasis, *Australian Journal of Biological Science*, vol. 13, pp. 150–162, 1960.

[142] Fraser, A., Simulation of genetic systems, *Journal of Theoretical Biology*, vol. 2, pp. 329–346, 1962.

[143] Freeman, J., *Simulating Neural Networks with Mathematics*, Addison-Wesley, New York, 1994.

[144] French, S., *Sequencing and Scheduling: An Introduction to the Mathematics of the Job-Shop*, John Wiley & Sons, New York, 1982.

[145] Fyffe, D., W. Hines, and N. Lee, System reliability allocation and a computational algorithm, *IEEE Transactions on Reliability*, vol. R-17, pp. 64–69, 1968.

[146] Gabriete, D. and K. Ragsdell, Large scale nonlinear programming using the generalized reduced gradient method, *ASME Journal of Mechanical Design*, vol. 102, pp. 566–573, 1980.

[147] Garey, M. and D. Johnson, Strong NP- completeness results: motivation, examples and implications, *Journal of ACM*, vol. 25, pp. 499–508, 1978.

[148] Garey, M., D. Johnson, and R. Sethi, The complexity of flowshop and jobshop scheduling, *Mathematics of Operations Research*, vol. 1, pp. 117–129, 1976.

[149] Gen, M., Reliability opimization by 0-1 programming for a system with several failure modes, *IEEE Transactions on Reliability*, vol. R-24, pp. 206–210, 1975.

[150] Gen, M. and R. Cheng, Optimal design of system reliability under uncertainty using interval programming and genetic algorithm, Technical report, ISE94-6, Ashikaga Institute of Technology, Ashikaga, Japan, 1994.

[151] Gen, M. and R. Cheng, Interval programming using genetic algorithms, in *Proceeding of First International Symposium on Soft Computing for Industry*, 1996.

[152] Gen, M. and R. Cheng, A survey of penalty techniques in genetic algorithms, in Fogel [129], pp. 804–809.

[153] Gen, M. and K. Ida, *Linear Programming and Goal Programming using BASIC*, Denki-Shoin Publisher, Tokyo, 1984 (in Japanese).

[154] Gen, M. and K. Ida, *Goal Programming Using Turbo C*, HBJ Publishers, Tokyo, 1993 (in Japanese).

[155] Gen, M., K. Ida, E. Kono, and Y. Li, Solving bicriteria solid transportation problem by genetic algorithm, in Gen and Kobayashi [160], pp. 572–575.

[156] Gen, M., K. Ida, and J. Lee, Optimal selection and allocation of a system availability using 0-1 linear programming with GUB structure, *Transactions of Institute of Electronics, Information and Communication Engineers*, vol J71-D, pp. 2140–2147, 1988 (in Japanese).

[157] Gen, M., K. Ida, and Y. Li, Solving Bicriteria Solid Transportation Problem with Fuzzy Numbers by Genetic Algorithm, *International Journal of Computers and Industrial Engineering*, vol. 29, pp. 537–543, 1995.

[158] Gen, M., K. Ida, M. Sasaki, and J. Lee, Algorithm for solving large-scale 0-1 goal programming and its application to reliability optimization problem, *International Journal of Computers and Industrial Engineering*, vol. 17, pp. 525–530, 1989.

[159] Gen, M., K. Ida, and T. Taguchi, Reliability optimization problems: a novel

genetic algorithm approach, Technical report, ISE93-5, Ashikaga Institute of Technology, Ashikaga, Japan, 1993.

[160] Gen, M. and T. Kobayashi, *Proceedings of the 16th International Conference on Computers and Industrial Engineering*, Ashikaga, Japan, 1994.

[161] Gen, M. and B. Liu, Evolution program for production plan problem, *Engineering Design and Automation*, vol. 1, no. 3, pp. 199–204, 1995.

[162] Gen, M. and B. Liu, A genetic algorithm for nonlinear goal programming, Technical report, ISE95-5, Ashikaga Institute of Technology, Ashikaga, Japan, 1995.

[163] Gen, M., B. Liu, and K. Ida, Evolution program for deterministic and stochastic optimizations, *European Journal of Operational Research*, 1996 (in press).

[164] Gen, M., B. Liu, K. Ida, and D. Zheng, Evolution program for constrained nonlinear optimization, in Gen and Kobayashi [160], pp. 576–579.

[165] Gen, M., Y. Tsujimura, and E. Kubota, Solving job-shop scheduling problem using genetic algorithms, in Gen and Kobayashi [160], pp. 576–579.

[166] Gen, M and G. Zhou, An approach to the degree-constrained minimum spanning tree problem using genetic algorithm, Technical report ISE95-3, Ashikaga Institute of Technology, Ashikaga, Japan, 1995.

[167] Giffler, B. and G. Thompson, Algorithms for solving production scheduling problems, *Operations Research*, vol. 8, no. 4, pp. 487–503, 1960.

[168] Gilbert, E., Minimum cost communication networks, *Journal of Bell Systems Technology*, vol. 9, pp. 2209–2227, 1967.

[169] Gillies, A., Machine learning procedures for generating image domain feature detectors, Ph. D. thesis, University of Michigan, Ann Arbor, 1985.

[170] Glover, F. and H. Greenberg, New approaches for heuristic search: a bilateral linkage with artificial intelligence, *European Journal of Operational Research*, vol. 39, pp. 119–130, 1989.

[171] Goldberg, D., *Genetic Algorithms in Search, Optimization and Machine Learning*, Addison-Wesley, Reading, MA, 1989.

[172] Goldberg, D. and K. Deb, A comparative analysis of selection schemes used in genetic algorithms, in Rawlins [347], pp. 69–93, 1991

[173] Goldberg, D., B. Korb, and K. Deb, Messy genetic algorithms: motivation, analysis, and first results, *Complex Systems*, vol. 3, pp. 493–530, 1989.

[174] Goldberg, D. and R. Lingle, Alleles, loci and the traveling salesman problem, in Grefenstette [186], pp. 154–159.

[175] Gong, D., M. Gen, W. Xu, and G. Yamazaki, Evolutionary strategy for obstacle location-allocation problem, in Zimmermann, H., editor. *Proceedings of the Third European Congress on Intelligent Techniques and Soft Computing*, pp. 426–433, Aachen, Germany, 1995.

[176] Gong, D., M. Gen, W. Xu, and G. Yamazaki, Hybrid evolutionary method for obstacle location-allocation problem, *International Journal of Computers and Industrial Engineering*, vol. 29, no. 1–4, pp. 525–530, 1995.

[177] Gong, D., M. Gen, G. Yamazaki, and W. Xu, Hybrid evolutionary method for capacitated location-allocation, *Engineering Design and Automation*, 1997 (to appear).

[178] Gong D., W. Xu, and M. Gen, Obstacle location-allocation in oil field, *Proceedings of 2nd Symposium of CIMS of Asian Countries*, Tokyo, 1994.

[179] Gordon, V. and D. Whitney, Serial and parallel genetic algorithms as functions optimizers, in Forrest [137], pp. 177–183.

[180] Gorges-Schleuter, M., ASPARAGOS: an asynchronous parallel genetic optimization strategy, in Schaffer [370], pp. 422–427.

[181] Graham, R. and P. Hell, On the history of the minimum spanning tree problem, *Annals of the History of Computing*, vol. 7, pp. 43–57, 1985.

[182] Graham, R., E. Lawler, J. Lenstra, and A. Rinnooy Kan, Optimization and approximation in deterministic sequencing and scheduling theory: a survey, *Annals of Discrete Mathematics*, vol. 5, pp. 287–326, 1979

[183] Graves, S., A review of production scheduling, *Operations Research*, vol. 29, pp. 646–675, 1981.

[184] Greenberg, H., A branch-and-bound solution to the general scheduling problem, *Operations Research*, vol. 16, pp. 353–361, 1968.

[185] Grefenstette, J., R. Gopal, B. Rosmaita, and D. Gucht, Genetic algorithms for the traveling salesman problem, in Grefenstette [186], pp. 160–168.

[186] Grefenstette, J., editor, *Proceedings of the First International Conference on Genetic Algorithms*, Lawrence Erlbaum Associates, Hillsdale, NJ, 1985.

[187] Grefenstette, J., Optimization of control parameters for genetic algorithms, *IEEE Transactions on Systems, Man, and Cybernetics*, vol. 16, pp. 122–128, 1986.

[188] Grefenstette, J., Incorporating problem specific knowledge into genetic algorithms, in Davis [102].

[189] Grefenstette, J., editor, *Proceedings of the Second International Conference on Genetic Algorithms*, Lawrence Erlbaum Associates, Hillsdale, NJ, 1987.

[190] Grefenstette, J., Lamarkian learning in multi-agent environment, in Belew and Booker [30], pp. 303–310.

[191] Grefenstette, J., *Genetic Algorithms for Machine Learning*, Kluwer Academic Publishers, Norwell, MA, 1994.

[192] Grefenstette, J. and J. Baker, How genetic algorithms work: a critical look at implicit parallelism, in Schaffer [370], pp. 20–27.

[193] Grötschel, M., On the symmetric traveling salesman problem: solution of a 120-city problem, *Mathematical Programming Studies*, vol. 12, pp. 61–77, 1980.

[194] Grötschel, M. and O. Holland, Solution of large-scale symmetric traveling salesman problems, *Mathematical Programming*, vol. 51, pp. 141–202, 1991.

[195] Gupta, J., A functional heuristic algorithm for the flow shop scheduling problem, *Operations Research Quarterly*, vol. 22, pp. 39–47, 1971.

[196] Gupta, M., Y. Gupta, and A. Kumar, Minimizing flow time variance in a single machine system using genetic algorithm, *European Journal of Operational Research*, vol. 70, pp. 289–303, 1993.

[197] Gupta, S. and J. Kyparisis, Single machine scheduling research, *Omega*, vol. 15, pp. 207–227, 1987.

[198] Hadley, G., *Linear Programming*, Addison-Werley, Palo Alto, CA, 1962.

[199] Hakimi, S., Optimum distribution of switching centers in a communication network and some related graph theoretic problems, *Operations Research*, vol. 13, pp. 462–475, 1964.

[200] Hall, N., Single and multi-processor models for minimizing completion time variance, *Naval Research Logistics Quarterly*, vol. 33, pp. 49–54, 1986.

[201] Hall, N., W. Kubiak, and S. Sethi, Earliness-tardiness scheduling problems II: deviation of completion times about a restrictive common due date, *Operations Research*, vol. 39, pp. 847–856, 1991.

[202] Hall, N. and M. Posner, Earliness-tardiness scheduling problems I: weighted deviation of completion times about a common due date, *Operations Research*, vol. 39, pp. 836–846, 1991.

[203] Halton, J., A retrospective and prospective survey of the Monte Carlo method *SIAM Review*, vol. 12, pp. 1–63, 1970.

[204] Hamacher, H. and S. Nickel, Restricted planar location problems and applications, *Naval Research Logistics*, vol. 42, pp. 967–992, 1995.

[205] Hammersley, J. and D. Handscomb, *Monte Carlo Methods*, Methuen, London, 1964.

[206] Hancock, P., An Empirical comparison of selection methods in evolutionary algorithms, in Fogarty [128], pp. 80–95.

[207] Hansen, E., *Global Optimization Using Interval Analysis*, Marcel Dekker, New York, 1992.

[208] Hardy, G., J. Littlewood, and G. Polya, *Inequalities*, Cambridge University Press, London, 1967.

[209] Harris, B., A. Farhi, and J. Dutour, Aspects of a problem in clustering, Technical report, University of Pennsylvania, Philadelphia, 1972.

[210] Haupt, R., A survey of priority rule based scheduling problem, *OR Spektrum*, vol. 11, pp. 3–16, 1989.

[211] Heragu, S. and A. Kusiak, Machine layout problem in flexible manufacturing systems, *Operations Research*, vol. 36, pp. 258–268, 1988.

[212] Hesser, J., R. Männer, and O. Stucky, Optimization of Steiner trees using genetic algorithms, in Schaffer [370], pp. 231–236.

[213] Hillier, F. and M. Connors, Quadratic assignment problem algorithms and the location of indivisible facilities, *Management Science*, vol. 13, pp. 42–57, 1966.

[214] Hillier, F. and G. Lieberman, *Introduction to Mathematical Programming*, McGraw-Hill, New York, 1991.

[215] Himmelblau, M., *Applied Nonlinear Programming*, McGraw-Hill, New York, 1972.

[216] Hinterding, R., Mapping, order-independent genes and the knapsack problem, in Fogel [132], pp. 13–17.

[217] Hitchcock, F., The distribution of a product from several sources to numerous locations, *Journal of Mathematical Physics*, vol. 20, pp. 224–230, 1941.

[218] Ho, J. and Y. Chang, A new heuristic for the n-job m-machine flow shop problem, *European Journal of Operational Research*, vol. 52, pp. 194–202, 1991

[219] Hock, W. and K. Schittkowski, *Test Examples for Nonlinear Programming Codes*, Springer-Verlag, New York, 1981.

[220] Holland, J., *Adaptation in Natural and Artificial Systems*, University of Michigan Press, Ann Arbor, 1975.

[221] Holsapple, C., V. Jacob, R. Pakath, and J. Zaveri, A genetics-based hybrid scheduler for generating static schedules in flexible manufacturing contexts, *IEEE Transactions on Systems, Man, and Cybernetics*, vol. 23, pp. 953–971, 1993.

[222] Homaifar, A., C. Qi, and S. Lai, Constrained optimization via genetic algorithms, *Simulation*, vol. 62, no. 4, pp. 242–254, 1994.

[223] Horn, J., N. Nafpliotis, and D. Goldberg, A niched Pareto genetic algorithm for multiobjective optimization, in Fogel [132], pp. 82–87.

[224] Hwang, C. and A. Masud, *Multiple Objective Decision Making Methods and Applications*, Springer, Berlin, 1979.

[225] Hwang, C. and K. Yoon, *Multiple Attribute Decision Making: Methods and Applications*, Springer-Verlag, Berlin, 1981.

[226] Ida, K., M. Gen, and T. Yokota, System reliability optimization with several failure modes by genetic algorithm, in Gen and Kobayashi [160], pp. 349–352.

[227] Ignall, E. and L. Schrage, Application of the branch and bound technique to some flow shop scheduling problems, *Operations Research*, vol. 13, pp. 400–412, 1965.

[228] Ignizio, J. P., *Goal Programming and Extensions*, Heath, Lexington, MA, 1976.

[229] Ignizio, J., *Linear Programming in Single & Multiple-Objective System*, Prentice Hall, Englewood Cliffs, NJ, 1982.

[230] Ijiri, Y., *Management Goals and Accounting for Control*, North-Holland, Amsterdam, 1965.

[231] Ishbuchi, H. and H. Tanaka, Formulation and analysis of linear programming problem with interval coefficients, *Journal of Japan Industrial Management Association*, vol. 40, no. 5, 320–329, 1989 (in Japanese).

[232] Ishibuchi, H. and H. Tanaka, Multiobjective programming in optimization of the interval objective function, *European Journal of Operational Research*, vol. 48, pp. 219–225, 1990.

[233] Ishibuchi, H., N. Yamamoto, T. Murata, and H. Tanaka, Genetic algorithms and neighborhood search algorithms for fuzzy flowshop scheduling problems, *Fuzzy Sets and Systems*, vol. 67, pp. 81–100, 1994.

[234] Janilow, C. and Z. Michalewicz, An experimental comparison of binary and floating point representations in genetic algorithms, in Belew and Booker [30], pp. 31–36.

[235] Jeffcoat, D. E. and R. Bulfin, Simulated annealing for resource constrained scheduling, *European Journal of Operational Research*, vol. 70, pp. 43–51, 1993.

[236] Johnson, S., Optimal two-and-three stage production schedules with setup times included, *Naval Research Logistics Quarterly*, vol. 1, pp. 61–68, 1954.

[237] Joines, J. and C. Houck, On the use of non-stationary penalty functions to solve nonlinear constrained optimization problems with GAs, in Fogel [132], pp. 579–584.

[238] Kaku, B. and G. Thompson, An exact algorithm for the general quadratic assignment problem, *European Journal of Operational Research*, vol. 23, pp. 383–390, 1986.

[239] Kall, P. and S. W. Wallace, *Stochastic Programming*, John Wiley & Sons, Chichester, 1994.

[240] Kanet, J., Minimizing the average deviation of job completion times about a common due date, *Naval Research Logistics Quarterly*, vol. 28, pp. 643–651, 1981.

[241] Kanet, J. and V. Sridharan, PROGENITOR: a genetic algorithm for production scheduling, *Wirtschaftsinformatik*, vol. 10, pp. 332–336, 1991.

[242] Katz, I. and L. Cooper, Facility location in the presence of forbidden regions; I. formulation and the case of the Euclidean distance with one forbidden circle, *European Journal of Operational Research*, vol. 6, pp. 166–173, 1981.

[243] Kaufmann, A. and M. Gupta, *Fuzzy Mathematical Models in Engineering and Management Science*, North-Holland, Amsterdam, 1988.

[244] Kaufmann, A. and M. Gupta, *Fuzzy Mathematical Models in Engineering and Management Science*, 2nd ed., North-Holland, Amsterdam, 1991.

[245] Kelley, J., The critical path method: resource planning and scheduling, in Muth and Thompson [307], pp. 347–365, 1963.

[246] Kennedy, S., Five ways to a smarter genetic algorithm, *AI Expert*, pp. 35–38, 1993.

[247] Kershenbaum, A., *Telecommunications Network Design Algorithms*, McGraw-Hill, New York, 1993.

[248] Kim, J. H., and H. Myung, A two-phase evolutionary programming for general constrained optimization problem, in *Proc. of the Fifth Annual Conference on Evolutionary Programming*, San Diego, 1996.

[249] Kobayashi, S., I. Ono, and M. Yamamura, An efficient genetic algorithm for job shop scheduling programs, in Eshelman [120], pp. 506–511.

[250] Koopmans, T. and M. Beckman, Assignment problems and the location of economic activities, *Econometrica*, vol. 25, pp. 53–76, 1957.

[251] Koulamas, C., The total tardiness problem: review and extensions, *Operations Research*, vol. 42, no. 6, pp. 1025–1044, 1994.

[252] Kouvelis, P., A. Kurawarwala, and G. Gutierrez, Algorithms for robust single and multiple period layout models for manufacturing systems, *European Journal of Operational Research*, vol. 63, pp. 287–303, 1992.

[253] Koza, John R., *Genetic Programming*, MIT Press, Cambridge, MA, 1992.

[254] Koza, John R., *Genetic Programming II*, MIT Press, Cambridge, MA, 1994.

[255] Kruskal, J., Jr., On the shortest spanning subtree of graph and the traveling salesman problem, *Proc. ACM*, vol. 7, no. 1, pp. 48–50, 1956.

[256] Kubota, A., Study on Optimal Scheduling for Manufacturing System by Genetic Algorithms, Master's thesis, Ashikaga Institute of Technology, Ashikaga, Japan, 1995.

[257] Kuo, T. and S. Hwang, A genetic algorithm with disruptive selection, in Forrest [137], pp. 65–69.

[258] Kusiak, A., *Intelligent Manufacturing System*, Prentice-Hall, Englewood Cliffs, NJ, 1990.

[259] Kusiak, A. and S. Heragu, The facility layout problem, *European Journal of Operational Research*, vol. 29, pp. 229–251, 1987.

[260] Lapaugh, A., Algorithms for Integrated Circuit Layout: Analytic Approach, Ph.D. thesis, Massachusetts Institute of Technology, Cambridge, MA, 1980.

[261] Larson, R. and G. Sadiq, Facility location with the manhattan metric in the presence of barriers to travel, *Operations Research*, vol. 31, pp. 652–669, 1983.

[262] Law, A. M. and W. D. Kelton, *Simulation Modeling and Analysis*, 2nd. ed., McGraw-Hill, New York, 1991.

[263] Lawer, E., J. Lenstra, A. Rinnooy Kan, and D. Shmoys, editors, *The Traveling Salesman Problem*, John Wiley & Sons, Chichester, 1985.

[264] Lawton, G., *A Practical Guide to Algorithms in C++*, John Wiley & Sons, New York, 1996.

[265] Lee, C., S. Danusaputro, and C. Lim, Minimizing weighted number of tardy jobs and weighted earliness-tardiness penalties about a common due date, *Computers and Operations Research*, vol. 18, pp. 379–389, 1991.

[266] Lee, C. and S. Kim, Parallel genetic algorithms for the tardiness job scheduling problem with general penalty weights, *International Journal of Computers and Industrial Engineering*, vol. 28, pp. 231–243, 1995.

[267] Lee, E. and R. Li, Comparison of fuzzy numbers based on the probability measure of fuzzy events, *Operations Research*, vol. 15, pp. 887–896, 1988.

[268] Lee, S., *Goal Programming for Decision Analysis*, Auerbach, PA, 1972.

[269] Lee, S. and D. Olson, A gradient algorithm for chance constrained nonlinear goal programming, *European Journal of Operational Research*, vol. 22, pp. 359–369, 1985.

[270] Lenstra, J., A. Rinnooy Kan, and P. Brucker, Complexity of machine scheduling problems, *Annals of Discrete Mathematics*, vol. 1, pp. 343–362, 1977.

[271] Li, C. and T. Cheng, The parallel machine min-max weighted absolute lateness scheduling problem, *Naval Research Logistics*, vol. 41, pp. 33–46, 1993.

[272] Leipins, G. and M. Hilliard, Genetic algorithm: foundations and applications, *Annals of Operations Research*, vol. 21, pp. 31–58, 1989.

[273] Liepins, G., M. Hilliard, M. Pallmer, and M. Morrow, Greedy genetics, in Grefenstette [189], pp. 90–99.

[274] Liepins, G., M. Hilliard, J. Richardson, and M. Pallmer, Genetic algorithm application to set covering and traveling salesman problems, in Brown, editor, *OR/AI: The Integration of Problem Solving Strategies*, 1990

[275] Liepins, G. and W. Potter, A genetic algorithm approach to multiple fault diagnosis, in Davis [101], pp. 237–250.

[276] Liou, T. and M. Wang, Ranking fuzzy numbers with integral value, *Fuzzy Sets and Systems*, vol. 50, pp. 247–255, 1992.

[277] Liu, L., (*s, S*) continuous review models for inventory with random life-times, *Operations Research Letters*, vol. 12, no. 3, pp. 161–167, 1990.

[278] Lozano-Perez, T. and M. Wesley, An algorithm for planning collision-free paths amongst polyhedral obstacles, Technical report, IBM Thomas J. Watson Research Center, 1978.

[279] Luenberger, D., *Linear and Nonlinear Programming*, 2nd. ed., Addison-Wesley, Reading, MA, 1984.

[280] Luss, H., Operations research and capacity expansion problems: a survey, *Operations Research*, vol. 30, no. 5, pp. 904–947, 1982.

[281] Manly, B. *The Statistics of Natural Selection on Animal Populations*, Chapman & Hall, London, 1984

[282] Martello, S. and P. Toth, *Knapsack Problems: Algorithms and Computer Implementations*, John Wiley & Sons, Chichester, 1990.

[283] Maza, M. and B. Tidor, An analysis of selection procedures with particular attention paid to proportional and Boltzmann selection, in Forrest [137], pp. 124–130.

[284] McDonnell, J., R. Reynolds, and D. Fogel, editors, *Evolutionary Programming IV*, MIT Press, Cambridge, MA, 1995.

[285] Männer, R. and B. Manderick, editors, *Parallel Problem Solving from Nature: PPSN II*, Elsevier Science Publishers, North-Holland, 1992.

[286] Merten, A. and M. Muller, Variance minimization in single machine sequencing problem, *Management Science*, vol. 18, pp. 518–528, 1972.

[287] Michalewicz, Z., *Genetic Algorithm + Data Structure = Evolution Programs*, 2nd ed., Springer-Verlag, New York, 1994.

[288] Michalewicz, Z., Genetic algorithms, numerical optimization, and constraints, in Eshelman [120], pp. 151–158.

[289] Michalewicz, Z., A survey of constraint handling techniques in evolutionary computation methods, in McDonnell et al. [284], pp. 135–155.

[290] Michalewicz, Z. and N. Attia, Evolutionary optimization of constrained problems, in Sebald and Fogel [375], pp. 98–108.

[291] Michalewicz, Z., D. Dasgupta, R. G. Le Riche, and Schoenauer, Evolutionary algorithms for Industrial Engineering problems, *International Journal of Computers and Industrial Engineering*, vol. 30, no. 4, 1996.

[292] Michalewicz, Z., T. Logan, and S. Swaminathan, Evolutionary operations for continuous convex parameter spaces, in Sebald and Fogel [375], pp. 84–97.

[293] Michalewicz, Z., and M. Schoenauer, Evolutionary algorithms for constrained parameter optimization problems, *Evolutionary Computation*, vol. 4, no. 1, 1996 (in press),

[294] Michalewicz, Z., G. A. Vignaux and M. Hobbs, A non-Standard Genetic Algorithm for the Nonlinear Transportation Problems, *ORSA Journal on Computing*, vol. 3, no. 4, pp. 307–316, 1991.

[295] Miller, J., W. Potter, R. Gandham; and C. Lapena, An evaluation of local improvement operators for genetic algorithms, *IEEE Transaction on Systems, Man, and Cybernetics*, vol. 23, no. 5, pp. 1340–1351, 1993.

[296] Mitchell, M., *An Introduction to Genetic Algorithms*, MIT Press, Cambridge, MA, 1996.

[297] Moinzadeh, K. and S. Nahmias, A continuous review model for an inventory system with two supply modes, *Management Science*, vol. 34, pp. 761–773, 1988.

[298] Morgan, B., *Elements of Simulation*, Chapman & Hall, London, 1984.

[299] Morton, T. and D. Pentico, *Heuristic Scheduling Systems-With Applications to a Production Systems and Project Management*, John Wiley & Sons, New York, 1993.

[300] Moscato, P. and M. Norman, A memetic approach for the traveling salesman problem: implementation of a computational ecology for combinatorial optimization on message-passing systems, in *Proceedings of the International Conference on Parallel Computing and Transportation Applications*, Amsterdam, 1992.

[301] Mühlenbein, H., How genetic algorithms really work: part I. mutation and hill-climbing, in Männer and Manderick [285], pp. 15–26.

[302] Mühlenbein, H. and D. Schlierkamp-Voosen, Predictive Models for the breeder genetic algorithm I. continuous parameter optimization, *Evolutionary Computation*, vol. 1, pp. 25–49, 1993.

[303] Mühlenbein, H., M. Schomisch, and J. Born, The parallel genetic algorithm as function optimizer, in Belew and Booker [30], pp. 271–278.

[304] Murtagh, B. and S. Niwattisyawong, An efficient method for the multi-depot location-allocation problem, *Journal of Operations Research Society*, vol. 33, pp. 629–634, 1982.

[305] Murty, K, *Linear and Combinatorial Programming*, John Wiley & Sons, New York, 1976.

[306] Muselli, M., and S. Ridella, Global optimization of functions with the genetic algorithm, *Complex Systems*, vol. 6, pp. 193–212, 1992.

[307] Muth, J. and G. Thompson, editors, *Industrial Scheduling*, Prentice Hall, Englewood Cliffs, NJ, 1963.

[308] Myung, H., and J. H. Kim, Hybrid evolutionary programming for heavily constrained problems, *Bio-Systems*, vol. 38, pp. 29–43, 1996.

[309] Myung, Y., C. Lee, and D. Tcha, On generalized minimum spanning tree problem, *Networks*, vol. 26, pp. 231–241, 1995.

[310] Nakagawa, Y. and S. Miyazaki, Surrogate constraints algorithm for reliability optimization problems with two constraints, *IEEE Transactions on Reliability*, vol. 30, no. 2, pp. 175–180, 1981.

[311] Nakahara, Y., M. Sasaki, and M. Gen, On the linear programming with interval coefficients, *International Journal of Computers and Engineering*, vol. 23, pp. 301–304, 1992.

[312] Nakahara, Y., M. Sasaki, K. Ida, and M. Gen, A method for solving 0-1 linear programming problem with interval coefficients, *Journal of Japan Industrial Management Association*, vol. 42, no. 5, pp. 345–351, 1991 (in Japanese).

[313] Nakano, R. and T. Yamada, Conventional genetic algorithms for job-shop problems, in Belew and Booker [30], pp. 477–479.

[314] Narula, S. and C. Ho, Degree-constrained minimum spanning tree, *Computers and Operations Research*, vol. 7, pp. 239–249, 1980.

[315] Nawaz, M., E. Enscore, and I. Ham, A heuristic algorithm for the m-machine n-job flow shop sequencing problem, *Omega*, vol. 11, pp. 11–95, 1983.

[316] Nemhanser, G. and L. Wolsey, *Integer and Combinatorial Optimization*, Wiley-Interscience, New York, 1989.

[317] Norman, B. and J. Bean, Random keys genetic algorithm for job-shop scheduling: unabridged version, Technical report, University of Michigan, Ann Arbor, 1995.

[318] Norman, B. and J. Bean, Random keys genetic algorithm for scheduling, Technical report, University of Michigan, Ann Arbor, 1995.

[319] Ogbu, F. and D. Smith, Simulated annealing for the flow-shop problem, *Omega*, vol. 19, pp. 64–67, 1991.

[320] Oliver, I., D. Smith, and J. Holland, A study of permutation crossover operators on the traveling salesman problem, in Grefenstette [89], pp. 224–230.

[321] Olsen, A., Penalty functions and the knapsack problem, in Fogel [132], pp. 554–558.

[322] Orvosh, D. and L. Davis, Using a genetic algorithm to optimize problems with feasibility constraints, in Fogel [132], pp. 548–552.

[323] Ostresh, L. Jr., An efficient algorithm for solving the two center location-allocation problem, *Journal of Regional Science*, vol. 15, pp. 209–216, 1975.

[324] Osyczka, A. and S. Kundu, A new method to solve generalized multicriteria optimization problems using genetic algorithm, *Structural Optimization*, vol. 10, no. 2, pp. 94–99, 1995.

[325] Otto, K. and E. Antonsson, Modeling imprecision in product design, in *Proceedings of the Third IEEE International Conference on Fuzzy Systems*, pp. 346–356, Yokohama, Japan, 1994.

[326] Padberg, M. and G. Rinaldi, Optimization of a 532-city symmetric traveling salesman problem by branch and cut, *Operations Research Letters*, vol. 6, pp. 1–7, 1987.

[327] Padberg, M. and G. Rinaldi, A branch and cut algorithm for the resolution of large scale symmetric traveling salesman problems, *SIAM Review*, vol. 33, pp. 60–100, 1991.

[328] Palmer, C., An Approach to a Problem in Network Design Using Genetic Algorithms, Ph.D. thesis, Polytechnic University, 1994.

[329] Palmer, C. and A. Kershenbaum, An approach to a problem in network design using genetic algorithms, *Networks*, vol. 26, pp. 151–163, 1995.

[330] Palmer, D. Sequencing jobs through a multi-stage process in the minimum total time—a quick method of obtaining a near optimum, *Operations Research Quarterly*, vol. 16, pp. 101–107, 1965.

[331] Panwalkar, S. and W. Iskander, A survey of scheduling rules, *Operations Research*, vol. 25, pp. 45–61, 1977.

[332] Paredis, J., Exploiting constraints as background knowledge for genetic algorithms: a case-study for scheduling, in Männer and Manderick [285], pp. 281–290.

[333] Patterson, J. and G. Roth, Scheduling a project under multiple resource constraints: a 0-1 programming approach, *AIIE Transactions*, vol. 8, pp. 449–455, 1976.

[334] Patterson, J., F. Talbot, R. Slowinski, and J. Weglaz, Computational experience with a backtracking algorithm for solving a general class of precedence and resource constrained scheduling problems, *European Journal of Operational Research*, vol. 49, pp. 68–79, 1990.

[335] Pearson, D., N. Steele, and R. Albrecht, editors, *Artificial Neural Nets and Genetic Algorithms*, Springer-Verlag, New York, 1995.

[336] Pedrycz, W., Fuzzy modeling: methodology, algorithms and practice, in Zurada et. al. [433], Section 2.3, 1994.

[337] Picone, C. and W. Wilhelm, Perturbation scheme to improve Hillier's solution to the facilities layout problem, *Management Science*, vol. 30, pp. 1238–1249, 1984.

[338] Piggott, P. and F. Suraweera, Encoding graphs for genetic algorithms: an investigation using the minimum spanning tree problem, in Yao [463], pp. 305–314.

[339] Potvin, J. and D. Dube, Improving a vehicle routing heuristic through genetic search, in Fogel [132], pp. 194–199.

[340] Prim, R., Shortest connection networks and some generalizations, *Journal of Bell Systems Technology*, vol. 36, pp. 1389–1401, 1957.

[341] Pritsker, A., L. Waters, and P. Wolfe, Multiproject scheduling with limited resources: a 0-1 approach, *Management Science*, vol. 16, pp. 93–108, 1969.

[342] Prüfer, H., Neuer beweis eines satzes über permutation, *Arch. Math. Phys.*, vol. 27, pp. 742–744, 1918.

[343] Radcliffe, N., Genetic Neural Networks on MIMD Computers, Ph.D. thesis, University of Edinburgh, UK, 1990.

[344] Radcliffe, N. and P. Surry, Formal memetic algorithms, in Fogarty [128], pp. 1–16, 1994.

[345] Rai, S. and D. Agrawal, editors, *Distributed Computing Network Reliability*, IEEE Computer Society Press, Los Alamitos, CA, 1990.

[346] Ramakumar, R., *Engineering Reliability: Fundamentals and Applications*, Prentice Hall, Englewood Cliffs, NJ, 1993.

[347] Rawlins, G., editor, *Foundations of Genetic Algorithms*, Morgan Kaufmann Publishers, San Mateo, CA, 1991.

[348] Reeves, C., Diversity and diversification in algorithms: some connections with tabu search, in Albrecht et al. [5], pp. 344–351.

[349] Reeves, C., Genetic algorithms and neighborhood search, in Fogarty [128], pp. 115–130.

[350] Reeves, C., A genetic algorithm for flow shop sequencing, *Computers and Operations Research*, vol. 22, pp. 5–13, 1995.

[351] Reeves, C. and C. Wright, An experimental design perspective on genetic algorithms, in Whitley and Vose [417], pp. 7–22.

[352] Reinelt, G., *The Traveling Salesman: Computational Solutions for TSP Application*, Springer-Verlag, Berlin, 1991.

[353] Renders, J. and H. Bersini, Hybridizing genetic algorithms with hill-climbing methods for global optimization: two possible ways, in Fogel [132], pp. 312–317.

[354] Richardson, J., M. Palmer, G. Liepins, and M. Hilliard, Some guidelines for genetic algorithms with penalty functions, in Schaffer [370], pp. 191–197.

[355] Ringuest, J. and D. Rinks, Interactive solutions for the linear multiobjective transportation problem, *European Journal of Operational Research*, vol. 32, pp. 96–106, 1987

[356] Rinnooy Kan, A., *Machine Scheduling Problem: Classification, Complexity Computation*, Martinus Nijhoff, The Hague, 1976.

[357] Rosenblatt, M., The dynamics of plant layout, *Management Science*, vol. 32, pp. 76–86, 1986.

[358] Rosenblatt, M. and H. Lee, A robustness approach to facilities design, *International Journal of Production Research*, vol. 25, pp. 479–486, 1987.

[359] Rosing, K., An optimal method for solving (generalized) multi-Weber problem, *European Journal of Operational Research*, vol. 58, pp. 414–426, 1992.

[360] Roy, B. and B. Sussmann, Les problemes d'ordonnancement avec constaintes disjonctives, Technical Report 9, SEMA, Note D.S., Paris, 1964.

[361] Rubinstein, R., *Simulation and the Monte Carlo Method*, John Wiley & Sons, New York, 1981.

[362] Saber, H. M. and A. Ravindran, Nonlinear goal programming theory and practice: a survey, *Computers and Operations Research*, vol. 20, no. 3, pp. 275–291, 1993.

[363] Sakawa, M., K. Kato, and T. Mori, Flexible scheduling in a machining center through genetic algorithms, *Computers and Industrial Engineering*, vol. 30, no. 4, pp. 931–940, 1996.

[364] Sakawa, M., K. Kato, and T. Shibano, Fuzzy programming for multiobjective 0-1 programming problems through revised genetic algorithms, *European Journal of Operational Research* (in press).

[365] Sampson, S. and E. Weiss, Local search techniques for the generalized resource constrained project scheduling problem, *Naval Research Logistics*, vol. 40, p. 665–675, 1993.

[366] Sannomiya, N. and H. Iima, Application of genetic algorithms to scheduling problems in manufacturing process, in Fogel [129], pp. 523–528, 1996.

[367] Sasaki, M., T. Yokota, and M. Gen, A method for solving fuzzy optimal reliability design problem by genetic algorithms, *Japanese Journal of Fuzzy Theory and Systems*, vol. 7, no. 5, pp. 1062–1072, 1995.

[368] Savelsbergh, M. and T. Volgenant, Edge exchanges in the degree-constrained spanning tree problem, *Computers and Operations Research*, vol. 12, pp. 341–348, 1985.

[369] Schaffer, J., Multiple objective optimization with vector evaluated genetic algorithms, in Grefenstette [186], pp. 93–100.

[370] Schaffer, J., editor, *Proceedings of the Third International Conference on Genetic Algorithms*, Morgan Kaufmann Publishers, San Mateo, CA, 1989.

[371] Schrage, L., Minimizing the time-in-system variance for a finite job set, *Management Science*, vol. 21, pp. 540–543, 1975.

[372] Schwefel, H., *Numerical Optimization of Computer Models*, John Wiley & Sons, Chichester, 1981.

[373] Schwefel, H., *Evolution and Optimum Seeking*, John Wiley & Sons, New York, 1994.

[374] Schwefel, H. and R. Männer, editors, *Parallel Problem Solving from Nature*, Springer, New York, 1990.

[375] Sebald, A. and L. Fogel, editors, *Proceedings of the Third Annual Conference on Evolutionary Programming*, World Scientific Publishing, River Edge, NJ, 1994.

[376] Sen, T. and S. Gupta, A state-of-art survey of static scheduling research involving due dates, *Omega*, vol. 12, pp. 63–76, 1984.

[377] Shaefer, C., The ARGOT strategy: adaptive representation genetic optimizer technique, in Grefenstette [189], pp. 50–55.

[378] Shi, G., H. Iima, and N. Sannomiya, A method for contructing genetic algorithm in job shop problems, *Proc. of 8th SICE Symposium Decentralized Autonomous System*, pp. 175–178, 1996

[379] Shiode, I., T. Nishida, and Y. Namasuya, Stochastic spanning tree problem, *Discrete Applied Mathematics*, vol. 3, pp. 263–273, 1981.

[380] Shore, R. and J. Tomkins, Flexible facilities design, *IIE Transactions*, vol. 12, pp. 200–205, 1980.

[381] Skiena, S., *Implementing Discrete Mathematics Combinatorics and Graph Theory with Mathematica*, Addison-Wesley, Reading, MA, 1990.

[382] Smith, A. and D. Tate, Genetic optimization using a penalty function, in Forrest [137], pp. 499–505.

[383] Sollin, M., Le trace de canalisation, in Berge, C. and A. Ghouilla-Houri, editors, *Programming, Games, and Transportation Networks*, John Wiley & Sons, New York, 1965.

[384] Solomon, M., The vehicle routing and scheduling problems with time windows constraints, *Operations Research*, vol. 35, pp. 254–265, 1987.

[385] Spears, W. and K. De Jong, On the virtues of parameterized uniform crossover, in Belew and Booker [30], pp. 230–236.

[386] Sprechcr, A., *Recourse-Constrained Project Scheduling: Exact Methods for the Multi-Model Case*, Springer-Verlag, Berlin, 1993.

[387] Steinberg, L., The backboard wiring problem: a placement algorithm, *SIAM Review*, vol. 3, pp. 37–50, 1961.

[388] Stinson, J., E. Davis, and B. Khumawala, Multiple resource constrained scheduling using branch and bound, *AIIE Transactions*, vol. 10, pp. 252–259, 1978.

[389] Storer, R., S. Wu, and R. Vaccari, New search spaces for sequencing problems with application to job shop scheduling, *Management Science*, vol. 38, no. 10, pp. 1495–1510, 1992.

[390] Sundararaghravan, P. and M. Ahmed, Minimizing the sum of absolute lateness in single-machine and multimachine scheduling, *Naval Research Logistics Quarterly*, vol. 31, pp. 325–333, 1984.

[391] Syswerda, G., Uniform crossover in genetic algorithms, in Schaffer [370], pp. 2–9.

[392] Syswerda, G., Scheduling optimization using genetic algorithms, in Davis [101], pp. 332–349.

[393] Taillard, E., Benchmarks for basic scheduling problems, *European Journal of Operational Research*, vol. 64, pp. 278–285, 1993.

[394] Talbot, F. and J. Patterson, An efficient integer programming algorithm with network cuts for solving resource-constraints scheduling problems, *Management Sciences*, vol. 24, pp. 1163–1174, 1978.

[395] Tam, K., Genetic algorithms, function optimization, facility layout design, *European Journal of Operational Research*, vol. 63, pp. 322–346, 1992.

[396] Tamaki, H., M. Mori, and M. Araki, Generation of a set of pareto-optimal solutions by genetic algorithms, *Transactions of the Society of Instrument and Control Engineers*, vol. 31, no. 8, pp. 1185–1192, 1995 (in Japanese).

[397] Tamaki, H. and Y. Nishikawa, A paralleled genetic algorithm based on a neighborhood model and its application to the jobshop scheduling, in Männer and Manderick [285], pp. 573–582.

[398] Tate, D. and A. Smith, A genetic approach to the quadratic assignment problem, *Computers and Operations Research*, vol. 22, pp. 73–83, 1995.

[399] Tate, D. and A. Smith, Unequal-area facility layout by genetic search, *IIE Transactions*, vol. 27, pp. 465–472, 1995.

[400] Thangiah, S. R. Vinayagamoorthy, and A. Guggi, Vehicle routing with time deadlines using genetic and local algorithms, in Forrest [137], pp. 506–513.

[401] Thierens, D. and D. Goldberg, Convergence models of genetic algorithm selection schemes, in Davidor et al. [95], pp. 119–129.

[402] Tillman, F., C. Hwang, and W. Kuo, *Optimization of Systems Reliability*, Marcel Dekker, New York, 1980.

[403] Tillman, F., Optimization by integer programming of constrained reliability problems with several modes of failure, *IEEE Transactions on Reliability*, vol. R-18, pp. 47–53, 1969.

[404] Tsujimura, Y., M. Gen, and R. Cheng, An efficient method for solving traveling salesman problems with advanced genetic algorithms, *Transactions of the Institute of Electronics, Information and Computer Engineering*, submitted for publication, 1997 (to appear).

[405] Tsujimura, Y., M. Gen, and E. Kubota, Flow-shop scheduling with fuzzy processing time using genetic algorithms, *The 11th Fuzzy Systems Symposium*, pp. 248–252, Okinawa, 1995 (in Japanese).

[406] Tsujimura, Y., M. Gen, and E. Kubota, Solving job-shop scheduling problem with fuzzy processing time using genetic algorithm, *Japanese Journal of Theory and Systems*, vol. 7, pp. 1073–1083, 1995.

[407] Tanino, T., M. Tanaka, and C. Hojo, An interactive multicriteria decision making method by using a genetic algorithm, in *Proceedings of International Conference on System Science and System Engineering*, pp. 381–386, 1993.

[408] Van de Panne, C. and W. Popp, Minimum cost cattle feed under probabilistic protein constraints, *Management Science*, vol. 9, pp. 405-430, 1963.

[409] Van Laarhoven, P., E. Aarts, and J. Lenstra, Job shop scheduling by simulated annealing, *Operations Research*, vol. 40, no. 1, pp. 113–125, 1992.

[410] Vignaux, G. A. and Z. Michalewicz, A Genetic Algorithm for the Linear Transportation Problem, *IEEE Transactions on Systems, Man, and Cybernetics*, vol. 21, pp. 445–452, 1991.

[411] Wang, P., G. Wilson, and N. Odrey, An on-line controller for production system with seasonal demands, *International Journal of Computers and Industrial Engineering*, vol. 27, pp. 565–574, 1994.

[412] Weistroffer, H., An interactive goal programming method for nonlinear multiple-criteria decision-making problems, *Computers and Operations Research*, vol. 10, no. 4, pp. 311–320, 1983.

[413] Wetzel, A., Evaluation of the effectiveness of genetic algorithms in combinatorial optimization, Technical report, University of Pittsburgh, 1983.

[414] Whitley, D., GENITOR: a different genetic algorithm, in *Proceedings of the Rocky Mountain Conference on Artificial Intelligence*, Denver, 1989.

[415] Whitley, D., V. Gordan, and K. Mathias, Lamarckian evolution, the Baldwin effect and function optimization, in Davidor et al. [95], pp. 6–15.

[416] Whitley, L., editor. *Foundations of Genetic Algorithms 2*, Morgan Kaufmann Publishers, San Mateo, CA, 1993.

[417] Whitley, L. and M. Vose, editors, *Foundations of Genetic Algorithms 3*, Morgan Kauffmann Publishers, San Mateo, CA, 1995.

[418] Winter, G., et al, *Genetic Algorithms in Engineering and Computer Science*, John Wiley & Sons, New York, 1996.

[419] Winston, W., *Operations Research: Appliciations and Algorithms*, 3rd ed., Duxbury Press, Belmont, CA, 1994

[420] Wong, D. and C. Liu, A new algorithm for floorplan design, in Wizard, V. and M. Yannakakis, editors, *Proceedings of the 23rd ACM-IEEE Design Automation Conference*, pp. 101–107, Boston, 1986.

[421] Wright, A., Genetic algorithms for real parameter optimization, in Rawlins [347], Chapter 4.

[422] Xu, W., On the quadratic minimum spanning tree problem, in Gen, M. and W. Xu, editors, *Proceedings of 1995 Japan-China International Workshops on Information Systems*, pp. 141–148, Ashikaga, Japan, 1995.

[423] Yamada, T. and R. Nakano, A genetic algorithm applicable to large-scale job-shop problems, in Männer and Manderick [285], pp. 281–290.

[424] Yamada, T. and R. Nakano, Genetic algorithm and job-shop scheduling problem, *Systems, Control and Information*, vol. 37, no. 8, pp. 484–489, 1993 (in Japanese).

[425] Yamamura, M., T. Ono, and S. Kobayashi, Character-preserving genetic algorithms for traveling salesman problem, *Journal of Japan Society for Artificial Intelligence*, vol. 6, pp. 1049–1059, 1992.

[426] Yang, X. and M. Gen, Evolution program for bicriteria transportation problem, in Gen and Kobayashi [160], pp. 451–454.

[427] Yokota, T., M. Gen, and K. Ida, System reliability of optimization problems with

several failure modes by genetic algorithm, *Japanese Journal of Fuzzy Theory and Systems*, vol. 7, no. 1, pp. 117–185, 1995.

[428] Yokota, T., M. Gen, K. Ida, and T. Taguchi, Optimal design of system relibility by an approved genetic algorithm, *Transactions of Institute of Electronics, Information and Communication Engineers*, vol. J78A, no. 6, pp. 702–709, 1995.

[429] Zadeh, L., Fuzzy sets as a basis for a theory of possibility, *Fuzzy Sets and Systems*, vol. 1, pp. 3–28, 1978.

[430] Zhang, L. and W. Zheng, Remodeling the film-copy deliverer problem, in *Proceedings of IEEE International Conference on Systems, Man, and Cybernetics*, pp. 543–547, San Antonio, 1994.

[431] Zhang, L. and W. Zheng, The extended traveling salesman problem and film copy deliver problem, *The Sixth Annual International Symposium on Algorithms and Computation*, Callaghan, Australia, 1995.

[432] Zimmerman, H., *Fuzzy Set Theory and Its Applications*, 2nd ed., Academic Publishers, Norwell, MA, 1991.

[433] Zurada, J., R. Marks II, and C. Robinson, editors, *Computational Intelligence: Imitating Life*, IEEE Press, New York, 1994.

[434] Alliot, J. M., E. Lutton, E. Ronald, M. Schoenauer, and D. Snyders, editors, *Artificial Evolution: European Conference. AE'95, Brest*, Springer-Verlag, Berlin, 1996.

[435] Fögel, L., P. J. Angeline, and T. Bäck, editors, *Proceedings of the 5th Annual Conference on Evolutionary Programming*, MIT Press, Cambridge, MA, 1996.

[436] Angeline, P. and K. E. Kinnerar, Jr., editors, *Advances in Genetic Programming*, Vol. 2, MIT Press, Cambridge, MA, 1996.

[437] Bäck, T. and H. Schwefel, Evolutionary computation: an overview, in Fogel [129], pp. 20–29.

[438] Ebeling, W. and H.-M. Voigt, editors, *Proceedings of the 4th Conference on Parallel Problem Solving from Nature*, Springer-Verlag, Berlin, 1996.

[439] Fogel, D. and A. Ghozeil, Using fitness distributions to design more efficient evolutionary computations, in Fogel [129], pp. 11–19.

[440] Furuhashi, T., editor, *Advances in Fuzzy Logic, Neural Networks and Genetic Algorithms*, Springer-Verlag, Berlin, 1995.

[441] Gen, M. and K. Ida, editors, *Proceedings of Mini-Symposium on Genetic Algorithms and Engineering Design*, Ashikaga, Japan, 1996.

[442] Gen, M. and Y. Tsujimura, editors, *Evolutionary Computations and Intelligent Systems*, Gordon & Breach Publishers, NJ, 1997.

[443] Gen, M. and R. Cheng, A hybrid search for machine scheduling problems, in Zimmermann [469], pp. 378–383.

[444] Gen, M. and B. Liu, Evolution program for optimal capacity expansion, *Journal of Operations Research of Japan*, 1997, in press.

[445] Gong, D., G. Yamazaki, and M. Gen, Evolutionary program for optimal design of material distribution system, in Fogel [129], pp. 139–143.

[446] Gong, D., M. Gen, and G. Yamazaki, Evolutionary method for facility location problem, in Zimmermann [469], pp. 373–377.

[447] Grierson, D. and P. Hajela, editors, *Emergent Computing Methods in Engineering Design: Applications of Genetic Algorithms and Neural Networks*, Springer-Verlag, Berlin, 1996.

[448] Herrera, H. and J. L. Verdegay, editors, *Genetic Algorithms and Soft Computing*, Physica-Verlag, Heidelberg, 1996.

[449] Koza, J. R., editor, *Genetic Programming: Proceedings of the First Annual Conference*, MIT Press, Cambridge, MA, 1996.

[450] Langton, C. and T. Shimohara, editors, *Artificial Life V: the Fifth International Workshop on the Synthesis and Simulation of Living Systems*, Nara, 1996.

[451] Li, Y., M. Gen, and K. Ida, Evolutionary computation for multicriteria solid transportation problem with fuzzy numbers, in Fogel [129], pp. 596–601.

[452] Michalewicz, Z. and M. Schoenauer, Evolutionary algorithms for constrained parameter optimization problems, *Evolutionary Computation* (to appear).

[453] Michalewicz, Z., Evolutionary computation: practical issues, in Fogel [129], pp. 30–39.

[454] Myung, H. and J. Kim, Constrained optimization using two-phase evolutionary programming, in Fogel [129], pp. 262–267.

[455] Sakawa, M., *Fuzzy Sets and Interactive Multiobjective Optimization*, Plenum Press, New York, 1993.

[456] Sakawa, M. and M. Tanaka, *Genetic Algorithms*, Asakura Syoten, Tokyo, 1995 (in Japanese).

[457] Schniederjans, M., *Goal Programming*, Kluwer Academic, Dordrecht, Germany, 1995.

[458] Singh, N., *Systems Approach to Computer-Integrated Design and Manufacturing*, John Wiley & Sons, New York, 1996.

[459] Tamaki, H., H. Kita, and S. Kobayashi, Multi-objective optimization by genetic algorithms: a review, in Fogel [129], pp. 517–522.

[460] Tsujimura, Y. and M. Gen, Genetic algorithms for solving multi-processor scheduling problems, in Yao, Kim, and Furuhashi [464].

[461] Yamamura, M., I. Ono, and S. Kobayashi, Emergent search on double circle TSPs using subtour exchange crossover, in Fogel [129], pp. 535–540.

[462] Yao, X., editor, *Progress in Evolutionary Computation*, Springer-Verlag, Berlin, 1995.

[463] Yao, X., editor, *Evolutionary Computation: Theory and Applications*, World Scientific Publishing, Singapore, 1996.

[464] Yao, X., J. H. Kim, and T. Furuhashi, editors, *Proceeding of the First Asia-Pacific Conference on Simulated Evolution and Learning*, Taejon, 1996.

[465] Yokota, T., M. Gen, and Y. X. Li, Genetic algorithms for nonlinear mixed integer programming problems and its applications, *Computers and Industrial Engineering*, vol. 30, no. 4, pp. 905–917, 1996.

[466] Zhao, L., Y. Tsujimura, and M. Gen, Genetic algorithms for fuzzy clustering, in Fogel [129], pp. 716–719.

[467] Zhou, G. and M. Gen, The genetic algorithms approach to the multicriteria minimum spanning tree problem, in Yao, Kim and Furuhashi [464].

[468] Zhou, G. and M. Gen, An effective genetic algorithm approach to the quadratic minimum spanning tree problem, in Yao, Kim and Furuhashi [464].

[469] Zimmerman, H., editor, *Proceedings of Fourth European Congress on Intelligent Techniques and Soft Computing*, Aachen, 1996.

[470] Hitomi, K. *Manufacturing Systems Engineering*, 2nd ed., Taylor & Francis, London, 1996.

[471] Higuchi, T., D. Mange, H. Kitano, and H. Iba, editors, *Proceedings of the First International Conference on Evolvable Systems: From Biology to Hardware*, Springer-Verlag, Berlin, 1996.

INDEX